To Laura, with all my heart – R.B.
To my family Gina, Justin, Maria, Alexis, and Roanna
for their enduring love and support – B.W.R.

Preface

Data science does not broker in abstractions or theories. One of the primary objectives of data science is to make sense of a large number of observations, and consequently to make sense of the real world. The type of analysis performed in data science is deeply rooted in the art of acquiring, wrangling, and visualizing information. Information either stored on servers, in the cloud, or in our brains can be said to have substance. Bits and bytes have weight both metaphorically and physically.

Our ability to decipher and filter large amounts of sensory data allows us to navigate successfully through our busy and sometimes dangerous world. Similarly, data science provides the methodology and tools to accurately interpret an increasing volume of incoming information in order to discern patterns, evaluate trends, and make the right decisions. The results of data science analysis provide real-world answers to real-world questions.

This book is about real-world questions and about arriving at the answers as expeditiously and precisely as possible. On the surface, these inquiries may appear to require simple, straightforward responses – yes or no. However, it is the depth and breadth of the question and the ramifications inherent in the answer that demand our focus and respect.

In one possible scenario, responding to the question in the affirmative may take us down a path involving significantly greater personal or financial commitments and resources than if the answer were a succinct and definitive "no." As with any true scientific investigation, answers should lead to further questions. The bottom line is that within the data science realm, answers and questions are equivalent in importance and the tools we use to derive the end results must rise to the level of proficiency and accuracy required by the analytical demands imposed by both question and answer.

Given a well-posed question, the topics in this book provide concise descriptions of the techniques and tools used in data science to generate viable answers. Data science as a discipline leverages strategies and technologies derived from computer science, statistics, and various business domains. Data science is the progeny of these fields but does not belong to any particular one.

The data scientist is an inter-disciplinarian, utilizing the core knowledge from several areas to make assessments of current data for the purposes of determining future directions in research and analysis. This book is intended for the beginning application-oriented data scientist who wishes to learn the essential methods necessary to extract meaning from numbers.

Is This a Textbook or a Practitioner's Book?

Yes – to both.

We believe that a modern textbook should cover the materials and applications that a student needs. Unlike traditional textbooks in which the student begins at "Chap. 1" and reads sequentially to the last chapter, we have made every effort to present and package the essential topics contained in this book so that the student is free to utilize select chapters based solely on their interests.

As a result, the chapters contained in this book are not necessarily meant to be read in the order in which they appear.

Programming Examples and Images

There are many Python example programs used throughout this book. We have done our best to make these examples readily available at a separate downloadable location. In addition, almost every image used in this book was created by Python code. For these plots and diagrams, the source code designed to create the image is cited and available. The current download location for the code included in this book is the URL: https://github.com/robertball/Beginners-Guide-Data-Science

There are three main reasons for posting the Python code at this separate location:

- There are many instances when having the code occupy unnecessary space in the book is neither meaningful nor instructive, so we have maintained the code separately to be perused, executed, and evaluated at your leisure.
- Python and its various associated libraries change. As a result, there may come a time when certain sample programs in the book no longer run efficiently or successfully on modern computers. Under these special circumstances we will respond by updating the relevant programs so they remain functional and informative. We cannot easily update the content of this book, but we can easily revise and refine downloadable code.
- Copying directly from this book to another resource such as a development environment often inadvertently transfers various formatting issues. Access to a download of the original code can avoid these formatting and program structure issues altogether.

Code Formatting
We differentiate code examples from the narrative text of the book. The following is an example of a code snippet:

```
File: an_example_file_name.py (does not actually exist - just an example)
print('This prints out text.')
print('This does too!')
```

Definitions
We highlight definitions in the book by *italicizing* the term being described.

Ship Names, Movie Names, Book Names, and Latin
Following modern conventions, we have also italicized ship names like the *HMS Titanic* (i.e., *The Titanic*), movie names like *Back to the Future*, book names like *Wizard of Oz*, and Latin words or phrases like *a priori*.

Emphasis
We designate special emphasis for specific points and statements by **bolding** them.

Ogden   Robert Ball
UT, USA

Ogden   Brian Rague
UT, USA

Contents

Chapter 1
Introduction to Data Science

***

The purpose of this book is very simple: to help you make money either directly or indirectly through data science. Before we provide our definition of data science, let us be clear about your potential motivations for reading this book.

If you wish to read a book about the theoretical foundations of data science (or big data/data analytics) then we recommend you do not waste your time with this book. This book at its core is a practical book that will help you make money. There are many other academic books on data science that are filled with mathematical symbols and expressions that explain the principles and algorithms behind data science in greater detail than this book. We do in fact use mathematical symbols and expressions from time to time; however, they are intended to help further clarify the topic under consideration, but you may ignore them without compromising your understanding of the material.

This book will help you make money directly if you are an entrepreneur seeking to produce a product such as a recommendation engine to move your business forward. In addition, this book will help you make money indirectly by launching your career into the expanding field of data science where you will assist your organization (e.g., business, government, or charity) to increase their revenue and you will consequently maintain a stable, growing income by retaining a prestigious position with a rewarding salary.

Regardless of which path you choose, you will want to read this book if you are a motivated person willing to learn a range of different skills to generate revenue, directly or indirectly, in data science. The path is not always obvious, but for the motivated person, it is well worth it.

If you wish to make money directly then the possibility exists that millions to billions of dollars can be earned by learning and effectively applying the concepts of data science. For example, consider Amazon (the company). The core part of their business is twofold: (1) recommending the right products to people by leveraging a recommendation system so that people buy what they want when they want and, (2) determining the best methods to reduce costs by shipping the products that people purchase in a fast and efficient way through optimized logistics and warehousing governed by processes related to supply chain management. These two main business strategies either originate from or are heavily influenced by data science and analytics.

If you wish to make money indirectly then data science *can* help your organization in making well founded decisions and predictions that cover all facets of business operations. The more valuable *you* are to your organization in designing and distributing new products, reducing costs, and discovering new markets, for example, the more secure and extensive your employment opportunities will be and the more money you will be able to earn.

Either way you choose, this book will help you succeed in a practical way by combining business (domain) knowledge and common sense with the reliable foundations of statistics and computer science.

However, whereas most data science books focus primarily on statistics and computer science, we realize that without fully considering and appreciating the business aspect of data science your end result is primarily of theoretical, academic value. In other words, without serious examination of a practical business purpose, your efforts and results will belong to the

R. Ball, B. Rague, *The Beginner's Guide to Data Science*, https://doi.org/10.1007/978-3-031-07865-1_1

nominal category of "Oh! That is neat! I am sure your mother is very proud of what you did." In contrast, our hope is that your data science results will produce the following reaction, "Wait... Are you saying that if we do that (or build that) that we can make that much money!! Woah... This is huge! Let me call the VP and see how fast we can get your results and recommendations into action right now."

Regardless of your level of interest in making money, it is important to have a revenue-generating mindset. No matter what kind of project you undertake, you will need money to operate. Your project might be especially noble, altruistic, and perfectly aligned with your inner values and priorities, and we fully support and congratulate those objectives. However, you will need money to keep the lights on in the building, to buy food, to hire additional people for the project, and in general to keep the motivational fire blazing. The more your results help finance either yourself or your organization the more likely both your short-term and long-term goals will be realized.

1.1 Superpowers

If you could have a superpower, what would it be?

Would you like the power to fly like Superman or Peter Pan? What about the capacity to predict the future? Alternatively, would you prefer the ability to walk through walls?

To many people data science is a superpower. For example, techniques and strategies related to data science allow you to predict the future with some level of confidence. Data science grants you the ability to project who may die and who may live and allows you greater insights about investing into the stock market.

Data science enables you to figure out who wrote a particular "anonymous" book based on word frequencies and helps you determine if the author of a book is male or female and the country where they most likely lived during their developing years as a child and teenager.

The principles and practices of data science allow you to detect fraud, scams, and other deceptive practices.

Data science enables you to peer into the far future and to see how many people will populate the earth during any given future year.

The analysis associated with data science also allows you to accomplish more mundane tasks such as understanding why students fail or succeed in their chosen major. It answers the questions of why someone was elected to office and why someone else was not.

Data science practice delves deeper and inspects the fundamental reasons why one house sells immediately at a given location while the exact same house would never sell somewhere else.

What about our ongoing fascination with past events? Data Science allows us to investigate historical data such as the passengers of the *Titanic*, a large ship that sank in the Atlantic Ocean in 1912 and allows us to determine with over 95% accuracy if a particular passenger was destined to live or die. More importantly, data science techniques clarify *why* one passenger died and *why* another passenger survived.

1.2 What Is Data Science?

If you invite 100 data scientists into a room and ask them to define "data science" then you will likely hear 100 different definitions.

For the purposes of this book, we will define *data science* as an inter-disciplinary field of investigation that is concerned with obtaining accurate and reliable insights about data. This insight is often obtained by using domain knowledge (i.e., the business side), statistics, linear algebra, machine learning, visualization, programming, cluster computing, and creativity. In more straightforward terms, data science is a field that involves primarily the domain (what the topic is about), statistics, and computer science.

Another way to view data science is as an exciting and expanding field of endeavor where we do whatever is necessary to fully comprehend and gain insight into data.

Insight, the ability to gain an accurate and intuitive understanding of data, is the key to success for modern businesses. Insight into what customers want and why they want it defines the profitable and rewarding pathways into the foreseeable future for many businesses.

Actionable insight enables people and businesses to react to what is going on around them in the invisible, data-driven world. The methodologies and techniques developed to discover "what we know we don't know" are vital to the success of a business.

Although we can sense the many physical, tangible properties of the world around us, such as smell, color, sound, and temperature, there are many things that we cannot detect. For example, we cannot observe with our eyes what the probability will be for snow next week nor see how traffic patterns are related to economies nor sense so many other millions of patterns that are hidden inside complex data.

Due to the complex nature of life and the dynamic processes associated with daily experience, data science itself is complex and involves many diverse topics that can never be fully covered in a single book. After you read this book, we encourage you to continue your lifelong journey of further exploring data science topics.

Data science is *interdisciplinary*, which by definition means that data science does not belong sequestered and siloed within a single field. By its very nature data science does not exist in isolation or separate from the human experience.

Does anything really exist in isolation? No subject exists by itself, even the seemingly independent axioms and theorems that define mathematics. Can a person study mathematics without enlisting their brain in the analysis of abstract concepts? A brain is made up of cells whose study is part of biology. Also, the study of the mind and behavior is psychology. Also, mathematicians write to describe their work. A mathematician cannot write without pencils, pens, chalk, or some other writing instrument. If mathematics could simply be thought of and not transcribed to paper, then it would have to be communicated through sound which involves physics.

To use the field of geography as a resource, the first law of geography states, "All things are related. The closer they are the more related they are to each other." That is true with all fields and all disciplines in life.

The main thing to remember about data science is that both statistics and computer science are **tools**. For example, programming languages such as R and Python are **used for a specific purpose** when applied to data science problems. A data scientist usually does not take a compiler design course to create programming languages. Similarly, although machine learning is **utilized in and integral** to data science, data scientists typically do not create new machine learning algorithms.

This conceptual overlapping organization resembles how e-commerce (buying and selling products online) **uses** advanced cryptography algorithms to ensure that sensitive information such as credit card numbers are not stolen during an online transaction. Although advanced mathematics was used to create the algorithms it is not necessary to have an advanced degree in mathematics to **employ** them.

Using a more widely familiar example, we all recognize that we do not need to know how to build a car to drive one. Automotive engineers require years of experience in mathematics and engineering to design and manufacture a car. However, to <u>drive</u> a car, you only need to complete a driver's education class and obtain your driver's license. Although it can be argued that knowing more about the detailed electrical and mechanical functions of your car can be a significant advantage when traveling from place to place, it is neither a necessary nor sufficient condition for driving the vehicle.

Figure 1.1 displays a Venn diagram of where data science resides in relation to computer science, statistics, and business, a visual confirmation that data science is truly an interdisciplinary endeavor.

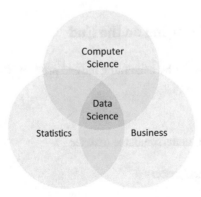

Fig. 1.1 The interdisciplinary nature of data science. In reality, nothing exists in isolation

1.3 Predicting the Future

Think about the temperature of air outside right now at your current location. What will be the recorded temperature at your location tomorrow? Let us be even more specific and state that we only want to know what tomorrow's low temperature will be.

This particular fact-finding task is framed as a simple question but figuring out the answer can be very challenging. *Meteorology*, the branch of science that is concerned with the atmosphere and forecasting of weather, is a complex field, covering metrics and topics such as pressure, temperature, jet streams, moisture, atmospheric layers, mesoscale processes, pollution, volcanic activity, and energetics.

Do you need a degree or advanced knowledge in meteorology to forecast tomorrow's temperature at your location? The answer is no.

If you allow a prediction within a specific range, then we could state that tomorrow's low temperature will be today's low temperature plus or minus (+/−) 50° Fahrenheit (F). For example, if it is currently 25° outside then tomorrow will most likely be between −25° and 75°. How would you assess the quality of this prediction? If you wait until tomorrow this guess will most likely be 100% correct. Our prediction is so broad and inclusive that it does not even matter if we measure temperature using the Fahrenheit or Celsius scale, as either outcome will likely fall within the predicted range.

However, can we narrow down this range of temperatures? Of course! We can predict that tomorrow's low temperature will most likely be today's low temperature +/− 10° F. We might not achieve 100% accuracy, but we will still be in close proximity to the actual temperature.

What level of accuracy can be achieved? Predicting tomorrow's temperature is not that hard because even a modest level of data analysis indicates that it is rare for tomorrow's temperature to be extremely different from today's temperature, although this may occur on rare occasions.

Let us construct a more complex query: What will the temperature at your current location be 100 days from now? More abstractly, what will the temperature be at your current location X days from now where X is any natural number? For example, X could be 120 or 200 or even 1000.

This question is essentially about a forecast, i.e. predicting the future. If we could accurately estimate the temperature and general weather (e.g., rain, snow, wind) of any location on earth for any day in the future to 100% accuracy then we will easily be the source of information for most everyone on the planet for all future weather forecasts. Even if we were only 95% accurate then we could open a weather forecasting business and effectively shut down all other weather predicting businesses across the globe.

In other words, predicting the future, more commonly known as forecasting, ultimately reduces to the essentials of business, providing value over a sustained period of time.

Data science helps businesses become and/or stay relevant, provides customers what they want in a timely manner, and most importantly, seeks to understand <u>why</u> the customer wants what they want even if they do not entirely understand it themselves.

Data science can be used in any area where data exists such as health care, finance, law, government, education, sports, and other areas of work or play. If data are generated in that field, then data science can be effectively utilized. In fact, what current field or topic does not utilize data in some form?

1.4 Understand the Process by Focusing on the End

One way to understand what data scientists do is by examining the end product. The following is a list of items and objectives that most data science projects include:

- Provide actionable insights
- Tell stories with data
- Communicate complex results in a clear, understandable manner
- Create consumable predictive products
- Align business goals with the data science process

1.4.1   Actionable Insights

One of the most important results of any data science project is to formulate *actionable insights*. For example, after analyzing publicly available campaign donations, key factors of donors might be found that shed light on the industries typically associated with each political party.

For example, if a person is employed in finance then it has been generally demonstrated they are more likely to vote for a Republican candidate in the United States. For most people, an insight that links careers and political parties may barely register. However, for potential political candidates (e.g., people trying to get elected or reelected), such information could establish a good return on investment regarding donor identification.

To illustrate, a political candidate canvassing a random cross-section of people for donations would likely receive a low overall ROI for their time and effort. A *return on investment (ROI)* is a key performance metric for determining the effectiveness in expending resources toward a specific business objective. For example, given the amount you invested in this book, what did you actually gain from reading it and applying the principles described?

Regarding the ROI for a political candidate's general survey of a random cross-section of people, did the hours of requesting financial backing justify the few donations received? Conversely, if the candidate knew exactly the industry and individuals to pursue for donations, like the finance sector mentioned previously, then a much higher ROI would be achieved by targeting only those people that evince a high probability of donating.

Another example of effectively applying actionable insights could involve the type of information discussed in a presentation about climate change. Given the same data and results, one presentation might produce actionable results while another does not.

For example, one presentation might emphasize only how humans have affected the global climate. The audience might find that presentation interesting, but not constructively respond by altering any behavior after the presentation has concluded. Another presentation based on the same information might instead offer actionable results and provide a list of measures the audience can perform to counteract any negative influence on climate. Specifically, the presentation might illustrate how switching from incandescent to LED light bulbs in a household is an impactful, actionable item for the audience to consider.

1.4.2   Tell Stories with Data

One important skillset that data scientists usually learn very early is that telling a story is instrumental in crystallizing the purpose of data analysis and the insights to be gained by exploring the data. Questions that should be addressed when working with data sets include:

- What does the data mean?
- Where did the data come from?
- What current insights are available from the data?
- What future predictions are available from the data?

For example, given all the data from the standardized test scores of elementary students in a given state, people have found that elementary students who take the standardized tests earlier in the day *before* lunch attain higher scores than peers in the same school who are administered the test in the afternoon.

An actionable insight would be to recommend that the elementary schools only administer the standardized tests in the morning. This approach would likely result in overall higher scores for the elementary school, which often translates into additional funding for these schools.

The descriptive storyline of these results and recommendations is vitally important though. Numerous psychological studies have found that people gravitate and respond to stories. People love stories and will pay attention if you relate an abstract idea as a story. However, simply listing facts will likely result in your audience becoming distracted and thinking about other pressing concerns, like their next meal or family responsibilities.

Consequently, a data scientist should present the insights gained from studying the standardized tests by recounting the personal stories and experiences of either actual or fictional students.

For example, which do you find more interesting: (1) listening to someone lecture at length about concepts related to circadian rhythm (the reason why students perform better on tests in the morning) or (2) listening to a person who introduces Sally and Sarah, identical twins from a loving home with well-adjusted parents in a middle-class neighborhood.

Sally and Sarah have similar IQ's, and on a particular weekday both ate the same thing for breakfast, and both rode their bikes to school. However, Sally earned a statistically higher score on a standardized test administered at school that day when compared with her twin sister, Sarah. Why? Now that the data scientist has engaged the audience's attention actionable insights can be more effectively shared and understood.

Without the audience's full attention, important information and explanations may be ignored or misinterpreted.

Stories provide a unique opportunity to bring the data to life! The data scientist learns there are many visualization techniques, some static and some interactive, that can provide significant insights into the data. You, the data scientist, will be able to literally show through storytelling how the data interacts on different dimensions and what it means in the broader context of the question.

Staring at spreadsheets or tables filled with thousands or millions of rows of data can simultaneously overwhelm and demotivate most individuals. Creating visualization examples and live animations of that same data can edify and energize those same people.

There are few things more exciting in life than sharing insights that you invested months to discover and then witnessing the metaphoric lightbulbs illuminate people's expressions as you lay out your data story. An isolated table of data is boring. Giving the numbers within that table a background, a present existence, and a potential future is a storywriter's dream. The crucial difference is the story behind your data will be firmly rooted in reality and not fiction.

1.4.3 Communicate Complex Results

One of the biggest challenges with data science projects is the results can be *very* complex. For example, we will find in Chap. 5 that multiple factors interact with each other to best predict the survival or death of passengers during the sinking of the *HMS Titanic* in the early morning hours of April 15, 1912. However, no matter the level of complexity of the relationships among the different factors, if you cannot effectively communicate the results to your target audience then you have wasted your efforts.

Some people believe that complex ideas cannot be communicated in simple ways. For example, the concept of integrals in calculus can be difficult to understand, since it involves understanding areas, curves, and infinitesimal intervals. Although integration can appear complex on the surface, elementary school children are able to sufficiently grasp the essential concept when taught carefully and appropriately. Although integrals are a form of advanced mathematics, they are merely an extension of a simple idea.

For elementary school children, if the area of a fence is represented by the number of dogs that can comfortably be enclosed within that fence then we can leverage that proposition to understand integrals. For example, we can state the premise that a one-foot by one-foot fence has a total area of one square foot and only one dog can be placed comfortably in that enclosure. We can extend this concept to understand rectangular areas. For instance, six dogs can be comfortably situated in a two-foot by three-foot fence. But what about a fence with a curved shape enclosing a specified area? By visualizing a large fence with a curved shape, we can demonstrate that an integral is simply adding up small dog-sized rectangles that fit within the area of interest -- we are really just counting dogs.

In other words, if elementary school children can be taught complex ideas using simple, relatable explanations then you can similarly communicate fundamental concepts to reasonably intelligent stakeholders who may not necessarily understand (or *wish* to understand) the intricate details.

The best teachers, communicators, and data scientists are not the ones who appear clever and intelligent but are those who express the most complex ideas in terms their audience can comprehend. It is easy to create material that appears complicated and is not well understood. What is often difficult and takes special effort is to present problematical ideas in a way that encourages and elevates the awareness and understanding of the target audience or stakeholders.

1.4.4 Create Consumable Predictive Products

The end product of a data science project depends significantly on the data and the project.

While the results of one data science project might simply be a presentation to stakeholders, another result might include the design, development, or application of wholesale predictive software systems. For example, one project might explain why a company should move from one location to another to cut costs. Conversely, a second project might create results in software that help people regulate their insulin levels.

Some examples of software utilized in data science projects could be any of the following:

- A suite of software that performs real-time analysis of news feeds and stock feeds to predict which stocks to buy or sell immediately based on current events.
- NLP (Natural Language Processing) software that analyzes all the current tweets from Twitter to create a summary of trends of what is currently taking place in the world. This information could be used by politicians, businesspeople, or the media.
- A recommendation engine that suggests what you should eat on any given day.
- A News feed that automatically provides local, state, and national news based on the user's current location.

Data science is a broad subject generating results for a wide variety of applications.

1.4.5   Aligning Business Goals with the Data Science Process

Whether you present new business opportunities or investigate ways to optimize current business endeavors, your focus is to maintain and grow your business.

One important metric in business is the *key performance indicator (KPI)*. Although we will not go through a long list of all the KPI's, we will bring them to your attention. For example, we previously mentioned ROI (Return on Investment).

The core measure behind ROI is that if you or your organization invest something, such as time or money, what will be gained in return? For example, if your organization compensates you for a period of two years to uncover actionable insights about when and how children write letters to Santa Claus, what do your employers expect to receive in return? Your time is valuable both to you and to your organization. Will this project help bring in more money or more customers? In other words, regardless of the project focus, how will your efforts ultimately help your organization achieve its goals?

We list just of few examples of KPI's to help you think about business goals in general:

- Customer lifetime value
- Customer acquisition cost
- Net profit margin
- Number of new contracts signed per period
- Average time for conversion
- Click-through rate
- Operational cash flow
- Inventory turnover
- Monthly website traffic

In the end, if your project moves your organization in the direction of its business goals then you will be retained and your organization will generate revenue. However, if your project is simply an interesting but ineffective endeavor and does not align with your organization's goals then eventually either your job, your project, or both will be reassessed.

Ultimately, aligning your data science projects with the business goals of your organization through various KPI's will help both you and your organization. If you do not align your data science projects with the goals of your organization then you will most likely not be involved with the organization very much longer.

1.5   It Is All About the Question!

In any data science project we pursue there must be some goal or key question we are trying to answer. Without a guiding question we seek to resolve we often get distracted by other interesting peripheral questions that may be noteworthy but will ultimately detract from our objective. Throughout the project duration, staying on target and remaining focused on the main objective is of primary importance.

This strategy does not preclude recognizing and recording other salient insights and avenues of investigation along the way. However, if your goal is to ultimately help your organization be profitable, then diversions that lead away from that goal should be carefully evaluated.

For example, Machine Learning (ML) is often used in many data science projects, which benefit from selecting and implementing correct and efficient ML algorithms. However, if your focus is to invest most of your time tweaking ML algorithms to run slightly faster then you are inadvertently creating a distraction whereas your primary goal should be to obtain the answer or result. After all, an answer to a question, even if generated by an inefficient algorithm, is better than never obtaining that answer because you are spending all your time fine-tuning the algorithm. The answer to your question is ultimately more important than the efficiency of the algorithm used to secure that answer.

There are many types of questions that you might answer for your projects. Regardless of the *type* of question, the question itself is the most important thing because it guides and supports us in our investigation.

The question should be specific and measurable. A purposely vague question like "How do people generally feel about the economy?" is hard to understand and it will be difficult to know when you have truly arrived at the answer to the question. However, a specific and measurable question like "How do people regard taxes on the day that taxes are due?" is an improvement.

1.5.1   Classification Questions

One of the common challenges that data scientists confront is classification. *Classification* problems are those that label data. For example, given an email, is it spam or a legitimate email? Or, given a set of books and 5 possible authors, who wrote which book?

Generally, you can think of classification problems as answering the following fundamental question: **Is this A or B?** Like the above email example, this choice involves a simple *binary answer*: spam or legitimate email. In some cases the response when attempting to assign an observations to a single category may be either positive (*yes*) or negative (*no*).

The solution to a classification problem, such as possible authors, might be a *multinomial answer*, an answer that has more than two possible choices. Instead of restricting the question to A or B, a multinomial question asks: **Is this A or B or … or Z?**

Whether your classification problem is binary or multinomial, classification problems are fairly common and may be expressed through various forms. For example, the following are different types of classification problems:

- Will this tire fail in the next 1000 miles: Yes or no?
- Which brings in more customers: a $5 coupon or a 25% discount?
- If there is a sale at a store, will people spend more at an in-person store (also known as a brick and mortar store) or an online store?
- The classic machine learning (ML) problem: Is the image a cat or a dog?
- Of the three possible ad types, which one are people more likely to click on to improve the click through rate (CTR)? (*Click through rate (CTR)* is a measure that many online businesses use to measure ad effectiveness.)
- Which of the types of marketing provides more customer loyalty?
- Given two texts (e.g., a tweet, blogs, articles, or book.) which one is more positive in terms of tone?

Humans are wired to classify things. We classify students based on grades. We classify teachers based on how well they teach. We classify our children, our pets, our neighbors, the areas of the world, and so on. Whether we do it consciously or unconsciously we are always classifying the world around us. In data science we get paid to do what everyone around us is doing anyway.

1.5.2   Anomaly Detection

"Is this weird?"…that is the basic question behind detecting anomalies. *Anomaly detection* can be applied in situations such as plagiarism, stock market trends, population trends, price spikes, and election results.

Your question might be one in which the new thing you just observed is clearly outside what would be considered normal or expected. If this is the case, is a new trend indicated or is this simply an outlier? An *outlier* is a data point or observation that is outside or at the extreme ends of the expected range.

To answer questions of this nature, we should know how to measure the similarities and differences between data points. Chapters 4 and 5 will focus heavily on these ideas and strategies.

An example of anomaly detection is to determine the best salesperson in your company. You might do this to identify the top salesperson to both reward that person and to understand their methods so that other salespeople can recognize and apply better sales techniques.

Another use for anomaly detection is to investigate election patterns to establish an occurrence of election fraud. For example, if a state traditionally reports approximately 40% of the eligible population voting every election and then suddenly this measure jumps to over 75% in the current election cycle, can this increase be attributed to election fraud or simply a quite successful "get-out-the-vote" campaign?

In another related example, assume you have data from a sports organization with statistics about all the players. If there are a few players noticeably and abnormally better in specific athletic categories (e.g., stronger, faster.), is it because these players practiced more or instead ingesting prohibited performance enhancing substances like steroids?

Regardless of the context, all anomaly detection questions highlight unexpected singular items or events that are substantially different from the remainder of the data.

1.5.3   Prediction/Forecasting

Prediction is one of the essential tasks for most data scientists. *Prediction* is the process of learning from a set of data, called training data, then estimating the outcome based on different factors. Prediction often answers the question **how much** or **how many?**

For example, if you receive approval by a bank for a loan then the bank researched your application and financials and formulated a prediction from the data that you will be able to pay back the loan. The factors included data from your financial history and the resulting prediction is that you will be able to pay back the loan. This is an example of classification.

For a prediction example that does not involve classification, consider a company that projects their sales amount for the next quarter. (A *quarter* is a business term that divides the year into four equal parts: January – March, April – June, July – September, and October – December.) The prediction of sales for the next quarter is not a classification prediction, but a regression problem.

A *regression* prediction is one that predicts continuous numerical values instead of discrete classifications. For example, examples of regression include predicting the population of people for a country in a given year, predicting the sales of a company, or predicting the number of goals a hockey team will tally over a season.

Predictions based on a time component are called *forecasting*. For example, all weather predictions are forecasts because the answer depends on a future point in time – the forecast next Tuesday might be sunny, but on Wednesday the forecast might be rainy.

1.5.4   Clustering

"To which group does this belong?" is the basic question addressed by clustering. *Clustering* is a form of classification that assigns the individual data points to specific categories.

There are two general types of clustering depending on whether the classification labels are known beforehand or not. For example, given a dataset of purebred dogs and dog breed labels (e.g., great Dane, golden retriever, chihuahua) you could place the dogs into clusters – the precise dog group – based on their breed physical characteristics such as weight, height, size, and shedding. However, what if you were given a set of dogs that were not purebred – a mix of different breeds – and thus did not possess distinctive labels? Mixed breeds do not have existing published statistics, so we need to cluster them based on similar weight and height without breed labels.

Since we do not have an established number of labels we might decide arbitrarily on three resulting clusters: big, medium, and small dogs. Alternatively, we might target a different number of groups such as four or five clusters based on other distinguishing characteristics. Regardless of the final number of groups, the clustering process is the same.

Clustering can also provide clarity on the following question: **How is this organized?** Sometimes you want to understand the underlying structure of a data set. For this question, there are no pre-established outcomes or labels associated with the observations in the dataset. The following provides example questions for this type of clustering problem:

- Which viewers like the same types of movies?
- Which printer models fail the same way?
- What categories of buyers do we have?
- How many distinct categories can we find in the (Republican/Democrat/Libertarian/etc.) party?
- Are most of our buyers early adapters, early majority, late majority, or laggards?

Clustering your dataset into recognizable categories can provide a significant degree of insight into the structure and meaning of the data.

1.5.5   Recommendations

Recommendations, generated from algorithms called recommendation engines, are a form of prediction. Fundamentally, a *recommendation engine* is a prediction engine that appraises and estimates people's opinions and preferences, answering the question **which option should be taken?** Chapter 7 explains the fundamentals of modern recommendation engines.

If you have ever utilized a search engine on the Internet then you have used a recommendation engine. In this case, a search engine's job is to predict the most preferred websites based on a few keywords.

For instance, if you simply entered "cat" into a search engine it has to assess if you would like to know more about the small mammal many people have as a pet, the UNIX operating system command "cat" (as in concatenate), a Broadway play named "Cats," or the company Caterpillar that produces construction equipment with stock symbol CAT. There are also dozens of other related but less widely-known concepts that "cat" could represent.

In a way, the software must look into a proverbial crystal ball and parse the dozens of potential meanings for the target keywords entered into the search field.

Recommendation engines are also closely connected with online stores. Virtually all major online retailers now have recommendation engines that assist the user/consumer. Familiar online shopping pitches such as "other people that bought this product also bought…" originate from recommendation engines.

There are many different types of recommendation engines but they all have the same core objective: Which option among many should be recommended to ensure the highest probability that the user will remain engaged? A recommendation engine adheres to the same operational philosophy as click through rate (CTR) discussed above.

1.5.6   Data Science Project Examples

There are many different data science projects that can be pursued and investigated. We provide the following brief list to help you understand the focus for many professionals in this discipline:

- We are losing approximately 5% of our customers per month. Why?
- What is the real cost, which involves all the parts of the company, of complying with a new regulation?
- Where should we set up another store, ATM, new distribution center, etc.?
- There is a cost associated with hiring and training new employees. It seems that once employees are trained sufficiently they tend to leave for higher salaries.
  - When do most of these employees leave? Why?
  - How can we increase loyalty to our company and retain them?
  - What type of employee, based on background, leaves most frequently? Can we pre-screen for this type of employee and not hire them in the first place?
- If we increase/decrease the quality of our product and also adjust our selling price higher/lower, how will that affect our total sales?
- Based on our products and our most loyal customers, is our marketing team targeting and offering our products to the right people?

The following is also a general list of questions that you might ask for any given data science project to ensure that the goals of the project are business oriented:

- Does the project support the core business or is it peripheral to the business mission?
- Why are we doing this?
- Can I explain the situation and context to a non-expert?
- Do I understand the business requirements to grow the company?
- Are my models shiny and fast or goal-oriented for the business?
- What are the concrete and actionable results?
- Is there an enthusiastic champion in upper management who will act on the insights provided by our results?
- Will the stakeholders accept the change(s) we propose?
- Once the project is finished will we have subject matter experts available to support the finished product or results?

There is an abundance of data science projects any company can undertake. Once the project is identified, the key question then becomes, **should you do it?** If the results of the effort ultimately build business value then you generally should pursue the project. However, it is very easy to become consumed and distracted by the exciting peripherals of the data science milieu which could result in costing your organization money with no appreciable, measurable return.

1.6   Understanding vs Specific Tools

Understanding the essential concepts of data science is more important than learning a specific tool.

For example, Python is currently the most widely used programming language in data science followed by R and VBA (the programming language used in Microsoft Excel). In Sect. 3.3 we explain how to use dataframes with Pandas, currently a popular library that is employed extensively by many data scientists. The Pandas tool is very useful and we have many examples demonstrating how this library will save you time with statistics in Chaps. 4 and 5 as well as with time series analysis in Chap. 10.

However, if you only learn how to use the **tool** then you will be doing yourself a great disservice. For example, in Chap. 10 it is much more important to learn the concepts related to smoothing and moving averages than to learn how to perform these operations specifically in Pandas.

We have found that it is relatively easy to learn a novel way of writing and implementing software if you have the core foundational programming knowledge. Twenty years from now most people may have moved away from Python to another programming language that does not yet exist. If you only know how to accomplish your tasks based on rote memorized functions from Python and the Pandas library, then you will have a hard time transitioning to the new language. Alternatively, if you have focused on the core foundational programming knowledge, then learning how to calculate moving averages in the new language will be a trivial adaptation in the future.

In summary, if you focus on the key principles of data science then you will soon discover that they never change. The current tools, such as Microsoft Excel, Python, and Pandas, will change, but not the principles upon which they are founded. Focus on the principles and you will always have a job. Focus only on the tools and you will find that you may be left behind relentlessly trying to catch up with the latest technology.

1.7   Data Science Life Cycle

As with any established methodology, data science has a predominant life cycle.

The beginning of every data science project defines the question that you hope to answer. This is both the most important and likely the most difficult step. Knowing what question to ask typically involves experience and domain-specific knowledge.

Once the question is identified and validated you then need to gather the data. Gathering the data comes in many forms and is explained in detail in Chap. 2. Whether you get data directly from surveys or indirectly from databases or downloads, you cannot do any type of analysis without the data. As in all research projects, data may be quantitative, qualitative, or a combination of both.

However, once you have the data you may discover that your guiding question is no longer relevant and needs to be modified. Realizing that your question requires revision because of the data under investigation is often as important as coming up with the initial question itself.

The process of data exploration is paramount in the life cycle. You must explore the data to understand the meaning and purpose embodied by this information. However, the data is often not clean. *Clean data* is data that is ready to be used immediately for exploration and analysis. To clean your data you will often need to wrangle, transform, or reconfigure your data, which is the topic of Chap. 3.

Once the data has been wrangled into a usable or clean form then you are ready to engage in a full exploratory analysis to achieve greater insights about the data. We describe the statistics necessary to examine our data in Chaps. 4 and 5.

Part of the exploration process is to visualize the data. *Data visualization* is the vital step of displaying or viewing the data in a form that enables a more comprehensive understanding of the context and trends inherent in the data science problem. In data science, the phrase "A picture is worth a thousand words" may not be accurate. Sometimes a picture or visualization provides instantaneous insight so much greater than simply looking at textual or numerical data that it may easily be worth far more than a thousand words.

Visualization is such an important aspect of the process that it is intertwined throughout the discussions in this book. You will come upon different methods of visualizations frequently throughout this text. Reading a data visualization book will be well worth your time.

You will find that several parts of the life cycle are iterative. Often the process of transforming or wrangling the data followed by visualization will lead to additional questions that bring about additional transformations and visualizations. This part of the data science life cycle is aptly described as "not knowing what you don't know." In other words, many questions arise from exploring and understanding the data that could not be conceived before the process started.

Data reduction is integral to the process. *Data reduction* is the transformation of the data from one form to another for the purposes of simplifying and decreasing the volume of data while simultaneously preserving the essential informational content. For example, given a date timestamp that includes the year, month, day, minute, and second of an event, it may be sufficient to reduce these data to simply the month when identifying quarterly business trends.

The end product of any data science life cycle is either a model that can be used for prediction, a presentation to clarify the meaning and future research of the data, or both. For example, you might develop a recommendation engine for your organization. After completing the software a presentation of the results to stakeholders and interested people will likely follow.

As emphasized above, the presentation is about telling the story of the data, about bringing the data to life, then sharing actionable insights with the audience. No matter how much behind-the-scenes work you put into your project, properly communicating its impact and influence is vitally important.

1.8   Python vs R

Python and R are both the most common programming languages used in data science and often listed as the top languages in job postings. R followed SAS (an older programming language) and emerged initially as the language utilized primarily for statistics and ML (Machine Learning). R is rarely used outside of statistics and data science. R is generally easier to learn for people that have no previous experience with programming languages.

Python followed later and is a general programming language that is one of the most popular in the world. Python is considered easier to learn for anyone that already has a background in programming languages. Also, Python is often one of the first programming languages introduced in university computer science programs.

All the code examples in this book are written in Python for one simple reason: according to years of extensive surveys and studies, Python is gradually gaining popularity over R in data science circles. In a nutshell, the older a person is the more likely they are to use R in industry, unless they are even older and use SAS. Conversely, the younger a person is the more likely they are to use Python in industry. Of course, familiarity with both Python and R provides advantages for individuals engaged in data science.

Industry is a term intended to mean working environments outside of a conventional educational setting. For example, employment or consulting with Apple, Google, or Walmart would be an example of working in industry. If a person does not work in industry, then they likely work in *Academia* or affiliated research institutions. Academia is a term intended to mean the workplace is predominantly an educational or research-oriented setting, such as a university, college, policy think tank, or FFRDC (federally funded research and development center).

1.9   Big Data, Data Analytics, and Data Science

Big data, data analytics, and data science are all terms used interchangeably by some and more precisely by others.

In our experience, universities and professors are more careful about defining the differences and overlapping concepts between the three terms. In contrast, employers of data scientists in the field usually are not overly concerned about these differences. People in industry usually care more about obtaining actionable results to their data science questions to make their organizations more profitable and impactful. They usually are not concerned about whether a "data analyst" or "data scientist" performs these tasks.

Consider the scenario in which you are being interviewed for a position and the interviewer asks if you can perform predictive forecasting work with linear regressions. You can safely assume the interviewer does not really care if you call yourself a "data analyst," a "data scientist," or even a "pink balloon." (Although they might be confused about why you call yourself a "pink balloon." However, some companies appreciate having eccentric employees, so that strategy might get you hired.)

In this book we distinguish between a data analyst and a data scientist by examining the level of involvement during the lifespan of the project. Data analysts typically identify trends and patterns, create visualizations, and help people make sense of the existing datasets to help them make decisions. These professionals focus on the things we know that we don't know.

On the other hand, the data scientist typically considers the larger context of the existing problem. This includes culling out new problems and creating additional projects with specific and measurable questions, gathering data that will help answer these questions, analyzing data (which includes identifying trends and patterns, and creating visualizations), writing software for new software prototypes, and providing presentations. Data scientists are generally interested in the things we don't know that we don't know.

However, identifying the differences between the fields of data analytics and data science is very difficult because few people agree on exactly what each area encompasses. In practical terms, instead of arguing semantics it does not matter what the difference is between data analytics and data science because what is more important is generating the actionable insights described earlier that help your organization become profitable.

Big data is more of a marketing term than a scientific term, similar in spirit to how non-technical people talk about "cloud computing" instead of the more technical term, "distributed computing." "Cloud computing" conjures up ideas of computing taking place mysteriously and magically in the ether while "distributed computing," although more technically accurate, sometimes confuses non-technical people. After all, what does "distributed" suggest to most people?

Big data is a term fundamentally implying that to make sense of the extremely large volumes of data currently available for consumption because of technological advances, visualizations and algorithms are necessary. In other words, to understand the vast amounts of data at our fingertips professionals should apply tools and techniques such as transformations, statistics, visualizations, and many other analytical algorithms to these data to discover relevant patterns and insights.

Many academics refer to the *"four V's" of big data*: volume, variety, velocity, and veracity.

Volume simply refers to the size of your data, with the commonly used modifier "big" implying an inordinate, unmanageable amount. However, in practical relative terms, how much is a lot? For some organizations big data is any data set that is larger than a gigabyte (10^9 bytes). For some, it is greater than a terabyte (10^{12} bytes). For others it might be the next level of a petabyte (10^{15} bytes). The amount of volume or size of a dataset will vary from project to project.

Variety means that the data is packaged in various formats. Your data might be clean and easy to use and derives directly from a database, or it might be unstructured and fragmented, meaning that you derive it from a variety of sources. For example, you might have a project that includes sources of data from video, sound, and text. In contrast, your data might be only stock symbols and numbers. The exact type of data that you use can be drastically different from project to project or it might always be sales data year over year. The data source and structure depend on the project and the kind of data your organization uses.

Velocity is the frequency at which new data is generated. For example, is your updated weather data available at 5-minute increments or daily increments? When analyzing stock data, are you only looking at historical data? For this particular case, the data never changes. However, if you are performing *real-time* stock analysis you might have to work with stock data that changes every second.

Veracity is the level of trust you place in your data. How much do you trust your data? For example, if you are using census data do you blindly accept the precision of all the information that you obtain from the government? Putting aside any conspiracy theories about the government and the census, can we expect the data to be 100% accurate? Tens of thousands of people work on every census. Did any of those hard-working people make a mistake? To accept the census data as 100% accurate is likely both naïve and wrong. We should assume people are fallible and make mistakes, and simple logic compels us to conclude we cannot abide 100% certainty for most data sets.

From a practical standpoint, "big data" is simply data that is too large for meaningful insights to be gained without methodically applying algorithms and expert analysis. To reduce this concept even further, big data is the name given to the datasets you examine in your projects as a data scientist.

Exercises

1. What is the difference between insight and actionable insight?
2. Explain how using stories in presentations can engage your audience.
3. What is the importance of explaining complex things in simple ways?
4. Why do KPI's matter to a data scientist?
5. In your own words, what is a data scientist? How does that differ from a computer scientist, mathematician, or statistician?
6. Is detecting credit card fraud an example of classification, anomaly detection, or both?
7. What is the point of clustering?
8. How does the size of your data impact your analysis?
9. There are many types of recommendation engines. However, it often appears that search engines are different from online retail stores. Compare and contrast general purpose search engines to recommendation engines utilized by online retail stores.

Chapter 2
Data Collection

***

Although there may be many points of contention regarding data science, one thing universally understood about the field is that the practice of data science is impossible without data. You may be given data, or you may have to go out and collect it. The bottom line is that any data science project begins with data.

As discussed in the previous chapter, the difference between a "data scientist" versus a "data analyst" is confusing at best. However, some people would claim that the main difference is a "data analyst" primarily analyzes pre-packaged data whereas a "data scientist" needs to identify, seek out, and gather the data as an important first step.

There are generally two ways to acquire the data of interest:

- Data Creation – Generate raw observations
- Data Gathering – Obtain it from an accessible source

2.1 Data Creation

Data creation is by far the harder approach when compared to obtaining the data from an immediately available and accessible resource.

There are many fields of study and research that collect data. To simplify, we will separate these fields into two types: those that collect information involving people and those that collect information about everything else in the universe.

The following fields are a sample of the many disciplines that collect information about people: psychology, anthropology, sociology, economics, political science, and human-computer interaction (HCI). These fields focus on the study of human behavior. For example, psychology researches the human mind, economics studies how people manage money, and HCI studies how people interact with computers and machines. These fields are interested in specific aspects of human nature and collect information about the motivations and theories that sufficiently describe human **behavior**.

There are other fields that also produce data based on measuring and representing human **activity**. For example, literature, art, theater, and sports, all produce their own data, but are different from the other fields in the kind of data they produce.

There are many other fields that are interested in the natural world and are not specifically concerned with large scale human systems and behavior, such as physics, biology, meteorology, geology, seismology and chemistry. These fields are often more concerned with **natural processes**, such as how forces interact, how weather works, and how the earth is formed and evolves.

The type of data you acquire from different fields will need to be studied and analyzed in different ways. For example, a sociology project analyzing migration patterns of immigrants will necessarily involve time series data, which we discuss in

R. Ball, B. Rague, *The Beginner's Guide to Data Science*, https://doi.org/10.1007/978-3-031-07865-1_2

Chap. 10. Other examples of time series data are analyzing the progression of stock market data, population trends, and election results based on year.

However, analyzing other kinds of data such as literature requires a completely different approach. Analyzing literature will most likely include natural language processing (NLP), a methodology we discuss in Chap. 9. Other examples of natural language processing include analyzing blogs, Twitter data, and emails.

Other data applications, like analyzing mineral deposits or predicting weather patterns, also demand different approaches based on the data. So, it is important to understand what kind of data you have before settling on the proper analysis strategy. Possessing or accessing some context and domain-specific knowledge (i.e., the kind of business you are dealing with) is vital to achieving any meaningful actionable insight.

Context is the area, domain, or situation that explains the data. For example, is it major league baseball data? Weather data? Astronomy data?

Unless you are an expert in a particular field then you most likely will not be gathering your own data. If you are interested in examining climate change then you will likely not start collecting your own dataset of barometric pressure, wind speeds, and precipitation amounts. You will probably work with experts in the area that have already completed some level of data acquisition from established weather stations around the world or at least have easy access to the data from these weather stations.

Examples of data creation include measuring interesting events, like how much rain fell yesterday, how much stress can be put on a pipe before leaks occur, how fast people memorize parts in a play, and other observable phenomena. Anything that is measurable and recorded can be captured as data.

2.2 IRB Approval

Generally, whenever you acquire data related to observations of **people** then you will be expected to secure IRB approval. The *IRB (Institutional Review Board)* is a group of people (or board) that evaluates the ethics of data collection involving people as research subjects. Any type of direct data retrieval from individuals, whether in the form of surveys, psychology tests, blood tests, or other means must be carefully examined and approved by the IRB associated with your organization **before** you begin your data collection. Even surveys, which may appear harmless on the surface but still are designed to extract personal information or opinions, must be approved.

The point of IRB approval is to protect people in the role of research subjects. The IRB is intended to protect people's privacy, their well-being, their identity, and their lives. The origins of the IRB dates back to when psychologists would apply electric shocks to people during experiments. The Stanford Prison experiment is a famous example of how a seemingly simple experiment conducted at Stanford University in 1971 got out of hand. In the experimental design, student volunteers were assigned roles as either fictional prisoners or fictional guards. The purpose and tenor of the experiment were derailed badly when the "guards" took their premise of authority too far. This famous experiment and other similar precarious research investigations led to the federal requirement that studies involving people need to be evaluated and approved by an objective oversight board prior to any experiment taking place.

Your organization will either have its own Institutional Review Board or will have defined a clear and consistent method to obtain approval from a review board that is external to your organization.

However, if you are collecting data that does not directly or indirectly affect people then you do not need to seek IRB approval. Some examples of this type of data collection include measuring the amount of rain, monitoring wind speed, or recording the speed of birds. In some instances, conducting analysis on anonymized data about people may not require IRB approval. However, we strongly recommend that you contact your IRB representative before conducting any type of analysis that involves the use of people as research subjects.

2.3 HCI: A Case Study

The myriad of details of data creation are beyond the scope of this book. However, you may find that reading a book on a particular field of study will be sufficient to enhance your knowledge of how best to accomplish your analysis. In the specific case of HCI (Human-Computer Interaction: the field where people are studied on how they interact with computers), you may find you need to gather additional information about how people navigate a particular website or piece of software to acquire a full understanding of the domain.

The following is a quick case study on how you might create your own data. Suppose that your project involved the question of understanding how the entries in Wikipedia affect political outcomes. You might be pleasantly surprised to find that many data points particular to your project have already been collected. For example, you might be primarily interested in a single election for a given year and all voting results are currently available. You can then cross reference that data with polling data and other demographic information. Additionally, you could use natural language processing (Chap. 9) to understand the Wikipedia page bias (or non-bias) with respect to the political candidates.

However, to completely understand the data, you might decide that you need additional insight into how people actually create and edit Wikipedia articles. We will not outline every method possible to create this data, but only highlight some of the more common approaches. The following are widely used methods to gather data in the field of HCI:

- Surveys
- Focus groups
- Lab-controlled experiments.
- Logs and diaries
- Longitudinal studies

After filling out the required IRB forms and following all subsequent IRB procedures you might conduct a survey on people you have identified as editors of Wikipedia pages. A *survey* is a type of questionnaire that targets a particular topic. Surveys are usually inexpensive and generally do not take a significant amount of time. However, response rates may be very low and how you advertise and promote your survey will directly influence the number of responses.

You could employ a focus group that convenes people who edit Wikipedia pages. A *focus group* is a form of group interview. The interaction with a focus group typically involves asking open-ended questions to the group and carefully observing the overall responses and reactions. Focus groups are an excellent method for acquiring more explicit details from individuals which may not have been otherwise obtained through surveys. However, focus groups require a strong leader (i.e. the researcher) who can guide the group discussion back to the main research objectives without getting too distracted by tangential topics.

Lab-controlled experiments are also common and involve inviting individuals, typically one at a time, into a laboratory setting where researchers observe the actions, behaviors, and original creations of the human subject. For example, the researcher might have the participant edit a Wikipedia page by performing a cognitive walkthrough. A *cognitive walkthrough* is an activity in which a participant speaks aloud or verbalizes what they are thinking. For instance, the participant might say, "Well, I usually start editing a Wikipedia page by first logging in. I do that by going to this page then clicking this button. Now, I ..."

Other common lab-controlled experiments in HCI include usability tests to determine the difficulty level for participants to complete a software task. During such tasks, participants are often observed using one-way mirrors, videotaped for later scrutiny, and/or their screen activity is recorded.

Logs and *diaries* are a form of writing where the participants write down their actions and events independently. They are valuable sources of information because they reveal what participants think when performing a task. This is a form of self-reporting and has been found to be valuable in many research studies. Note that HCI researchers have discovered that what people do and what they think they do are sometimes strikingly different. When human subjects document in written form what they think they are doing it is often very enlightening to both the participant and researcher.

Longitudinal studies are similar to lab-controlled studies except they monitor the participant over a longer period of time. In this case, instead of focusing on how a participant edits a Wikipedia article in a single session, a longitudinal study examines how a participant varies their editing strategies over time.

Although you may never be directly involved in HCI research, you may find that you require additional training to be able to successfully gather the data necessary to complete your analysis. In this case you or your organization may have to hire a specialist in the field or instead devote the time and effort to become sufficiently familiar with the domain such that you can gather the data yourself.

2.4   Data Gathering

Obtaining data from an existing source is typically much easier than creating your own data. This process may involve querying a database, downloading data from a website, or even acquiring an email attachment from a colleague with the data you need to analyze.

Assembling the appropriate dataset could also involve downloading data that is publicly available, for example, from the census bureau, CDC (Center for Disease Control and Prevention) or other data collection resource.

Gathering data may also be part of the ETL process. The *ETL (Extract, Transform, Load)* process involves extracting relevant information, often from multiple sources, transforming it (see Chap. 3 for more details), which most likely involves cleaning the data, and transforming or changing the data to fit into particular databases.

2.5   Databases

Your organization may already have the data you need. If so, this information is often stored in some version of database.

Invariably, as a data scientist, you will use databases at some point in your career. You may not need to be an expert on databases, but you will have to possess the skillsets that allow you to extract the data from these databases. Being able to navigate and utilize a database is one of the top requirements for most data scientist positions. Even data originating from various sources will inevitably end up in a database before you begin analysis.

Databases cover a wide spectrum from traditional relational databases to distributed databases such as Hadoop and Spark. A *relational database* is a highly structured database that stores data in tables. Each table comprises rows of entries, much like a spreadsheet. The different tables are then linked or related to each other to provide an environment that allows users to maintain large amounts of data with very fast insert and access speeds. The organizational design by which tables are interconnected is called a *schema*. Relational databases are currently the most common type of databases and reside on servers worldwide.

A relational database uses SQL to input and access data. *SQL (Structured Query Language)* is a declarative query language and allows you to programmatically access and control relational databases. There are many dialects of SQL (e.g., Oracle vs Microsoft SQL), but once you become expert in one then you can easily learn the syntax of another.

A *distributed database* or *cluster computing* is the concept of integrating multiple computers instead of one computer to store and process the data. Whereas a relational database is usually found on a single computer server, distributed databases like Hadoop and Spark may be used to harness the processing power of thousands of computers. However, unlike a relational database, a distributed database is unstructured, meaning that specific rules like tables and schemas do not always apply.

Hadoop leverages *NoSQL ("Not Only SQL")* distributed databases, which are unstructured in nature, trading off strict consistency requirements for speed and agility. NoSQL is based on the mechanism of distributed databases, where unstructured data may be stored across multiple processing nodes (an individual computer), and often across multiple servers. However, there are many addons for Hadoop and Spark such as HBase that allow you to use large tables like a relational database on top of the extensive multi-computer infrastructure.

Distributed databases are required when you have data sufficiently large that it does not fit on a single computer. For example, you might use Hadoop when your dataset is a size greater than a traditional database can handle, for example on the scale of terabytes currently. Distributed databases represent a form of cloud computing in the sense these data stores might be hidden from the user in that they are not fully aware how many computers constitute the distributed database.

In addition, databases vary dramatically in size. There are single-file databases (e.g., SQLite), enterprise-level databases that employ one or more servers devoted to the database (e.g., Oracle's database or Microsoft's SQL Server), or distributed databases that use dozens, hundreds, or even thousands of distributed computers (e.g., Hadoop).

As a data scientist you will have many reasons to use databases. You may be responsible for inserting data into a database, accessing data, merging data, or many other data-related duties.

Effectively utilizing a database is one of the top requirements for most data scientist positions. Even if you obtain data from various sources, the data will invariably end up in a database before you begin analysis.

Due to the extensive diversity of databases and their many idiosyncrasies and applications we will not explain database usage and operation in this book. There are many available resources, online tutorials, and documentation that you can find that will assist you along your journey.

However, with that said, as a data scientist it is vital to learn how to navigate databases. Basic database skills are taught in virtually all data science programs and are generally expected in data-related jobs.

A data warehouse is another tool you may encounter during your career. A *data warehouse* is a system that accumulates data from different sources within an organization for reporting and analysis purposes. A data warehouse stores historical data about your organization so you can gain insights. In general, a data warehouse does not store current information nor is it updated in real-time.

There are many tools associated with data warehousing. For example, Tableau is a powerful software system that allows the user to drag and drop queries for complex visualizations and reports. SAS is another widespread data management tool that incorporates powerful analytics and visualizations for effective reporting. There are many other reporting tools designed for businesses that also provide intuitive interfaces for complex reporting and visualizations.

2.6 Downloading Data

Data that does not currently exist in your organization's databases will need to be acquired. There are many cases when data are compiled by another organization and made available for convenient download.

Unfortunately, the data that you seek may reside across multiple locations. For example, a list of donors to politicians will provide some partial information about these supporters. You may need to obtain additional data about these donors, which are also available, but this process of gathering relevant data may now require merging the new dataset with the original dataset. This scenario lends importance to the techniques described in Chap. 3 which focus on wrangling the data into consumable chunks to make some sense of it.

There are vast amounts of information available for analysis. For example, the US Census Bureau has massive datasets that you can download at www.census.gov/data/datasets.html. This resource includes information about the population's race, ethnicity, religion, and age. The following is a short list of publicly available datasets to illustrate our point:

- The NOAA (National Oceanic and Atmospheric Administration) https://www.ncdc.noaa.gov/cdo-web/datasets
- The CDC (Center Disease Control and Prevention) https://www.cdc.gov/nchs/data_access/ftp_data.htm
- Maps: https://catalog.data.gov/dataset/u-s-state-boundaries
- UN (United Nations) world population data: https://population.un.org/wpp/Download/Standard/Population/
- National Snow & Ice Data Center: https://nsidc.org/data/smmr_ssmi
- NASA Planetary Data System: https://pds.nasa.gov/

In addition, many state and municipal data are also available. For instance, most counties in the US post the residential and commercial properties that are in arrears (delinquent in property tax payments). Although the period of delinquency varies from state to state if a property tax has not been paid within a certain amount of time (e.g., three years in Arizona, five years in Utah) then the county will foreclose on the owners and auction those properties off to the highest bidder.

In this situation, if you were working for an investment organization interested in identifying the best current tax auction possibilities in the country, you would likely want to start by downloading all the data from each county. According to the US Census,[1] there are 3,006 counties in the United States. Although downloading the list of delinquent tax properties might take a few weeks, this data still would not provide the complete story.

For every property of interest, you would need to merge its parcel number (the technical name for any property in a county) with its actual GIS boundary information and additional metadata about the type of property (e.g., residential, commercial, industrial), determine the worth of the property, determine if there are additional liens on the property, and so on. In other words, finding all the relevant information for your analysis begins with finding the delinquent properties then merging all the additional, supporting information you need into one consumable data file or database.

A similar data gathering example would be investing in a government guaranteed instrument called tax lien certificates. Tax lien certificates are only available in a few states and involve an investor paying one-year delinquent property taxes instead of the property owner. The investor pays the delinquent taxes and in return the property owner pays the investor (through the county clerk) the delinquent taxes with interest (15–18%, depending on the state). If the property owner does not pay the investor the delinquent taxes, then after a set amount of time (e.g., 3 years in Arizona, 4 years in Wyoming), the investor acquires the property (although in a few states the property will go up for auction).

Under these circumstances, your organization may be trying to determine which areas in the country have the best return on redeemable tax lien certificates. Regardless of the state and county from which you choose to acquire the data, you will be able to download a list of all tax delinquent properties for that county. However, only knowing where these properties are located will likely not be enough to allow for an informed decision. You probably only want to buy properties either zoned residential or zoned commercial (depending on your investment strategy). Unfortunately, that information is not included in

---

[1] https://www2.census.gov/geo/pdfs/reference/GARM/Ch4GARM.pdf

the downloadable information from the county treasurer's office where you obtained the list of available tax lien certificates for sale.

Augmenting your data with additional zoning data is essential before you can reasonably assess which state offers the best investment opportunity. Determining how a property is zoned may take a long time if you do it manually. For some perspective, there are tens of thousands of delinquent property taxes in Maricopa County, Arizona (the Phoenix metropolitan area) every year. Manually determining how each property is zoned is an insurmountable task. Considering that the list of delinquent properties is published only three weeks before posting the tax lien certificates for auction requires this task to be automated to adequately address the time crunch. As emphasized above, determining the zoning for tens of thousands of properties will take a long time by hand.

So, if the county has a website that lists how a property is zoned then you could augment your original downloaded dataset by automatically entering the property's address into the website and retrieving the appropriate zoning classification.

Lastly, we want to mention that even though there are troves of publicly available data ready for download, you may occasionally come across a paywall where a fee is required to download the target data. In that case, consumption cost must be figured into the analysis effort when compared to simply downloading the data for free.

2.7   Web Scraping

Clearly, the scenario in which the complete dataset is available by simply visiting a website, clicking a link, and performing a download is highly preferred. However, often the data are not always so easily accessible. Many times, the data are present at a website, but not in a conveniently downloadable format.

For example, you may be interested in obtaining a list of all the residential houses currently for sale in a given area. That specific collection of information may be freely available at realtor websites (e.g., realtor.com, zillow.com, mls.com), but typically those websites do not make the data available as a single download. In this case, you may have to resort to web scraping, one of the most common information gathering tasks for any data scientist.

Web scraping is the activity of gathering unstructured data from a webpage or webpages and transforming that data into a consumable form. A variety of job advertisements for prospective data scientists list web scraping as a common skill requirement for many applicants. The results from web scraping over 100,000 recipes and their accompanying reviews (over 4 million reviews) from the Internet to make recommendation engines are utilized in Chap. 7.

Before we look at code examples that demonstrate how to perform web scraping, we need to answer two important questions:

1. Why web scraping?
2. What does it really mean to perform web scraping?

2.8   Why Web Scraping?

The main reason we apply web scraping is to confront and address the common situation in which data are available somewhere on the Internet, but not always in an easily downloadable form. Alternatively, you may be able to download the data you want, but the data are woefully incomplete. For example, in Sect. 2.6 we shared several useful links to publicly available data. However, that data may only represent the initial samples of the full dataset, or it might not be available to download.

A simple example of web scraping is copying information from a browser and pasting the data into a document. As an example, assume you were interested in determining the 30 companies that constitute the Dow Industrial Average (a stock market index) and you could not easily locate a file that contained this list of information to download. In this case, you could visit a website that displays the requisite information in a table (e.g., Wikipedia), highlight and copy the table, then paste it to your document.

Thirty companies are relatively easy to copy and paste. However, the Russell 5000 is another stock market index which contains 5000 companies. Most websites that include this index list the 5000 companies across twenty-six individual webpages labeled A - Z. Gathering the complete dataset for this index is tedious at best because you would need to visit twenty-six pages and perform a copy and paste for each page. Nevertheless, this approach may still be considered viable.

However, many working data scientists would not regard the repetitive action of manually copying and pasting information from a website to a separate document to accurately represent "web scraping." Since the web scraping task usually involves a significant transfer of data it is much more efficient to write a script or program that automatically extracts the information for you.

For example, as part of your efforts to obtain the list of residential homes for sale, you visit one of the above realtor websites (e.g., realtor.com) and enter in your search criteria. Let's say you are interested in all the residential houses for sale in the greater New York City area. After submitting your specifications, the realtor website shows there are tens of thousands of results. Writing down the results by hand or even copying and pasting all the information into a spreadsheet is out of the question, so you create a script, or small program, to programmatically scan each page of your results and store the target information in your database or files.

To reiterate one of the key points mentioned above, the purpose of web scraping is to acquire data available on websites but not currently accessible in a single downloadable form. You would then proceed to extract that information – web scrape the information – from the website. Once you have all the data then you can apply appropriate analysis and make intelligent actionable insights based on that analysis. Whether you web scrape to obtain the complete dataset or to augment existing data, the process would be the same. Chapter 3 addresses how to efficiently merge data from different sources, but now we answer the second important question.

2.9   What Does It Really Mean to Perform Web Scraping?

The fundamentals of web scraping can be characterized in two steps:

1. Download a webpage
2. Parse the webpage

2.9.1   Download a Webpage

Downloading a webpage is nothing more than obtaining the HTML, CSS, and JavaScript components that are the building blocks of a modern webpage. As mentioned earlier, one of our favorite websites is data.census.gov. If you navigate to data.census.gov using a modern web browser then the HTML, CSS, and JavaScript elements will be rendered visually to you by that browser. You will most likely see images, text located and formatted in certain ways, a variety of colors, and a selection of interactive buttons. Also, if the URL takes you to a commercial website then you should expect to see advertisements.

However, if you want to view the raw source code the browser processes, then you can use your browser to display this source. For example, with Firefox or Google Chrome (or any other major browser) you can readily view the raw HTML, CSS, and JavaScript components by right-clicking on the webpage and selecting "View Page Source" or something similar. (Please note that software is constantly changing and there may in fact be different steps required to view the source. However, these steps should be similar to the ones described above and easily determined.)

Please note that it is beyond the scope of this book to explain the design principles and coding techniques associated with web programming (HTML, CSS, and JavaScript). However, there are many good books available that can teach you the basics of this important discipline.

Even so, we have great news! You only need a rudimentary knowledge of web programming to succeed in the endeavor of web scraping. If you understand the most basic concept that all programming and markup languages comprise a series of characters called strings, then the essentials of web scraping reduce to downloading a website's source and then applying a series of string comparisons and string manipulations (parsing) to the acquired source code.

An additional interesting thing to know is you can access and download a webpage *without a* browser. For example, wget is an open-source utility program often associated with the Linux operating system that allows for the direct download of a single webpage or even entire websites.

To illustrate web scraping, we will focus on capturing a random recipe. Due to the ever-changing nature of the Internet we have created an example HTML page for this chapter named *example_recipe_page.html* which can be found with the files for this chapter. Our original example from an existing "live" site on the Internet was altered before we completed the book so we created this example page for consistency.

If the webpage resided on a server then instead of employing our browser, we could first use wget to download the webpage. wget is a powerful utility and has many useful features, such as generating mirrors (exact duplicates) of websites, converting links, and using proxies. The official documentation for wget is located at: https://www.gnu.org/software/wget/manual/wget.html.

To use wget, it must first be installed on your computer. The wget software usually comes preinstalled with most Linux distributions, but will typically need to be installed separately on Windows and Macs. However, before installing wget please finish reading this section where we also demonstrate how to achieve similar web scraping functionality in Python.

We could download a webpage using wget from the command prompt as follows:

```
wget "https://icarus.cs.weber.edu/~rball/book_files/example_recipe_page.html"
```

Although the above action performs a download operation, there may be instances in which the incorrect webpage is retrieved. Sometimes wget returns different results when compared to a browser. The reason is that some websites attempt to rebuff direct data access attempts by employing defensive strategies that make it more difficult to web scrape the website's information.

To guarantee the webpage source code is identical to the code we would retrieve from a browser we instruct wget to simulate a web browser. To accomplish this, we provide something known as a user agent. Every browser has an associated *user agent* – a software identification mechanism and client intermediary that web servers use to communicate appropriate responses. For example, "Mozilla/5.0 (Macintosh; Intel Mac OS X 10.15; rv:84.0) Gecko/20100101 Firefox/84.0" is a user agent for Firefox version 84 used on a Mac.

A second important consideration is that websites often have downloading restrictions specified in a file called "robot.txt" and wget normally adheres to these customized access rules. Since in this case we are only trying to show an example of how to download a website without a browser, we will instruct wget *not* to follow the rules in the "robot.txt" file by passing in the "--execute robots=off" option. However, in general we encourage you to follow the rules included in the "robot.txt" files whenever possible and ignore these restrictions only on very rare occasions.

The ethics of web scraping can be complicated at best. However, we encourage you to be a good net citizen, or netizen, by carefully following website guidelines. If you do not abide by website regulations, then your IP address may end up blacklisted by the website resource. To be *blacklisted* means that the server where the webpage is found may refuse to respond to your specific IP address. More to the point, if you abuse a website's guidelines as outlined in the "robot.txt" file you may discover that the website has prohibited you from any future content download attempts from their servers which may include resources beyond the single website of current interest.

Adding a user agent and setting the option *not* to follow the rules in the "robot.txt" file is accomplished by entering the following command:

```
wget   -e   robots=off   --user-agent="Mozilla/5.0   (Macintosh;   Intel   Mac   OS   X   10.15;
rv:84.0) Gecko/20100101 Firefox/84.0"
"https://icarus.cs.weber.edu/~rball/book_files/example_recipe_page.html"
```

The resulting downloaded file is an html file, but it will not be stored on your machine with the appropriate file extension. To view the file on your computer using a browser, you may have to add the ".html" extension to the downloaded file. Once the command successfully executes, you should now have a local copy (a file on your computer instead of the web) of the target web page. By navigating to the website location and viewing the page source you will discover that the local file is an exact copy of the page resident on the website.

Downloading websites can be done in a similar fashion using Python. The following is Python code that performs the same operation as `wget` described above but instead saves the file as "output.html":

```
File: retreive_html_page_example1.py:
import requests
link = 'https://icarus.cs.weber.edu/~rball/book_files/example_recipe_page.html'
response = requests.get(link, headers={'User-Agent':'Mozilla/5.0 (Macintosh;
Intel Mac OS X 10.15; rv:84.0) Gecko/20100101 Firefox/84.0'}) # get the webpage
    with open("output.html", "w") as f:  # save the response as a file
        f.write(response.text)
```

2.9.2   *Parse the Webpage*

Now that you know how to retrieve a copy of the raw web page source to your machine with either Python or wget it is time to extract the information of interest. The following Python program is identical to the prior code section above, except it displays the raw source for viewing by using the print statement:

```
File: retreive_html_page_example2.py:
import requests
link = 'https://www.website-name.com/website.html'
response = requests.get(link, headers={'User-Agent':'Mozilla/5.0 (Macintosh;
Intel Mac OS X 10.15; rv:84.0) Gecko/20100101 Firefox/84.0'}) # get the webpage
print(response.text)
```

The code produces a large amount of text – the entire contents of the raw website. You may initially know very little about the contents of a webpage, so how do you effectively extract data from this resource? The good news is that prior knowledge about the source of a webpage is not necessary to parse the data on the page, although a little background about the information and its organization would certainly be an advantage. Distilling the targeted information from the page source really only requires knowledge about how to parse strings.

The first thing we will do is try to make the website content more manageable by arranging the source code into lines separated by a carriage return, which in Python is expressed as the single escape character: '\n.'

To extract the ingredients from the website, we will introduce a few basic Python string manipulation functions. The first one is the *split* function.

The *split* function will divide a string into a list based on the separator character provided as the input parameter. For example, if a string is defined as "1-2-3" and you want to split the string to retrieve only the numbers contained in the string then the following Python code can be applied:

```
line = "1-2-3"
print(line.split('-'))

['1', '2', '3']
```

We will use the same approach with our web page source code. Instead of printing out the entirety of the source code as one lengthy string, we will separate the content into individual lines and store them as a list using the following Python code:

```
lines = response.text.split('\n')
```

Now the entire website can be saved or printed out as smaller, more manageable units of information, essentially source code lines organized as a list. However, some of the lines will include a substantial amount of white space before and after the text. To get rid of this whitespace we use the *strip* string manipulation function. The following code illustrates this approach:

```
example = '    Once upon a time...        '
print(example.strip())

'Once upon a time...'
```

To capture and understand the data housed within the HTML file, we "inspect" the source code of the page. In other words, we highlight text in a webpage and view its corresponding HTML in another window. Chrome and Firefox have similar capabilities when inspecting HTML. Typically you right-click the text in the browser and select "inspect" or a similar action from the context menu. Figure 2.1 shows how this is accomplished in Firefox.

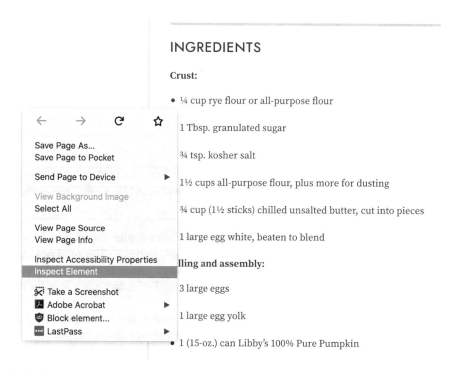

Fig. 2.1 First step to invoke the inspect tool in a browser

Figure 2.2 shows the result when you hover over the text in the browser on the left and the corresponding HTML is highlighted on the right. Specifically, in this example we notice that the recipe ingredient is enclosed in "span" tags with the class attribute "ingredients-item-name."

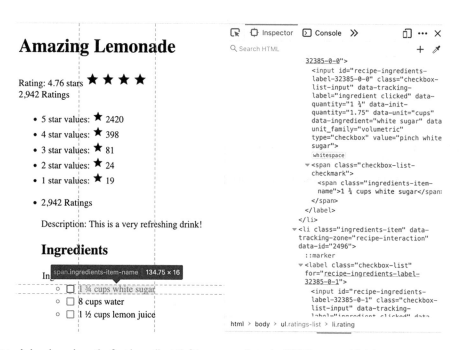

Fig. 2.2 The inspect tool showing where the first ingredient (left) corresponds to the HTML source (right)

However, for our purposes, the specific meaning and directives of the HTML source do not matter (although some knowledge of this popular markup language certainly helps). Our primary objective is to know where the specific ingredient information is stored. In Python there are various ways to parse a string. However, the key to effectively parsing strings is clearly defining **what** you want to extract. In this case we seek the ingredients. By employing the inspection tool (the right side of Fig. 2.2), we have discovered that the ingredient is embedded in the string `<span class="ingredients-item-name">1¾ cups white sugar </span>`.

To print only those lines with ingredients, we add a for-loop to navigate through every line in the list (`for line in lines`) and an if-statement to select what is printed out (`if '<span class="ingredients-item-name">' in line:`). The following Python code downloads the webpage and prints out any line that contains `<span class="ingredients-item-name">`:

```
File: retreive_html_page_example3.py:
import requests
link = 'https://icarus.cs.weber.edu/~rball/book_files/example_recipe_page.html'
response = requests.get(link, headers={'User-Agent':'Mozilla/5.0 (Macintosh;
Intel Mac OS X 10.15; rv:84.0) Gecko/20100101 Firefox/84.0'}) # get the webpage

lines = response.text.split('\n')  # divide the text into a list
for line in lines:
    line = line.strip()  # strip out the white space
    if '<span class="ingredients-item-name">' in line:
        print(line)
```

The result is a large block of text with various HTML tags and the target ingredients all occurring together:

```
<li class="ingredients-item" data-tracking-zone="recipe-interaction" data-id="1526">
<label class="checkbox-list" for="recipe-ingredients-label-32385-0-0"> <input
class="checkbox-list-input" data-tracking-label="ingredient clicked" data-
quantity="1¾" data-init-quantity="1.75" data-unit="cups" data-ingredient="white sugar"
data-unit_family="volumetric" type="checkbox" value="pinch white sugar" id="recipe-
ingredients-label-32385-0-0"> <span class="checkbox-list-checkmark"> <span
class="ingredients-item-name">1¾ cups white sugar </span> </span> </label> </li>
<li class="ingredients-item" data-tracking-zone="recipe-interaction" data-id="2496">
<label class="checkbox-list" for="recipe-ingredients-label-32385-0-1"> <input
class="checkbox-list-input" data-tracking-label="ingredient clicked" data-quantity="8"
data-init-quantity="8" data-unit="cups" data-ingredient="water" data-
unit_family="volumetric" type="checkbox" value=" water" id="recipe-ingredients-label-
32385-0-1"> <span class="checkbox-list-checkmark"> <span class="ingredients-item-
name">8 cups water </span> </span> </label> </li>
<li class="ingredients-item" data-tracking-zone="recipe-interaction" data-id="5107">
<label class="checkbox-list" for="recipe-ingredients-label-32385-0-2"> <input
class="checkbox-list-input" data-tracking-label="ingredient clicked" data-
quantity="1½" data-init-quantity="1.5" data-unit="cups" data-ingredient="lemon juice"
data-unit_family="volumetric" type="checkbox" value=" lemon juice" id="recipe-
ingredients-label-32385-0-2"> <span class="checkbox-list-checkmark" <span
class="ingredients-item-name">1½ cups lemon juice </span> </span> </label> </li> </ul>
```

Note the above block is composed of three separate lines (one ingredient per line) extracted from the page source. We know the text for every ingredient is preceded by `<span class="ingredients-item-name">`, so we can split the text for a given line with this string as the separator using:

```
ingredient = line.split('<span class="ingredients-item-name">')
```

The statement above produces a list called ingredient with extraneous HTML markup in the zeroth position in the list (prior to the occurrence of the `<span class="ingredients-item-name">` string) and the targeted ingredient text and closing tags in the first position in the list. We seek only the ingredient text and so we need to remove the straggling closing tags. If we print out the second item in the list (`print(ingredient[1])` we obtain the following:

```
1¾ cups white sugar </span> </span> </label> </li>
```

To retrieve only the ingredient substrings we use the powerful bracket notation available in Python. To summarize, use brackets (i.e. [and]) with a colon (:) to identify part of a string, as illustrated in the following Python code:

```
example = "I love my mother and father!"
print(example[2:10])


'love my '
```

The first number within the brackets (2 in this example) specifies the starting location in the string. In Python, and most programming languages, indices begin with 0. Consequently, [2:10] means start with the third character up to but not including the eleventh character. In Python both the first and second number in the brackets are optional. If you omit the first number preceding the colon the default is 0 and if you omit the second number after the colon the default is to include all remaining characters in the string.

To determine the substring location correctly and obtain only the text prior to " </span> </span> </label> </li>" we need the starting position of " </span> </span> </label> </li>" in the string. The *find* function and *index* function (you can use either one) returns the index of the starting location of a substring. The following Python code provides an example of the *find* function:

```
example = "I love my mother and father!"
print(example.find('love'))


2
```

Incorporating all these concepts together creates the following Python parsing code:

```
File: retreive_html_page_example4.py:
import requests
link = 'https://icarus.cs.weber.edu/~rball/book_files/example_recipe_page.html'
response = requests.get(link, headers={'User-Agent':'Mozilla/5.0 (Macintosh;
Intel Mac OS X 10.15; rv:84.0) Gecko/20100101 Firefox/84.0'})  # get the webpage

lines = response.text.split('\n')  # divide the text into a list
for line in lines:
    line = line.strip()  # strip out the white space
    if '<span class="ingredients-item-name">' in line:
        ingredient = line.split('<span class="ingredients-item-name">')
        ingredient_text = ingredient[1]
        end_of_ingredient_text = ingredient_text.find(' </span>')
        ingredient_text = ingredient_text[:end_of_ingredient_text]
        print(ingredient_text)
```

The above code produces the following result:

```
1¾ cups white sugar
8 cups water
1½ cups lemon juice
```

There are many introductory books and websites that cover the basics of string parsing in Python. To effectively web scrape you will need to be proficient in Python and learn the basics of web programming, particularly HTML. The more you learn about and apply coding practices, the better you become.

In addition, we have described in detail only one example. Every website will be structured differently and you may find your first attempt at web scraping requires hours to procure the data you seek. Most likely subsequent efforts to scrape a web page will be easier. Using the inspection tools in web browsers may initially appear confusing, but the more familiarity you have with these tools, the more streamlined the process becomes.

One important final observation is that web scraping with string manipulation does not represent the most efficient way to accomplish this time intensive task. In the next section we show you how to perform web scraping using a Python library called BeautifulSoup instead of string parsing.

2.10   Web Scraping with BeautifulSoup and Selenium

String parsing may prove challenging for some individuals but not for others. Given this variability in experience with string processing, it's important to consider more efficient ways to web scrape in Python. In particular we will leverage the BeautifulSoup library to help make our data gathering task easier.

BeautifulSoup is one of many Python libraries and frameworks designed for web scraping. Other related libraries are available such as lxml, selenium, and scrapy. How do you choose the "best" library that meets your processing requirements?

We reemphasize a key notion mentioned previously: Learning the conceptual foundations of a data science topic is the most important step; learning a particular customized tool is typically less important. With this as our guiding principle, we have shown an example of web scraping with string parsing to provide the context for this essential data gathering task. Now we will show a web scraping example with BeautifulSoup. The official documentation for BeautifulSoup is located at https://www.crummy.com/software/BeautifulSoup/bs4/doc/

When learning to perform any new task it is important to secure an overview of the main tools people generally use to accomplish the task and then devote some time to consider which strategy will work best for your specific context and problem. Sometimes the most important aspect about learning any new skill is simply to start doing it. After all, can you really learn how to swim just by reading about it? In our experience, you ultimately have to test the water.

For our implementation using BeautifulSoup , we are going to import the requests library again to download the website (`import requests`). Next, we will import BeautifulSoup (`from bs4 import BeautifulSoup`). BeautifulSoup is a powerful library that can be used for various purposes. For our application we will be using BeautifulSoup to primarily parse websites written in HTML, so we will pass the website content that we get from the requests library to the BeautifulSoup constructor with the option to parse HTML (`BeautifulSoup(response.text, 'html.parser')`). The resulting Python code looks like this:

```
import requests
from bs4 import BeautifulSoup

link = 'https://icarus.cs.weber.edu/~rball/book_files/example_recipe_page.html'
soup = BeautifulSoup(response.text, 'html.parser')
```

Although BeautifulSoup is more efficient than string parsing, it requires more in-depth knowledge about web programming. If you know HTML attributes *class* and *id* as well as basic HTML tags, then you should be able to immediately utilize BeautifulSoup. However, if you are not familiar with basic HTML elements and structure then we recommend taking some time to learn the fundamentals of HTML before continuing with this chapter.

Given that you know the definition and usage of the *class* attribute in HTML, we can now look once again at the span tag that encloses all the ingredients of interest:

```
<span class="ingredients-item-name">
```

As described during our string parsing discussion above, you will notice that every ingredient is enclosed in a *span* tag with the "ingredients-item-name" class attribute. You will not always be so lucky to have access to explicit identifiers when you web scrape, but in most cases detailed descriptors such as these will be available.

BeautifulSoup possesses many powerful ways to parse a website. For example, you could accumulate all information associated with a particular type of tag: `soup.find_all('span')`. You can display HTML in a nice easy-to-read format: `soup.prettify()`. You can find HTML siblings and parents, insert additional HTML, and perform other useful data gathering operations. The BeautifulSoup documentation is well written and offers many examples.

The current plan of action is to use BeautifulSoup to acquire every span tag that specifies an "ingredients-item-name" class. We have already seen how to obtain every span tag by using `soup.find_all('span')`. However, that will retrieve every span item in the entire webpage including those unrelated to our ingredient search. We want to focus solely on the span tags that have the "ingredients-item-name" class, so we will use the following statement instead: `soup.find_all('span',` class_='ingredients-item-name'). Please note that BeautifulSoup denotes "class_" with an underscore because the word "class" is a keyword in Python.

Putting it all together at this point gives us the following Python code:

```
File: BeautifulSoup_example1.py:
import requests
from bs4 import BeautifulSoup
link = 'https://icarus.cs.weber.edu/~rball/book_files/example_recipe_page.html'
response = requests.get(link, headers={'User-Agent':'Mozilla/5.0 (Macintosh;
Intel Mac OS X 10.15; rv:84.0) Gecko/20100101 Firefox/84.0'})

soup = BeautifulSoup(response.text, 'html.parser')
ingredients = soup.find_all('span', class_='ingredients-item-name')
for ingredient in ingredients:
    print(ingredient)
```

The code above produces the following output:

```
<span class="ingredients-item-name">1¾ cups white sugar </span>
<span class="ingredients-item-name">8 cups water </span>
<span <span="" class="ingredients-item-name">1½ cups lemon juice </span>
```

You could use conventional string parsing at this point, leveraging the *find* or *index* functions and bracket notation described previously to obtain the requisite substrings. However, BeautifulSoup offers an easier way to extract the text from any tag by using the `text` attribute associated with each item in the list. So, instead of printing an individual complete ingredient span tag, we print `ingredient.text`, as shown here:

```
File: BeautifulSoup_example2.py:
import requests
from bs4 import BeautifulSoup
link = 'https://icarus.cs.weber.edu/~rball/book_files/example_recipe_page.html'
response = requests.get(link, headers={'User-Agent':'Mozilla/5.0 (Macintosh; Intel Mac OS X
10.15; rv:84.0) Gecko/20100101 Firefox/84.0'})

soup = BeautifulSoup(response.text, 'html.parser')
ingredients = soup.find_all('span', class_='ingredients-item-name')
for ingredient in ingredients:
    print(ingredient.text)
```

This produces the desired set of results:

```
1¾ cups white sugar
8 cups water
1½ cups lemon juice
```

BeautifulSoup reduces the overhead and effort associated with string parsing to just a few lines of code. However, the tradeoff when using BeautifulSoup is that more familiarity with HTML is necessary.

As exhibited above, BeautifulSoup often makes web scraping an easier task with cleaner code. However, BeautifulSoup can be insufficient if a webpage requires interactivity. For example, many webpages possess an "infinite scrolling" feature, a technique that loads a portion of the available data until the user scrolls to the bottom of the page at which point additional content is displayed.

For example, assume a fictional webpage that loads thirty recipes when the site is initially visited. As the user scrolls to the bottom of the existing list the webpage automatically loads an additional thirty recipes and continues to show the remaining recipes in increments of thirty until exhausting all available recipes. If BeautifulSoup were employed in this situation then you would only retrieve the first thirty recipes.

Selenium is a library that works in conjunction with a browser to mimic human interaction with simulated mouse clicks, scroll bar movements, and other user interface actions. Selenium will work with Python only after the installation of additional drivers. However, once selenium is properly installed you can automatically control the browser from Python, thus allowing you to replicate human interactions with the browser environment. Selenium can be programmed to successfully simulate any activities between a human user and a browser.

There are many tools and techniques to perform web scraping. Regardless which strategy you choose to collect data, web scraping is an essential skill that most employers expect from prospective data scientists. Take the time to master web scraping concepts and skillsets and you will be on your way to effectively acquiring and analyzing data.

Exercises
1. What is the main difference between a data scientist and a data analyst?
2. What does it mean to gather data?
3. Compare and contrast gathering data by survey, focus group, lab-controlled experiment, logs or diaries, and longitudinal studies.
4. What are the fundamental constraints of a relational database?
5. When would you use a distributed database over a relational database?
6. Why do you think servers sometimes blacklist IP addresses?
7. When might it be considered unethical to web scrape a webpage?
8. Is there ever a time when you would want to use string parsing over using a library like BeautifulSoup? If so, when?
9. Provide examples when BeautifulSoup is insufficient and it is necessary to use interactive tools like selenium.

Chapter 3
Data Wrangling

You will most likely reference this chapter quite often as you become more proficient with data wrangling. However, we suggest going carefully through this chapter at least once so that you are conversant in the fundamentals of reshaping and transforming data.

The procedures for conditioning data in preparation for analysis is akin to steps listed in an instruction manual about how to operate a complicated tool for carpentry. The tool in each case has many features and options, but you will not fully appreciate them until applying these capabilities in a real world context.

Data wrangling, also known as *data munging*, refers to specific methods applied to format and package the data for subsequent analysis. This might involve merging data from various resources, cleaning up the data, or creating new data from existing data sources.

In the end, properly organizing and preprocessing the raw data will lead to actionable insights and working knowledge about the key information contained in the data.

3.1 Data vs Information vs Knowledge

What exactly are data? Although you may have heard this word frequently used in conversation and writing, the word *data* can be very confusing. First, data is plural for datum, but that fact is more academic than anything else because most people simply refer to all data as "data," regardless of the quantity.

For practical purposes, we define *data* to be facts, definitions, figures, numerical quantities, survey results, and any other raw values that **can be** measured, reported, collected, analyzed or organized. The emphasis is on the words "can be" because a sizeable portion of potential data is just sitting there collecting proverbial dust, or stalled along the data processing pipeline.

Primarily for the purpose of overwhelming and impressing the general public, experts at the end of each year will typically share the estimated volume of new data that have been collected annually worldwide. For example, how many quintillions of bytes of new data have been created every day in the last year alone? We learn a new prefix every yearly cycle, from giga-bytes to tera-bytes, and increasing through peta-, exa-, zetta-, and yotta-bytes.

From a practical sense, it does not matter how much data are available, because the numbers and characters constituting data have no intrinsic or practical value until you process them. We cannot overemphasize this point: **Non-processed data is useless!**

Conversely, how do we make data useful? By processing it.

The steps involved in processing – measuring, reporting, collecting, analyzing, organizing, reshaping – are the critical actions that transform data into information. Data that is processed and presented clearly and concisely to an individual or group consequently becomes *information*.

© The Author(s), under exclusive license to Springer Nature Switzerland AG 2022
R. Ball, B. Rague, *The Beginner's Guide to Data Science*, https://doi.org/10.1007/978-3-031-07865-1_3

Data are effectively the raw material that once processed becomes information. Sometimes information is simply data that are given a relevant degree of context. For example, take the following data: 7 and 1. The numbers 7 and 1 are data in its most raw form, but if given context, these data can become information. For this case, transforming 7 and 1 to 7 days = 1 week provides a context which sublimates the raw data to the level of information.

The next step after gathering information is acquiring knowledge. *Knowledge* is the insight, the organizational patterns, and the experience that you gain from information. You can easily share information from one person to another, but knowledge, either directly inferred or indirectly acquired, is relatively difficult to share.

For example, assume we create a collection of information about the stock market by transforming raw stock market data into tables and graphs. We then proceed to share these details with a randomly selected person. It is likely they will not amass the same knowledge from this information as we do. Knowledge is a deeply personal shift in mindset that cannot be shared. It is an internal process that ultimately changes a person's perspective and behavior.

If knowledge were easily and effectively shared between people, then when all people view precisely the same information they would all discern roughly identical patterns, and make similar deductions based on those perceived patterns. However, if you ever follow highly charged reports about politics, religion, or sports, then you will observe quite quickly that many people are exposed to the same information but arrive at entirely different conclusions (knowledge).

Ultimately, the goal for data scientists is to transform data to knowledge as actionable insights.

3.2 From Data to Information

Chapter 2 introduced several techniques to collect data. Now that we have data, how do we elevate it to the level of information?

After this chapter the remainder of the book investigates and presents different ways to gain knowledge from data. However, before we can effectively obtain this knowledge, the data must be reshaped into information.

Although in the case of small datasets there are a myriad of trivial ways to transform data to information, our focus is extracting meaning from data that cannot be viewed via a brief glance. For example, you might have thousands or millions of rows of data you want to process.

This leads us to data wrangling – the purpose of this chapter. The term *data wrangling* heralds back to the time of herding, or wrangling, horses and cattle to a particular destination. Wrangling a large herd of cattle from an expansive field to a specified corral can be a tedious and time-consuming task. So it is with data wrangling.

Many professional data scientists anecdotally report that about 80% of their time is spent modifying, organizing, compressing, combining, and reshaping data – that is, data wrangling.

This important component of data science may be the least glamorous aspect of the job, but significant deficiencies in carrying out proper data wrangling will prevent the data scientist from experiencing the more rewarding and exciting parts of the project.

There is an expression that teachers teach for free but are paid to grade. This sentiment could be mapped analogously to the discipline of data science in which "teachers" is replaced by "data scientists", "teach" is replaced by "discover insights", and "grade" is replaced by "wrangle the data", or, after substituting, *Data scientists discover insights for free but are paid to wrangle the data*. In any large data science project, there is a significant amount of cleaning, changing, transforming, and creating to accomplish.

Data wrangling is an important component of the *ETL (Extract, Transform, Load)* process. Wrangling data is the 'T', the transform part in which the raw data from one source is processed by some means from one form to another ready for analysis.

There are many examples of data wrangling from simple to complex. For example, converting a spreadsheet column of measurements from the imperial system (inches and feet) to the metric system (meters and kilometers) is easy. However, merging purchasing data from various departments in the government (e.g., the Air Force and Department of Energy), determining the category of every transaction, extracting the complete description of every product purchased, and combining all the various pieces into one complete comprehensible dataset is much more difficult.

Sometimes the end product of the data wrangling process is a new database or data warehouse, and sometimes it is simply a spreadsheet.

3.3 Pandas

Pandas is the name of an open-source library in Python that is designed to facilitate the organization and manipulation of data. Pandas is a powerful tool and is frequently a requirement listed on data science and data analyst job postings.

Covering every function and capability of the Pandas library is beyond the scope of this text. However, we will address several of the most useful functions for data science beginners. After you become familiar with Pandas we highly recommend you explore the official documentation and user guide available online. Although it may only take you a week or two to become proficient with the basics of Pandas you will likely be learning more nuances of the library for years to come.

Pandas provides an optimized data structure known as a dataframe. A *dataframe* is simply a 2D matrix that holds data, much like a modern spreadsheet. As we would expect from a 2D layout, the fundamental organization of the data is in columns and rows.

The dataframe structure follows the three rules of *Tidy Data*: (1) each feature or variable is a column, (2) each observation or record is a row, and (3) each collection of related experimental data is contained in a single dataframe. Pandas is built on top of the Numpy library partly because of Numpy's built-in parallel processing with large arrays. In other words, Numpy arrays (and subsequently Pandas dataframes) run at faster speeds with large amounts of data compared to traditional Python lists.

As a beginner most of your data will initially come packaged as a CSV file. *CSV* stands for *comma-separated values*, which means that for a given observation values in different columns are separated by a comma (,). Why use CSV files over XLS or XLSX files? The main reason to use CSV files is because they are not tied to any particular software package, which allows for more flexibility within your organization. To read a CSV file with Pandas you use the `read_csv` function.

Alternatively, XLS and XLSX files are Excel files designed by Microsoft, Inc. to be used in Microsoft Excel. If you want to store your data in Excel format then you can read in Excel files with the Pandas `read_excel` function.

In fact, there are many functions that allow you to read from different file formats such as `read_clipboard`, `read_json`, `read_orc`, `read_sas`, and `read_spss`. In general, if there is a file format that is popular with data scientists then there is often a function available to read the structured data and populate a Pandas dataframe.

In this chapter we will primarily use the `royal_line.csv` file for demonstration purposes. The `royal_line.csv` file was created from public sources and contains family history information about Elizabeth II, Queen of England. Figure 3.1 shows a screenshot of the first four observations in `royal_line.csv` as displayed in Excel.

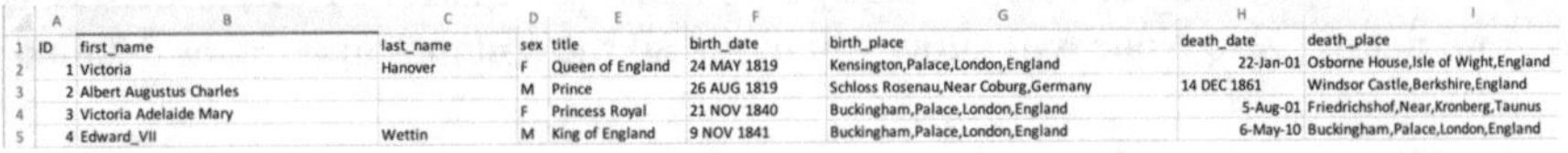

	A	B	C	D	E	F	G	H	I
1	ID	first_name	last_name	sex	title	birth_date	birth_place	death_date	death_place
2	1	Victoria	Hanover	F	Queen of England	24 MAY 1819	Kensington,Palace,London,England	22-Jan-01	Osborne House,Isle of Wight,England
3	2	Albert Augustus Charles		M	Prince	26 AUG 1819	Schloss Rosenau,Near Coburg,Germany	14 DEC 1861	Windsor Castle,Berkshire,England
4	3	Victoria Adelaide Mary		F	Princess Royal	21 NOV 1840	Buckingham,Palace,London,England	5-Aug-01	Friedrichshof,Near,Kronberg,Taunus
5	4	Edward_VII	Wettin	M	King of England	9 NOV 1841	Buckingham,Palace,London,England	6-May-10	Buckingham,Palace,London,England

Fig. 3.1 A screenshot of the first four entries of the royal_line.csv file, which contains Elizabeth II, Queen of England's family and relations

3.3.1 Series and Dataframe Basics

A Series is Pandas' version of a one-dimensional array. However, it is still optimized for large amounts of data and has an extensive number of associated functions.

A Dataframe is Pandas' version of a two-dimensional array and can be thought of as a matrix or a table also optimized for large amounts of data like a series.

There are various ways to access or display the data contained in a series or dataframe. The easiest way is to use the `print` command. For example, the following will show the first five and last five rows of the dataframe and provide summary information about how many columns and rows are in the dataframe:

```
File: import_csv_example.py:
import pandas as pd
df = pd.read_csv('royal_line.csv')
print(df)
```

There are many ways to create a dataframe from scratch. However, these modestly sized dataframes are typically only created for demonstration purposes because a dataset that can be viewed in a single glance is usually not a component of an enterprise data science project. The following example code shows simple examples of how to create a dataframe and a series:

```
File: dataframes_from_scratch.py:
import pandas as pd
import numpy as np

# Create a dataframe:
dictionary_data = {'Column1': ['Value1', 'Value2'],  # keep adding as many values as you like
    'Column2': ['Value1', 'Value2'],   # you can keep adding as many columns as you like too
    }
df = pd.DataFrame(dictionary_data)

# Create another dataframe with number 0-7 in two columns:
df = pd.DataFrame(np.arange(8).reshape(4, 2), columns=['column 1', 'column 2'])

# Create a series:
data = np.array(['g', 'e', 'e', 'k', 's'])
ser = pd.Series(data)
```

Why import Pandas as 'pd'? First, `import pandas as pd` is included in all the official Pandas documentation. Second, referring to *Pandas* as the abbreviated 'pd' is more efficient because you type less and therefore experience a concomitant decrease in keyboard mishaps within your code, thus making the 'pd' alias preferred. The same thing is true about the statement `import numpy as np`.

Why 'df'? The alias 'df' stands for dataframe. Similar to the use of 'pd' and 'np', 'df' for dataframe is employed in all official documentation and is also shorter. So, 'pd' for pandas, 'np' for numpy, and 'df' for dataframe are industry standard. We strongly recommend you follow this convention when composing your programs.

When we read in the CSV above the code only printed out some of the data. By default Pandas will only display an abridged version of the dataframe, omitting a majority of row and column entries. Fortunately, there are several display options that allow for some flexibility in how the data are listed. For example, we can view the data in a select number of columns through the command `pd.set_option('display.max_columns', X)` where X is the number of max columns you want to view. Setting X to None will allow you to see all the columns at once, as detailed below:

```
File: import_csv_example_max_columns.py:
import pandas as pd
pd.set_option('display.max_columns', None)
df = pd.read_csv('royal_line.csv')
print(df)
```

Unfortunately, employing this option alone often results in a confusing presentation format because Pandas wraps the data and distributes each row across separate display lines. This wrapping behavior can be suppressed by using another option (`pd.set_option('display.expand_frame_repr', False)`). Note that the `'display.expand_frame_repr'` option is True by default, so the second argument (`False`) in the `set_option` command above toggles this option off. For similar reasons the `display.max_rows` option can be very useful too. Here is an example with each of these three options specified that presents all columns and rows without any column wrapping:

```
File: import_csv_example_three_options.py:
import pandas as pd
pd.set_option('display.max_columns', None)
pd.set_option('display.max_rows', None)
pd.set_option('display.expand_frame_repr', False)
df = pd.read_csv('royal_line.csv')
print(df)
```

With large amounts of data, setting the `display.max_rows` option to `None` may not be a good idea. In our example above all 3009 rows are printed. To display a specific number of rows, the `head` function might be more convenient. The `head` function invoked with no argument lists the first five entries. Likewise, the default `tail` function displays the last five entries. Any valid integer can be passed to either the `head` or `tail` functions to indicate how many rows to display. The following demonstrates how to print the default first five entries with `head` and the last ten entries with `tail`:

```
Partial code: full program in head_and_tail_example.py:

print(df.head())
print(df.tail(10))
```

You may have noticed in the printed dataframe an additional column of numbers located to the far left of the data. For instance, without modifying the display options, `df.head()` will produce the following:

```
   ID  ...                        death_place
0   1  ...  Osborne House,Isle of Wight,England
1   2  ...       Windsor Castle,Berkshire,England
2   3  ...    Friedrichshof,Near,Kronberg,Taunus
3   4  ...     Buckingham,Palace,London,England
4   5  ...                   Darmstadt,,,Germany
```

The far left column containing the sequence 0 to 4 is the *index* that Pandas assigns to each row. An index is typically a unique value associated with each row in the dataframe that is leveraged to expedite searches and retrievals. However, since we already have a distinct ID column index (second column) included with the original data file this additional index introduced by Pandas is redundant. When reading the data, we indicate our existing index called ID using the following code:

```
File: set_index_in_read.py:
import pandas as pd
df = pd.read_csv('royal_line.csv', index_col='ID')
print(df.head())
```

Running the above code results in the following:

```
                 first_name  ...                        death_place
ID                           ...
1                  Victoria  ...  Osborne House,Isle of Wight,England
2   Albert Augustus Charles  ...       Windsor Castle,Berkshire,England
3    Victoria Adelaide Mary  ...    Friedrichshof,Near,Kronberg,Taunus
4                Edward_VII  ...     Buckingham,Palace,London,England
5           Alice Maud Mary  ...                   Darmstadt,,,Germany
```

Note the absence of the auto-generated index column in the above output. With Pandas dataframes you can also limit the information to certain columns. For instance, the following two print statements are equivalent and display only the first five data entries of the 'title' column:

```
Partial code: full program in print_only_title_column.py:

print(df['title'].head())
print(df.title.head())
```

Note that `df.title` only works in the above statement because the column name ('title') does not include any spaces. If the column were named 'column 1' then `df.column 1` would not work because the Python interpreter would not associate the '1' with 'column' due to the intervening space which violates the Python naming convention. For any column names with whitespaces the bracket syntax is required: `df['column 1']`.

Multiple columns may be viewed by passing in a list of column names. This syntax often confuses students because of the introduction of multiple brackets. Just remember that the outermost brackets are accessing the dataframe and the innermost brackets are defining a Python list. The following retrieves the first five data entries of both the `title` column and the `first_name` column:

```
Partial code: full program in print_only_title_and_first_name_columns.py:

print(df[['title', 'first_name']].head())
```

Whenever you need to view the specific column labels for a particular dataframe use `df.columns` as shown below:

```
Partial code: full program in print_columns.py:

print(df.columns)
```

The nice thing about `df.columns` is that it returns an index object that behaves as an iterable list. In other words, you can access the contents like a list (e.g., `df.columns[0]`) or with a for-loop.

Additionally, two common functions provide general information about the dataframe: `info` and `describe`. The following code uses both functions to reveal summary information:

```
Partial code: full program in use_info_and_describe.py:

print(df.info())
print('\n', df.describe())
```

The `info` function produces the following result, indicating under the 'Dtype' (datatype) heading that all columns are considered "object" type or strings:

```
<class 'pandas.core.frame.DataFrame'>
Int64Index: 3009 entries, 1 to 3009
Data columns (total 8 columns):
 #   Column        Non-Null Count    Dtype
---  ------        --------------    -----
 0   first_name    2984 non-null     object
 1   last_name     1142 non-null     object
 2   sex           2996 non-null     object
 3   title         1398 non-null     object
 4   birth_date    1734 non-null     object
 5   birth_place   486 non-null      object
 6   death_date    1692 non-null     object
 7   death_place   449 non-null      object
dtypes: object(8)
memory usage: 211.6+ KB
None
```

The results of the `describe` function are given below (with a reduced font to show all the information). This function provides summary information about the data in each of the columns:

	first_name	last_name	sex	title	birth_date	birth_place	death_date	death_place
count	2984	1142	2996	1398	1734	486	1692	449
unique	2006	421	2	308	1005	307	1084	315
top	John	Hanover	M	Prince	1872	St. James Palace,London,England	1958	Paris,France
freq	43	70	1685	118	11	19	9	15

Please note the `describe` function will generate different results for different datatypes. For example, to see how `describe` provides a statistical summary of numeric data, please refer to the end of Sect. 3.3.4.

3.3.2 *Dropping or Removing Data*

There are many reasons why you might wish to remove data. In some cases, you may not require the entire dataset to extract actionable insights, or you might remove rows that are missing information so that machine learning algorithms can function properly. Let's start with dropping two columns. The following code drops the `birth_place` and `death_place` columns:

```
Partial code: full program in drop_columns_fails.py:

df.drop(columns=['birth_place', 'death_place'])
print(df.head())
```

Or does it? If you run the above code, you will notice that the `birth_place` and `death_place` columns still remain. The reason is because most dataframe and series functions return a **new** dataframe or series. In the above code, `df.drop()` returns a new dataframe. You can either put the results in a new variable or you can indicate that you want 'inplace' changes to act upon the current dataframe or series. The following shows how to drop columns and return a new dataframe as well as how to drop the columns in the original dataframe by setting the `inplace` keyword argument to `True`:

```
Partial code: full program in drop_columns_succeeds.py:

# drop columns in a new dataframe (the old dataframe is not affected):
new_df = df.drop(columns=['birth_place', 'death_place'])

# drop columns in the current dataframe:
df.drop(columns=['birth_place', 'death_place'], inplace=True)
```

You can also drop a row by indicating the specific index. Note that our ID index adopted from the original datafile begins at 1. In contrast Pandas indexes start at 0 by default. In the following code, the first drop function drops the first row and the second drops multiple rows by passing a list of indexes:

```
Partial code: full program in drop_by_rows.py:

df.drop(index=1, inplace=True)   # drop a single row

df.drop(index=[4, 5, 6], inplace=True)   # drop multiple rows
```

Note that these operations act on the existing dataframe by setting `inplace = True`. Knowing the specific index value is not necessary when dropping a row using `df.index`, which avoids potential variations in index numbering and always references the first row starting with 0. For example, the following code consecutively drops the first row twice, effectively removing the first two rows, regardless of their assigned index value:

```
Partial code: full program in drop_first_index_twice.py:

df.drop(df.index[0], inplace=True)
df.drop(df.index[0], inplace=True)
```

A range of entries can also be removed by `df.index` using standard Python indexing and slicing notation.. The following example drops the first ten rows in the dataframe and the second example drops both specified rows and columns at the same time:

```
Partial code: full program in drop_rows_and_columns.py:

df.drop(df.index[:10], inplace=True)   # drop the first ten rows

df.drop(index=[11, 12, 13], columns=['birth_place'], inplace=True)   # drop rows and columns
```

Conditionals, explained in Sect. 3.3.5, can also be leveraged to drop rows and columns.

The `drop_duplicates` function drops rows that are duplicated. This can be helpful when more than one row contains identical data.

The `drop_na` function drops every row that includes at least one NA entry. Essentially, an NA is a blank indicating missing or omitted data in the dataframe or series, and represents a general form of NaN (not a number), NaT (not a time), or null. The `drop_na` function, like most Pandas functions also has many options. For example, you can specify which columns to drop with the `subset` parameter and a maximum number of rows to be dropped using the `thresh` parameter.

Be very careful with the `drop_na` function! If there is **any** missing information in any of the columns for a given row then the entire row is dropped. For the `royal_line.csv` example dataset we are using, most rows are missing some data. For instance, many royals do not have a last name and for many rows birthplace and deathplace are missing. The following code shows that before the `drop_na` function is invoked there are a total of 3009 rows and 8 columns. After the `drop_na` function is called there remains 77 rows and 8 columns, effectively removing 97% of the rows:

```
File: drop_nas.py:
import pandas as pd
df = pd.read_csv('royal_line.csv', index_col='ID')

print(df.shape)   # prints (3009, 8)
df.dropna(inplace=True)
print(df.shape)   # prints (77, 8)
```

Note that `df.shape` is an effective way to quickly determine the number of rows and columns in a dataframe or series by returning a tuple in the form of (rows, columns).

A useful strategy to understand the proportion of missing data is through visualization. The following code reads the data and visualizes the entries that are missing. Figure 3.2 shows the resulting heatmap produced by the following code, plotting a black line for each entry with data and a cream line for each entry with missing data. The first column is mostly black because a majority of royal's first names are known. On the other hand, the third to last column and the last column, which represent the birth and death places of the individuals respectively, are mostly missing information.

```
File: visualize_nas.py:
import seaborn as sns
import matplotlib.pyplot as plt
import pandas as pd

df = pd.read_csv('royal_line.csv', index_col='ID')
sns.heatmap(df.isnull(), yticklabels=False, cbar=False)
plt.show()
```

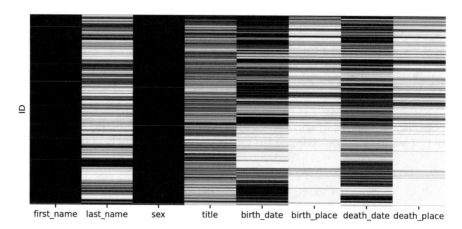

Fig. 3.2 Visual depiction of missing data in the royal_line.csv dataset. Black lines represent entries with data and cream lines represent missing data. Image created from visualize_nas.py

3.3.3   Adding, Modifying Data, and Mapping

Along with dropping or removing data, other common tasks applied to existing datasets include adding or modifying entries. These tasks are employed so frequently that there are a bewildering number of ways made available by Pandas to accomplish these actions.

The first method to address is the `fillna` function that, as its name implies, fills in NA's in the dataset. There are many options associated with the `fillna` function. The following is a sampling of available methods with comments explaining each usage:

```
File: fillna_examply.py:
import pandas as pd

df = pd.read_csv('royal_line.csv', index_col='ID')

# replace ALL NA entries with a fixed value:
df.fillna(0, inplace=True)

# replace the first 2 NA entries in each column with a fixed value:
df.fillna(0, limit=2, inplace=True)

# replace ALL NA first names with a fixed value:
df['first_name'].fillna('no first name', inplace=True)

# replace specific columns with specific values:
values = {'first_name': 'no_first_name', 'last_name': 'no_last_name', 'sex': 'no_sex', 'title':
'no_title', 'birth_date': 'no_birth_date', 'birth_place': 'no_birth_place', 'death_date':
'no_death_date', 'death_place': 'no_death_place'}
df.fillna(value=values, inplace=True)

# ffill and pad: from first row to last row, propagate the most recent row that
# is not an NA forward until next valid row
df.fillna(method='ffill', inplace=True)
# bfill and backfill: like ffill, except from last row to first row
df.fillna(method='bfill', inplace=True)
```

Note that the keyword values 'ffill' and 'pad' are synonymous as are the keyword values 'bfill' and 'backfill.'

One of the easiest ways to add or update data is to simply create or modify a column with a single value. The following code creates a column called `birth_year` with the value of 42 for every entry:

```
Partial code: full program in create_birth_year.py:

df['birth_year'] = 42
```

This approach is not very practical, but it does illustrate how easy it can be to create a new column and to set a value for every row. You can also set the values of a column to the corresponding values of another column as shown in the following code:

```
Partial code: full program in copy_column.py:

df['birth_year'] = df['birth_date']
```

Computations and constructions may be performed on existing columns to create new ones, such as concatenating the first name and last name to generate the individual's full name as coded here:

```
Partial code: full program in concatenate_columns.py:

df['full_name'] = df['first_name'] + ' ' + df['last_name']
print(df[['first_name', 'last_name', 'full_name']].head())
```

Which produces the following output:

```
ID                   first_name last_name          full_name
1                      Victoria   Hanover   Victoria Hanover
2    Albert Augustus Charles         NaN                NaN
3     Victoria Adelaide Mary         NaN                NaN
4                    Edward_VII    Wettin   Edward_VII Wettin
5          Alice Maud Mary         NaN                NaN
```

These results reveal a problem: NaN concatenated with any other string equals NaN. Ideally, the values of full_name should be valid even if only a first name or last name exists. To resolve this issue, we implement a function that operates on each row of data. The following code defines the create_full_name function and then utilizes the apply operation to invoke that function on each row which effectively creates another column called full_name whose entries are all valid names if values exist for either one or both of first_name or last_name.

```
File: create_full_name.py:
import pandas as pd
import numpy as np

def create_full_name(row):
# both first_name and last_name are strings
    if isinstance(row['first_name'], str) and isinstance(row['last_name'], str):
        result = row['first_name'] + ' ' + row['last_name']
    elif isinstance(row['first_name'], str):  # only first_name is a string
        result = row['first_name']
    elif isinstance(row['last_name'], str):  # only last_name is a string
        result = row['last_name']
    else:  # neither first_name nor last_name are strings, they are both NaN
        result = np.nan
    return result

df = pd.read_csv('royal_line.csv', index_col='ID')

df['full_name'] = df.apply(create_full_name, axis=1)
```

Note the apply operation in the above code example takes a function as the first parameter. Python lambda functions can also be used to create any required function argument inline or as needed. For example, consider the following statement that employs a lambda function to either calculate the length of each valid full name or return 0 if the full name is NaN (a float datatype):

```
Partial code: full program in lambda_example.py:
df['name_length'] = df.apply(lambda row: 0 if isinstance(row['last_name'], float) else
len(row['last_name']), axis=1)
```

Note the option `axis = 1` means to process the data row by row as opposed to column by column specified by `axis = 0`. For the default `axis = 0`, `apply` examines each individual column. For example, the following code creates a two-column dataframe where the `apply` function returns the maximum value for each column. Note that the data type of each column in a dataframe is important to know because the data type determines the functions that are available. For example, for numeric types statistical functions such as max and min may be applied as shown:

```
File: apply_max_example.py:
import pandas as pd
import numpy as np


df = pd.DataFrame(np.arange(12).reshape(6, 2), columns=['column 1', 'column 2'])
print(df)


new_df = df.apply(lambda column: column.max())
print(f'\nThe result =\n{new_df}')
```

Which produces the following result:

```
     column 1   column 2
0           0          1
1           2          3
2           4          5
3           6          7
4           8          9
5          10         11


The result =
column 1    10
column 2    11
dtype: int64
```

There are three main functions used to create or change data in dataframes: `apply`, `map`, and `applymap`. The following code creates a small example dataframe, displays it, then provides specific examples for each of the three functions. First, the results of passing the max function from the Numpy library (abbreviated as 'np') to the `apply` function with `axis = 0` and `axis = 1` are generated. Second, the code includes an example of the `map` function which modifies every individual cell in a series or column. Third, an example of the `applymap` function is demonstrated which updates every cell in the entire dataframe.

```
File: apply_map_and_applymap.py:
import pandas as pd
import numpy as np


df = pd.DataFrame(np.arange(8).reshape(4, 2), columns=['column 1', 'column 2'])
print(f"Original dataframe:\n{df}\n")
print(f"df.apply(np.max, axis=0):\n{df.apply(np.max, axis=0)}\n")
print(f"df.apply(np.max, axis=1):\n{df.apply(np.max, axis=1)}\n")
print(f"df['column 1'].map(lambda x: x*2):\n{df['column 1'].map(lambda x: x*2)}\n")
print(f"df.applymap(lambda x: x*2):\n{df.applymap(lambda x: x*2)}")
```

Which produces the following:

```
Original dataframe:
   column 1  column 2
0         0         1
1         2         3
2         4         5
3         6         7

df.apply(np.max, axis=0):
column 1    6
column 2    7
dtype: int64

df.apply(np.max, axis=1):
0    1
1    3
2    5
3    7
dtype: int64

df['column 1'].map(lambda x: x*2):
0     0
1     4
2     8
3    12
Name: column 1, dtype: int64

df.applymap(lambda x: x*2):
   column 1  column 2
0         0         2
1         4         6
2         8        10
3        12        14
```

Figure 3.3 visually renders the data coverage for each function utilized in the previous code.

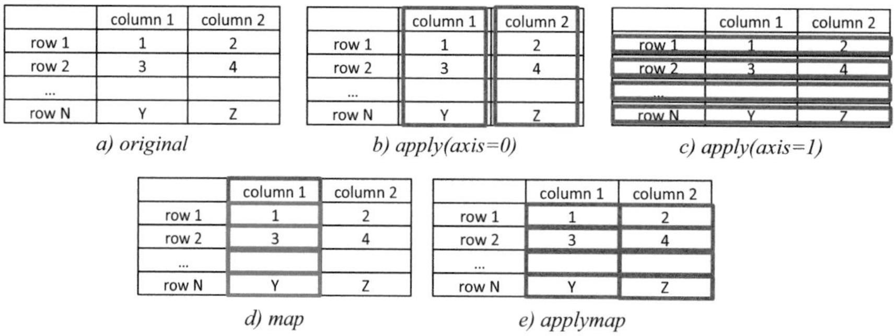

Fig. 3.3 (a) Original dataframe. (b) Visualization of apply(axis = 0) – processes the dataframe by columns or series. (c) Visualization of apply(axis = 1) – processes the dataframe by rows. (d) Visualization of map – processes one column of the dataframe cell by cell. (e) Visualization of applymap – processes each cell of the dataframe

Often data analysts are confused by the optimized, block-processing functions of `apply`, `map`, and `applymap`. In contrast, consider the `iterrows` function, one of the more intuitive and **slowest** techniques to iterate through the rows in a dataframe. The `iterrows` function is not nearly as optimized as `apply`, `map`, and `applymap`, but the use of this function adheres to traditional repetitive control structures common in imperative programming. Assuming the `full_name` column was created from the code above, the following shows an example of employing a for-loop with `iterrows` to print out the full name of all people with a valid birth date entry:

```
Partial code: full program in iterrows_example.py:

for index, row in df.iterrows():
    if isinstance(row['birth_date'], str):
        print(f"{index}: {row['full_name']} was born {row['birth_date']}.")
```

Table 3.1 provides a summary of these four major functions.

Table 3.1 Description of how apply, map, applymap, and iterrows process data. Compare to Fig. 3.3

Function	Description
apply	Process each column (if axis=0) or row (if axis=1)
map	Applicable for only a series (or individual column in a dataframe)
applymap	Process each cell in the entire dataframe
iterrows	Iterates through each row (similar to *apply(axis=1)*) and typically used in a for-loop. Slow compared to apply, map, and applymap

Functions similar to `apply`, `map`, and `applymap` include `transform` and `agg` (an alias for the `aggregate` function). The `itertuples` and `items` functions are similar to `iterrows`. Whereas `itertuples` returns tuples rather than rows, in our opinion the `items` function should be named "itercolumns" since this function returns the name of the column and the column data as a series.

3.3.4 Changing Datatypes of Series or Columns

Up to this point all the datatypes for each column in our examples has been 'object' because all columns contain string data. You can verify this using the *info* function or the *dtypes* attribute:

```
Partial code: full program in dyptes_example.py:

print(df.info())
print(df.dtypes)
```

There are various ways to convert columns from one data type to another. For example, consider the following example dataframe:

```
Partial code: full program in three_types.py:

df = pd.DataFrame({'ints': [1, 2, 3, 4],
                   'strings': ['a', 'b', 'c', 'd'],
                   'floats': [1.1, '2.2', '3.3', 4]})
print(df)
print(df.dtypes)
```

Which produces the following output:

```
    ints strings  floats
0      1       a     1.1
1      2       b     2.2
2      3       c     3.3
3      4       d       4
ints           int64
strings       object
floats        object
dtype: object
```

Pandas did a good job automatically figuring out the types of the columns based on the individual entries. However, the last column 'floats' is listed as type 'object' or string because some of the entries ('2.2' and '3.3') were entered as strings even though these strings contain numerical information and can be converted to floats.

To convert a column between different datatypes we can utilize one of the following:

```
Partial code: full program in three_types_converted.py:

# convert only one column:
df['floats'] = df['floats'].astype(float)

# convert all the columns based on a dictionary:
convert_dict = {'ints': int, 'strings': str, 'floats': float}
df = df.astype(convert_dict)
```

A word of caution: if there are NaN entries in the column to be transformed then errors will often occur when converting to other data types.

There are various datatypes in Pandas. The following represents a sample of these datatypes: int, int32, int64, float, float64, bool, and any valid object type. There are also several specific functions that convert a column to a specific type such as to_datetime, to_timedelta, and to_numeric.

The to_datetime function is generally quite useful, but for our particular example it falls short. For example, to convert the birth_date column from object to datetime we can try the following code:

```
Partial code: full program in failed_conversion_to_datetime.py:

# remove all NA rows in the birth_date column:
df['birth_date'].dropna(inplace=True)
df['birth_date'] = pd.to_datetime(df['birth_date'])   # this fails!
```

Which produces the following:

```
Out of bounds nanosecond timestamp: 1292-01-01 00:00:00
```

The above code generates errors due to several issues. First, there are entries in the dataset formatted like the following: ABT 751. This notation means that the family history experts believe the person was born about (ABT) 751. The second is related to an out of bounds nanosecond timestamp error above.

The datetime format used in Pandas can only support approximately 580 years in the range from around 1677 to 2262. To test this yourself consider the following code:

```
print(pd.Timestamp.min, pd.Timestamp.max)
```

Alternatively, we could use pd.to_datetime(x, errors = 'coerce'), but this transforms all attempts that generate errors into NaT (not a time) – a variant form of NA.

For this particular situation, there are a number of workarounds, such as storing the year, month, and day separately in three different columns. For demonstration purposes, we will extract the year from the `birth_date` column and assign this value to a new column named `birth_year`. The following code is rather involved and is included here to demonstrate that occasionally the built-in Pandas functions are not sufficient to convert between datatypes:

```
File: get_year_from_birth_date.py:
import pandas as pd
import numpy as np

def get_year(x):
    if pd.isna(x):
        year_result = np.nan  # if the birth_year is nan then return nan
    else:  # checking a number of edge cases in the data and stripping it out:
        if "ABT" in x:  # for example: ABT  1775
            x = x[3:]
            x = x.strip()
        if "/" in x:  #  For example: 1775/1776
            x = x[:x.find('/')]
        num_spaces = x.count(' ')
        if num_spaces == 0:  # only has the year
            year_result = int(x)
        elif num_spaces == 1:  # example: FEB 1337
            x = x[x.rfind(' ') + 1:]  # 'rfind' finds the last space. The 'r' stands for 'reverse.'
            if x.isnumeric():
                year_result = int(x)
            else:  # This could happen if there is only a day and month, like '10 JAN'
                year_result = np.nan
        elif num_spaces == 2:  # example: 16 FEB 1337
            x = x[x.rfind(' ') + 1:]  # 'rfind' finds the last space. The 'r' stands for 'reverse.'
            year_result = int(x)
        else:
            # There are a few other strange dates that aren't worth our time to fix,
            # so just return nan for those.
            year_result = np.nan

df = pd.read_csv('royal_line.csv', index_col='ID')
df['birth_year'] = df['birth_date'].map(get_year)

print(df['birth_year'])
```

Executing the above code will convert most birthdates into `birth_year` with the `float64` datatype assigned. (There are a few edge cases we did not address for the sake of brevity.) Why `float64` instead of `int` or `int64`? Because NaN is considered a `float` (or `float64`), any column that includes NaN's will have the `float64` datatype.

Now that we have some experience generating the `birth_year` column with numeric type `float64`, we can now observe that invoking the `describe` function will reflect different results based on the column datatype. The following displays the results from the `describe` function applied separately to a float column and a string column:

```
Partial code: full program in different_dtypes_for_describe.py:

print(df['birth_year'].describe())
print('\n', df['title'].describe())
```

Which produces the following results:

```
count    1727.000000
mean     1739.392588
std       253.463283
min       686.000000
25%      1675.000000
50%      1837.000000
75%      1907.000000
max      1991.000000
Name: birth_year, dtype: float64


count       1398
unique       308
top        Prince
freq         118
Name: title, dtype: object
```

Note that for a numerical value, the describe function generates descriptive statistics, whereas for a string value a frequency/mode analysis occurs.

3.3.5 Conditionals in Dataframes and Series

Conditionals are a very useful feature of Pandas which typically produce a Numpy array of Booleans or a Pandas Boolean series. For example, assuming the creation of the column birth_year from Sect. 3.3.4, consider the following code that produces a Boolean mask of True if the birth_year column is greater than or equal to 1990 and False otherwise:

```
Partial code: full program in birth_year_mask.py:

# Boolean masks:
df['birth_year'] = df.birth_date.map(get_year) # get_year function from previous code example
boolean_mask = (df.birth_year >= 1990)
print(f'Boolean mask = \n{boolean_mask}')
print('\n', df[['first_name', 'last_name', 'birth_year']][boolean_mask])
```

Which outputs the following results:

```
Boolean mask =
ID
1        False
2        False
3        False
4        False
5        False
       ...
3005     False
3006     False
3007     False
3008     False
3009     False
Name: birth_year, Length: 3009, dtype: bool
```

```
                    first_name last_name   birth_year
 ID
 2958  Eugenie Victoria Helena    Windsor       1990.0
 2961                      NaN     Mowatt       1990.0
 2963                    Kitty        NaN       1991.0
```

The last statement in the above code demonstrates how we can couple our original dataframe values with a Boolean mask to select and display only those rows for which the mask is True, in this case royals in the dataset with a `birth_year` of 1990 or later.

Boolean masks can be leveraged in a variety of ways. Also, different Boolean expressions may be combined using the Pandas logical operators shown in Table 3.2.

Table 3.2 Logical operators in Pandas for creating Boolean masks

Logical operator	Symbol in Pandas
and	&
or	|
not	~

The following is an example of the 'and' (&) operator in a dataframe conditional. Specifically, the code prints out all the people born in a year greater than or equal to 1500 and who held the title of "Queen:"

```
Partial code: full program in conditional_and_example.py:

print(df[(df.birth_year >= 1500) & (df.title.str.contains('Queen'))][['first_name', 'title',
'birth_year']])
```

In the above code the Boolean expression enclosed in the first set of square brackets [] acts as the Boolean mask that selects the appropriate rows from the dataframe. Also note that to use string functions (e.g. *contains*, *capitalize*, *find*) the column must be designated as an "object" datatype and the `str` attribute is applied. In the above example this functionality is demonstrated with the `title` column by using `df.title.str.contains`.

Consider the following Python example with standard strings in which the code checks if the variable `example` contains the word "Queen:"

```
File: standard_string_comparison.py:
example = 'Queen of England'
if 'Queen' in example:  # Preferred method
    print('contains example 1')

if example.__contains__('Queen'):  # Using the __ functions is frowned upon.
    print('contains example 2')
```

In both cases the Boolean expression returns True and prints the following results:

```
contains example 1
contains example 2
```

However, when checking if a column or series data contains a string then use the built-in `contains` function as illustrated in the following example:

```
Partial code: full program in dataframe_string_built_in_example.py:

# drops all rows with an NA in the title column:
df.dropna(inplace=True, subset=['title'])
print(df[df.title.str.contains('Queen')])  # without brackets to specify title
print(df[df['title'].str.contains('Queen')])   # with brackets to specify title
```

The following statement prints all the string functions available for a given series or column with string (object) datatype:

```
Partial code: full program in print_str_functions.py:

print(dir(df['title'].str))
```

3.3.6   loc and iloc Functions

One of the most common tasks for data scientists is filtering the information to more efficiently derive actionable insights about the data. The head and tail functions provide a quick, truncated view of the beginning or end, respectively, of the dataframe or series. However, what if you're interested in examining results that are not necessarily at the very beginning or end of the dataset?

For this purpose, the loc function is designed to access rows and columns by **label**. In contrast, the iloc function is used to access rows and columns by **integer value** – the 'i' in iloc stands for 'integer', or, more specifically, integer-location based indexing.

Regarding the labels referenced by the loc function, we are by now familiar with column labels, but rows can be labeled as well. The concept of row labels may be confusing initially until we understand that Pandas lets us name a row anything we want. For example, a row could be assigned an integer, category, or string. For more information about advanced indexes see Sect. 3.3.9.

In our royal line example, we have a custom index starting with 1 instead of the default 0 because the CSV file includes an ID column that begins numbering with 1. Also recall that when initially reading the CSV file the index_col option was set to the ID column:

```
df = pd.read_csv('royal_line.csv', index_col='ID')
```

Given this, to access row 1 (currently in position 0 unless the dataframe was sorted differently) we use the following code:

```
Partial code: full program in loc_example1.py:

print(df.loc[1])
```

Which produces the following output:

```
first_name                                    Victoria
last_name                                      Hanover
sex                                                  F
title                                Queen of England
birth_date                                24 MAY 1819
birth_place          Kensington,Palace,London,England
death_date                                22 JAN 1901
death_place      Osborne House,Isle of Wight,England
birth_year                                      1819.0
Name: 1, dtype: object
```

Recognizing that each individual row and individual column in a Pandas dataframe or series has a label is key to understanding the loc function. To clarify this concept further, the following two lines of code use the same lookup index, but the first statement returns an error and the second statement is successful:

```
Partial code: full program in loc_example2.py:

print(df.loc[0])   # causes an error because there is no row labeled 0
print(df.iloc[0])  # works because it returns the first (zeroeth) row
```

The comments in the code above provide an explanation for these outcomes, but rest assured that if you are still confused you are not alone. Although column labels are intuitive, row labels may take some practice. Consider the following code that creates a dataframe with variously labeled indexes:

```
Partial code: full program in index_example1.py:

df = pd.DataFrame(np.arange(10).reshape(5, 2), columns=['A', 'B'], index=['cat', 42, 'stone',
42, 12345])# Five rows each with an associated index
print(df)
```

Which produces the following output:

```
       A  B
cat    0  1
42     2  3
stone  4  5
42     6  7
12345  8  9
```

Although this indexing scheme appears arbitrary and even non-unique in the case of the duplicate use of 42, it is intended to illustrate that every row has a label. The following are some example statements that should solidify this point:

```
Partial code: full program in index_example2.py:

print(df.loc[12345])  # prints the last row: values 8 & 9
print(df.loc['stone'])  # prints the third row: values 4 & 5
print(df.loc[42])  # prints two rows (because two rows have the label 42)
print(df.loc['A'])  # produces an error. There is no row labeled 'A'
print(df.loc['cat':'stone'])  # prints the first three rows
print(df.loc[['cat', 'stone']])  # prints the first and third rows
print(df.loc['stone', 'B'])  # prints the value of 5: row 'stone', column 'B'
print(df.loc[df['A'] > 3])  # prints the last three rows


# using conditionals:
# row 12345, column 'A' == 8, so it changes the value in column 'B' to 120:
df.loc[df['A'] == 8, 'B'] = 120
print(df.loc[12345])  # prints 8 & 120, showing the change

print(df.iloc[0])  # print the first row with the values of 0 & 1
print(df.iloc[0:3])  # print the first three rows
print(df.iloc[[0, 2, 4]])  # prints 'cat,' 'stone,' and 1234 rows
print(df.iloc[0, 1])  # prints the first (zeroeth) row ('cat') for column 'B'
print(df.iloc[0:3, 1])  # prints the first 3 rows for column 'B'
```

Along with the filtering operations described in the previous section in Sect. 3.3.5, using conditional Boolean expressions with the loc and iloc functions makes many tasks with dataframes and series much easier.

Returning to the royal family history data as an example, let's create a new column named era. The era column signifies if a person was born in one of three distinct time periods: 'ancient,' 'middle_years,' and 'modern.' The following creates a new column and initially assigns the value 'unknown':

```
Partial code: full program in era1.py:

df['era'] = 'unknown'
```

That was easy enough. If you view the contents of the dataframe (e.g., using `head`, `tail`, etc.) you will see a new column labeled `era` filled with the 'unknown' string. Our next step is to decide how to divide the birth years. The following code will print the maximum and minimum values of the birth years to help us with this decision:

Partial code: full program in era2.py:

```
print(f"The earliest year = {df['birth_year'].min()} and the latest year = {df['birth_year'].
max()}.")
```

The results show 686 as the earliest year and 1991 as the latest year. Section 3.3.7 examines in detail how to automatically determine a range of numbers, but for now we will naïvely take the range of 1305 years (1991–686 = 1305 years) and divide it by three to obtain roughly 435 years per era.

For our example we will claim that royals born between 686–1121 arrived during the 'ancient' era (686 + 435 = 1121). Consequently, the period 1121–1555 is labeled as the 'middle_years' era and people born after 1555 arrived during the 'modern' era. The following code utilizes the `loc` function in conjunction with conditionals to establish these three eras:

Partial code: full program in era3.py:

```
df['era'] = 'unknown'
df.loc[df['birth_year'] < 1122, 'era'] = 'ancient'  # 686 - 1121
df.loc[(df['birth_year'] >= 1122) & (df['birth_year'] <= 1555), 'era'] = 'middle_years'
# 1122 - 1555
df.loc[df['birth_year'] > 1555, 'era'] = 'modern'  # after 1555
```

Note these same assignments could have been accomplished by a custom function used in conjunction with the `map` utility (for more information about the `map` function see Sect. 3.3.3).

Other helpful functions that support conditional selection operations include `at`, `xs`, `take`, `numpy.select`, and `numpy.where`.

3.3.7 Binning

To achieve accurate interpretations of our data, often numerical values need to be aggregated. Section 3.3.6 manually aggregated birth years to delineate specific eras when people were born.

The concept of binning is based fundamentally on ranges of numbers. For example, using our royal family history dataset, for all the people with valid birth years and death years we can easily determine their age when they died. This calculation would result in a large distribution of ages and would likely not be overly helpful in providing us meaningful insights. However, if we were to bin these lifetime lengths into a few descriptive categories, such as infancy, child, teenager, young adult, and so on, then we might better understand life spans in terms of these summary categories.

First, let's visualize the birth years of the people listed in the data. We will count how many people were born in each year and visualize the distribution of birth years using the following code:

Partial code: full program in visualize_birth_years.py:

```
import matplotlib.pyplot as plt

# gets rid of nan's and return a list:
just_birth_years = df.birth_year.dropna().tolist()
just_birth_years = sorted(just_birth_years)  # sort the birth years
years_count = {}  # create a dictionary of year to number of births
for year in np.arange(just_birth_years[0], just_birth_years[-1]):
    years_count[year] = 0  # set a default of zero births to every year
```

```
for year in just_birth_years:
    if year in years_count:
        years_count[year] += 1  # add 1 for every birth year found in the data

plt.bar(range(len(years_count)), list(years_count.values()))  # bar chart
plt.xticks([])  # remove the year labels - there are too many to show
plt.ylabel('Number of people born in year')
plt.xlabel('Year Born')
plt.title('Histogram of birth years')
plt.show()
```

The resulting bar chart is shown in Fig. 3.4. Because there are many individual years depicted from 686 to 1991, we removed year labels on the horizontal axis to reduce clutter. The easiest conclusion to derive from this data visualization is that most of the people represented in the dataset were born in a later era… but what's the best approach to manage this information?

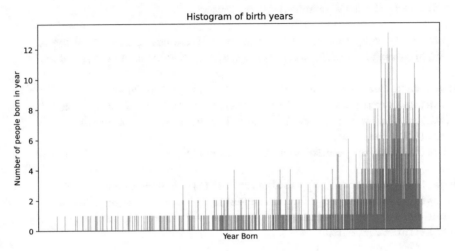

Fig. 3.4 Bar chart displaying the number of royals born each year. The taller the bar the more people born that year. (Image created from visualize_birth_years.py)

Binning the birth years will make the overall span of data easier to understand. Initially, let's acquire an overview of the birth years using the `describe` function:

```
Partial code: full program in describe_birth_years.py:

print(df['birth_year'].describe())
```

Which produces the following output:

```
count    1727.000000
mean     1739.392588
std       253.463283
min       686.000000
25%      1675.000000
50%      1837.000000
75%      1907.000000
max      1991.000000
Name: birth_year, dtype: float64
```

From this listing we note that from the 3009 total count of people in the dataset only 1727 possess a valid birth year between 686 and 1991. The quartile values indicate that most of the birth dates occurred later in this range of years. From this auspicious beginning, we now proceed to bin the data.

In Pandas there are two primary ways to bin data:

- `qcut` function: bins all data into quartiles (thus the 'q')
- `cut` function: bins all data based on equal intervals

The following code produces two bar charts, shown in Fig. 3.5, to illustrate the different results generated by these two functions:

```
Partial code: full program in bin_and_visualize_birth_years.py:

import matplotlib.pyplot as plt
df['year_q4'] = pd.qcut(df.birth_year, q=4)
df['year_c4'] = pd.cut(df.birth_year, bins=4)
q_results = df.year_q4.value_counts().sort_index()
c_results = df.year_c4.value_counts().sort_index()

fig, (ax1, ax2) = plt.subplots(1, 2)  # create two plots: 1 row, 2 columns
fig.suptitle('qcut example (left) with cut example (right) with 4 bins each')

ax1.bar(range(len(q_results)), q_results.tolist())  # bar chart
ax1.set_xticks(np.arange(len(q_results.index)))  # set the # of xticks
ax1.set_xticklabels(q_results.index)  # set the values of the xticks
ax1.set_xlabel('Interval Born')
ax1.set_ylabel("Number of people born in interval")

ax2.bar(range(len(c_results)), c_results.tolist())  # bar chart
ax2.set_xticks(np.arange(len(c_results.index)))  # set the # of xticks
ax2.set_xticklabels(c_results.index)  # set the values of the xticks
ax2.set_xlabel('Interval Born')

plt.show()
```

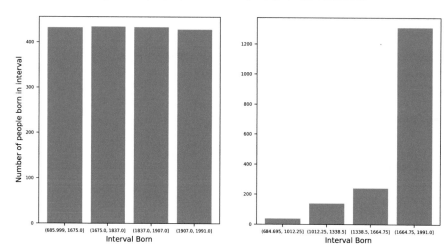

Fig. 3.5 Example of the difference between the qcut (left) and cut (right) functions. The qcut function bins all data into quartiles (thus the 'q') while the cut function bins based on equal intervals. (Image created from bin_and_visualize_birth_years.py)

Essentially, qcut divides the data so that each bin contains roughly equal data counts, whereas cut divides the range so that each bin covers an equal interval. We can also view the results of the binning with the following code:

```
print(q_results)
print('\n', c_results)
```

Produces the following results:

```
(685.999, 1675.0]      432
(1675.0, 1837.0]       434
(1837.0, 1907.0]       433
(1907.0, 1991.0]       428
Name: year_q4, dtype: int64

(684.695, 1012.25]      38
(1012.25, 1338.5]      139
(1338.5, 1664.75]      240
(1664.75, 1991.0]     1310
Name: year_c4, dtype: int64
```

To understand the interval index format (e.g. ((685.999, 1675.0]), consider Fig. 3.6. For any defined interval, a parentheses ('(' or ')') means to exclude the number and a bracket ('[' or ']') means to include the number. For example [1, 3] means 1 to 3 including both 1 and 3. Alternatively, [1, 3) means 1 to 3 including 1, but not 3, a range that would contain 2.999, but not 3.0. Consequently, (685.999, 1675.0] means greater than 685.999 (i.e. 686) up to and including 1675.

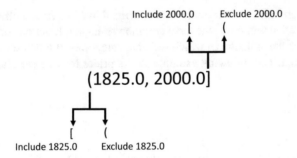

Fig. 3.6 Binning is based on the idea of mathematical ranges. The '[' designates number inclusion and the '(' designates number exclusion. The above '(1825.0, 2000.0]' reads, "From 1825 to 2000, excluding 1825 and including 2000." This notation defines an interval greater than 1825 up through and including 2000. Image created by Robert Ball

The qcut and cut functions allow the number of bins to be specified using the q and bins keyword parameters respectively.

In Sect. 3.3.6 we showed how to manually divide birth years into era bins using the loc function. The following shows how this binning operation can be achieved in one line with the cut function. The additional code in the file is used for printing out and visualizing the results as shown in Fig. 3.7:

```
Partial code: full program in bin_label_and_visualize_birth_years.py:

df['year_c3'] = pd.cut(df.birth_year, bins=3, labels=['ancient', 'middle_years', 'modern'])
```

The output follows:

```
ancient           67
middle_years     264
modern          1396
Name: year_c3, dtype: int64
```

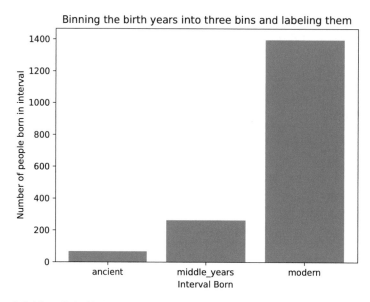

Fig. 3.7 Visual representation of dividing all the birth years into three bins labeled 'ancient,' middle_years,' and 'modern.' Compare to Fig. 3.4. (Image created from bin_label_and_visualize_birth_years.py)

3.3.8 Reshaping with Pivot, Pivot_Table, Groupby, Stack, Unstack, and Transpose

Frequently it is convenient or informative to restructure data contained in a dataframe, effectively organizing the data into a different shape or format. This section will cover the most common reshaping functions provided by Pandas.

The `pivot` function addresses the situation in which separate categories of a dataset feature are enumerated and highlighted using a cross tabular format. The following example shows prices for two car brands in both the 'new' and 'used' categories:

```
File: cross_tab1.py:
import pandas as pd
df = pd.DataFrame({'Car': [1, 1, 2, 2],
                   'Type': ['new', 'used', 'new', 'used'],
                   'Price': [10, 5, 12, 7]})
print(df)
```

Which produces the following output:

```
   Car  Type  Price
0    1   new     10
1    1  used      5
2    2   new     12
3    2  used      7
```

Calling the `pivot` function on this dataframe will rework the data into a more compact and usable format. Careful inspection of the dataframe above reveals that each car is designated by a number, but each car has two separate entries representing the 'new' and 'used' versions of that vehicle. The goal is to reshape the data such that each car brand is represented on a single row. A use case for this particular reorganization would be a car salesperson who needs to quickly view all the prices of a given car brand for the different 'Type' categories.

To use the `pivot` function the features in the original dataframe are used to specify the new index and new columns, as well as the values for each cell in the resulting table:

```
Partial code: full program in pivot1.py:

p = df.pivot(index='Car', columns='Type', values='Price')
print(p)
```

Which produces the following output:

```
Type   new  used
Car
1       10    5
2       12    7
```

The `pivot` function **only** works if there is either zero or one entries per cell in the result. Consider the following dataframe:

```
File: cross_tab2.py:
import pandas as pd
df = pd.DataFrame({'Car': [1, 1, 2, 2, 2],
                   'Type': ['new', 'used', 'new', 'used', 'used'],
                   'Price': [10, 5, 12, 7, 6]})
print(df)
```

Which produces the following output:

```
   Car  Type  Price
0    1   new     10
1    1  used      5
2    2   new     12
3    2  used      7
4    2  used      6
```

Invoking the `pivot` function on this dataframe (df.pivot(index = 'Car', columns = 'Type', values = 'Price')) will generate an error. The problem is that there are now two 'Price' entries (7 and 6) for car number 2 with a 'Type' category of 'used'. Which of the two values should go into the cell?

The solution to this problem is the `pivot_table` function. The `pivot_table` function resolves this issue by specifying how multiple entries are handled by applying an aggregation function. For example, the aggregation operation can indicate that a total count or a mean should be calculated for each cross-tabular value entry. The following code specifies the `mean` function from the Numpy class as the `aggfunc` keyword parameter to compute the average for each entry:

```
Partial code: full program in pivot2.py:

p = df.pivot_table(index='Car', columns='Type', values='Price', aggfunc=np.mean)
print(p)
```

Which produces the following output when applied to the most recent dataframe listed above:

```
Type   new  used
Car
1      10.0   5.0
2      12.0   6.5
```

Note that by employing the `pivot_table` function the value for Car 2 and Type 'used' is now 6.5, the average of 6 and 7 from the original dataframe. Pivot tables can result in immensely complex tabular formats with multiple indexes, multiple columns, and various aggregation functions specified. We demonstrate only the basic single-index, single-column case here to highlight the facility of pivot tables when applied to data analysis. The following code demonstrates another simple use case of a pivot table where the average birth year of people in our royal family history dataset are reported with title as the index and sex as the column:

```
Partial code: full program in pivot3.py:

df.dropna(inplace=True, subset=['title', 'sex', 'birth_year'])
p = df.pivot_table(index='title', columns='sex', values='birth_year', aggfunc='mean')
```

Which produces the following:

```
sex                  F        M
title
Admiral             NaN   1759.0
Admiral Sir         NaN   1881.0
Archduchess      1825.0      NaN
Archduke            NaN   1844.6
Baron               NaN   1901.0
...                 ...      ...
Tsarina          1859.5      NaN
Vicount Althorp     NaN   1964.0
Vicount Linley      NaN   1961.0
Viscount            NaN   1913.8
Viscount Hampden    NaN   1877.5
```

The following code provides two more examples. The first fills blank entries in the resulting pivot table after aggregation with 0 instead of NaN's and uses two aggregate functions, mean and count. The second is similar to the first, but instead declares two indexes and two columns producing a much more complicated, nested output result. The results are fairly extensive and the book cannot do justice showing them, but we encourage you to run the code yourself.

```
Partial code: full program in pivot4.py:

# example 1:
p = df.pivot_table(index='title', columns='sex', values='birth_year', aggfunc=[np.mean,
'count'], fill_value=0)

# example 2:
p = df.pivot_table(index=['title', 'first_name'], columns=['sex', 'last_name'], values='birth_
year', aggfunc=[np.mean, 'count'], fill_value=0)
```

The groupby function's recasting of information is very similar to that of the pivot_table function. In general, the main difference is how the resulting output is shaped. The most common novice mistake is to create a group object without specifying an aggregating function like mean, sum, and std. Consider the following code:

```
File: groupby1.py:
import pandas as pd
df = pd.DataFrame({'Car': [1, 1, 2, 2, 2],
                   'Type': ['new', 'used', 'new', 'used', 'used'],
                   'Price': [10, 5, 12, 7, 6]})

g = df.groupby(by='Car')
print(g)
```

The above code will print out something like <pandas.core.groupby.generic.DataFrameGroupBy object at 0x1179a6d60>, which is a result that fails to contribute to our data analysis goals. However, if we group the data by the 'Car' column and subsequently invoke the `mean` function on the resulting `DataFrameGroupBy` object then we will obtain more useful results, as illustrated by the following code:

```
File: groupby2.py:
import pandas as pd
import numpy as np
df = pd.DataFrame({'Car': [1, 1, 2, 2, 2],
                   'Type': ['new', 'used', 'new', 'used', 'used'],
                   'Price': [10, 5, 12, 7, 6]})

print(f'Original dataframe:\n{df}')

p = df.pivot_table(index='Car', columns='Type', values='Price', aggfunc=np.mean)
print(f'\nPivot table example:\n{p}')

g = df.groupby(by='Car').mean()
print(f'\nGroupby example:\n{g}')
```

Which produces the following output:

```
Original dataframe:
   Car  Type  Price
0    1   new     10
1    1  used      5
2    2   new     12
3    2  used      7
4    2  used      6

Pivot table example:
Type   new  used
Car
1     10.0   5.0
2     12.0   6.5

Groupby example:
        Price
Car
1    7.500000
2    8.333333
```

Note the mean aggregate function is only applied to columns with the appropriate datatype for calculating a mean value, which is why only the Price averages are displayed. In contrast, the count aggregate function can successfully calculate results for both numerical and string columns as verified by this next code example:

```
Partial code: full program in groupby3.py:

g = df.groupby(by='Car').count()
print(f'Groupby example:\n{g}')
```

Which produces the following output:

```
Groupby example:
     Type  Price
Car
1       2      2
2       3      3
```

The `groupby` function's various parameter options can produce results that are every bit as sophisticated as the results generated by the `pivot_table` function. The following example demonstrates the `groupby` function over two columns (Car and Type) and uses the `agg` (short for aggregate) function to produce two separate mean and sum calculations on the data:

```
Partial code: full program in groupby4.py:

print(df.groupby(by=['Car', 'Type']).agg(['mean', 'sum']))
```

Which produces the following output:

```
          Price
          mean sum
Car Type
1   new   10.0  10
    used   5.0   5
2   new   12.0  12
    used   6.5  13
```

The `groupby` function can be used in combination with many other functions covered in this chapter. For example, the following code effectively groups all royals by title and sex, counts the total in each title/sex category using the `value_counts` function, and finally filters the results so that only rows with a count of 50 or more are displayed:

```
Partial code: full program in groupby5.py:

print(df.groupby('title')['sex'].value_counts().loc[lambda x: x >= 50])
```

Which produces the following output:

```
title     sex
Duke      M       75
Lady      F       62
Prince    M      118
Princess  F       96
Name: sex, dtype: int64
```

The `stack` and `unstack` functions, as their naming implies, perform inverse operations. Based on the documentation describing these functions, stacking is analogous to taking a row of books arranged side by side horizontally as is typically done on a library shelf and ordering them vertically with the leftmost book now on top of the stack. For the purposes of our analogy, each feature for a given dataframe row constitutes a single book. The following code illustrates the results of applying this operation to a dataframe:

```
File: stack_example.py:
import pandas as pd
df = pd.DataFrame({'Car': [1, 1, 2, 2, 2],
                   'Type': ['new', 'used', 'new', 'used', 'used'],
                   'Price': [10, 5, 12, 7, 6]})
print(df)
s = df.stack()
print('\n', s)
print('\n', s.unstack())   # back to the original
```

Which produces the following output:

```
   Car  Type  Price
0   1   new    10
1   1   used    5
2   2   new    12
3   2   used    7
4   2   used    6

0  Car         1
   Type       new
   Price       10
1  Car         1
   Type      used
   Price        5
2  Car         2
   Type       new
   Price       12
3  Car         2
   Type      used
   Price        7
4  Car         2
   Type      used
   Price        6
dtype: object

   Car  Type Price
0   1   new    10
1   1   used    5
2   2   new    12
3   2   used    7
4   2   used    6
```

The stack function allows the programmer to specify which level or levels are stacked and whether to drop NA rows. For the unstack function the programmer can also specify which level or levels to unstack.

The transpose function (or simply T, with no parentheses), transposes a dataframe. For example, given the following code:

```
File: transpose_example.py:
import pandas as pd
import numpy as np
df = pd.DataFrame(np.arange(6).reshape(3, 2), columns=['A', 'B'])
print(f'Original\n{df}')

df = df.transpose()   # or df.T

print(f'\nTransposed:\n{df}')
```

produces the output:

```
Original
     A  B
0   0  1
1   2  3
2   4  5


Transposed:
      0  1  2
A   0  2  4
B   1  3  5
```

Essentially, transposing the dataframe will exchange the index and the columns. In other words, the axis for the columns and indexes of the dataframe will be swapped.

Table 3.3 provides a summary of the various reshaping functions.

Table 3.3 General description of the different reshaping functions

Function	Description
pivot	Create a new dataframe with the index, column(s), and values specified. Only ONE entry per cell is permitted.
pivot_table	Create a new dataframe with the index, column(s), and values specified. An aggregation function must be specified (e.g., count, sum, max, min, etc.).
groupby	Similar to *pivot_table*, but with less restrictions on the resulting shape. The result of the function returns a *DataFrameGroupBy* and so an aggregation function should be used.
stack	Change the dataframe from horizontal to vertically stacked.
unstack	Change the dataframe from vertically to horizontally stacked.
transpose (or *T*)	Swap the indexes and columns (rotate the dataframe 90 degrees).

3.3.9   Understanding Dataframe Indexes

A dataframe or series can be assigned any customized index different from the default zero-based index autogenerated by Pandas. When a new dataframe or series is imported or created, Pandas will by default set the first row index to 0, the next row index to 1, and continue this sequential numbering until N-1, where N is the total number of rows. If there are ten rows (N = 10) then the index will be 0 to 9 inclusive.

The following code produces a dataframe with six rows, indexed 0 through 5:

```
File: index_example3.py:
import pandas as pd
import numpy as np
df = pd.DataFrame(np.arange(12).reshape(6, 2), columns=['column 1', 'column 2'])
print(df)
```

Which produces the following output:

```
   column 1  column 2
0         0         1
1         2         3
2         4         5
3         6         7
4         8         9
5        10        11
```

However, you can redefine the indexing with strings as shown here where the index range is defined as 'a' through 'f':

```
File: index_example4.py:
import pandas as pd
import numpy as np
df = pd.DataFrame(np.arange(12).reshape(6, 2), columns=['column 1', 'column 2'], index=['a',
'b', 'c', 'd', 'e', 'f'])
print(df)
```

Which produces the following output:

```
   column 1   column 2
a         0          1
b         2          3
c         4          5
d         6          7
e         8          9
f        10         11
```

The index can follow any arbitrary numerical order as demonstrated by the following code:

```
File: index_example5.py:
import pandas as pd
import numpy as np
df = pd.DataFrame(np.arange(12).reshape(6, 2), columns=['column 1', 'column 2'], index=[4, 2,
1, 3, 5, 0])
print(df)
```

Which produces the following output:

```
   column 1   column 2
4         0          1
2         2          3
1         4          5
3         6          7
5         8          9
0        10         11
```

There are many built-in indexes available. For example, the RangeIndex lets you specify a range of values with a given increment, as demonstrated by the following code:

```
File: index_example6.py:
import pandas as pd
import numpy as np
df = pd.DataFrame(np.arange(12).reshape(6, 2), columns=['column 1', 'column 2'], index=[4, 2,
1, 3, 5, 0])
df.index = pd.RangeIndex(start=0, stop=30, step=5)
print(df)
```

Which produces the following output:

```
    column 1   column 2
0          0          1
5          2          3
10         4          5
15         6          7
20         8          9
25        10         11
```

Other options include but are not limited to `CategoricalIndex`, `MultiIndex`, `IntervalIndex`, `DateTimeIndex`, and `PeriodIndex`.

Useful attributes that characterize and test index structure are available such as `is_unique`, `is_monotonic` (meaning that the index stays the same or increases, but never decreases), and `size`. The following example examines the `is_unique` attribute:

```
File: index_example7.py:
import pandas as pd
import numpy as np

df = pd.DataFrame(np.arange(4).reshape(2, 2), columns=['column 1', 'column 2'])
print(f"First dataframe = \n{df}\nDoes it have an unique index: {df.index.is_unique}")
df = pd.DataFrame(np.arange(4).reshape(2, 2), columns=['column 1', 'column 2'], index=[1, 1])
print(f"Second dataframe = \n{df}\nDoes it have an unique index: {df.index.is_unique}")
```

Which results in the following output:

```
First dataframe =
    column 1   column 2
0          0          1
1          2          3
Does it have an unique index: True
Second dataframe =
    column 1   column 2
1          0          1
1          2          3
Does it have an unique index: False
```

One of the most practical index functions is `reset_index`. Because filtering and sorting operations preserve original indices for each of the rows the result after applying these operations will typically be a `non-sequential` index. For example, with our royal dataset of 3009 observations, filtering the rows with the title "Princess" will generate a result containing only 96 rows. After sorting the resulting dataframe we can utilize the `reset_index` function to reestablish a sequential numerical index from 0 to 95.

However, when first using the `reset_index` function, beginning data scientists are sometimes surprised to see the original index is preserved as an entirely new column as revealed in the following example:

```
File: index_example8.py:
import pandas as pd
import numpy as np

df = pd.DataFrame(np.arange(4).reshape(2, 2), columns=['column 1', 'column 2'], index=[1, 1])
df.reset_index(inplace=True)
print(df)
```

Which produces the following output:

```
    index   column 1   column 2
0     1         0          1
1     1         2          3
```

To delete the prior index, the *drop* parameter should be set to True as follows:

```
Partial code: full program in index_example9.py:

df.reset_index(drop=True, inplace=True)
```

Which produces the following output:

```
   column 1   column 2
0      0          1
1      2          3
```

3.3.10 Common Statistics Functions, Counting, and Sorting

There are many other helpful functions beyond the ones addressed in this chapter. In fact, expect to encounter new Pandas functions currently not available at the time this book is written. Given this reality, we highly recommend that an integral part of your lifetime pursuit of learning includes periodically exploring the Pandas documentation.

However, there exist a few remaining functions worth mentioning that do not fit neatly with the topics addressed in the other sections in this chapter. For example, the following functions are useful for basic descriptive statistics (see Chap. 4 for more information on descriptive statistics): min, max, idxmax (first occurrence of the maximum value), idxmin (first occurrence of the minimum value), sum, count, std. (standard deviation), median, and mode.

The unique function returns a list of the unique entries in a column. For example, running the unique function on the title column returns a list of the distinct titles contained in the dataframe observations. In a similar vein, the nunique function returns the total number of unique entries. However, be aware that nunique is NOT equivalent to calling len(unique) because the unique function also counts the occurrence of NA's whereas the nunique function does not. Consider the following illustrative example:

```
File: unique_and_nuniuqe.py:
import pandas as pd
df = pd.read_csv('royal_line.csv', index_col='ID')

# prints a list of the unique titles in the dataframe:
print(df['title'].unique())

print(df['title'].nunique())    # prints 308 (does NOT count NA's)
print(len(df['title'].unique()))    # prints 309 (does count NA's)

df.dropna(inplace=True, subset=['title'])    # get rid of NA's in title column
print(len(df['title'].unique()))    # now prints 308
```

The value_counts function utilized in one of the earlier code examples counts the number of a given set of entries in a column. The value_counts function is extremely versatile in that it can perform the following: return the total count, return the relative percentage of entries using the `normalize` parameter, include or not include NA's using the `dropna` parameter, and categorize the data with the `bin` parameter. The following demonstrates the various uses of the `value_counts` function when applied to the title column:

```
Partial code: full program in value_counts_example.py:

print(df['title'].value_counts())
print(df['title'].value_counts(normalize=True))
print(df['title'].value_counts(dropna=False))


# bins are only allowed for numerical values, so we use the birth_year column:
print(df['birth_year'].value_counts(bins=10))
```

Which produces the following output:

```
Prince               118
Princess              96
Duke                  76
Lady                  62
King of France        43
                     ...
Earl Lennox 6th        1
duc d' Alencon         1
Earl Mountbatten       1
Earl of Fife           1
Earl Dartmouth         1
Name: title, Length: 308, dtype: int64
Prince            0.084406
Princess          0.068670
Duke              0.054363
Lady              0.044349
King of France    0.030758
                    ...
Earl Lennox 6th   0.000715
duc d' Alencon    0.000715
Earl Mountbatten  0.000715
Earl of Fife      0.000715
Earl Dartmouth    0.000715
Name: title, Length: 308, dtype: float64
NaN               1611
Prince             118
Princess            96
Duke                76
Lady                62
                   ...
Holy Roman Empr      1
Prince Ilynsky       1
Countess of Kent     1
Earl of Bedford      1
Earl Dartmouth       1
Name: title, Length: 309, dtype: int64
(1860.5, 1991.0]          744
(1730.0, 1860.5]          444
(1599.5, 1730.0]          189
(1469.0, 1599.5]          103
```

```
(1208.0, 1338.5]                          78
(1338.5, 1469.0]                          70
(1077.5, 1208.0]                          44
(947.0, 1077.5]                           27
(816.5, 947.0]                            19
(684.6940000000001, 816.5]                 9
Name: birth_year, dtype: int64
```

The first statement above reveals there are more princes than any other title (118) followed by princesses (96). The second command normalizes the title data and shows that princes constitute 8.4% of the data and princesses account for 6.9%. The third line of code is informative because it uncovers the fact that there are 1611 people without any title.

The last statement with the `bin` parameter can only be applied to numeric data, so the `birth_year` column was processed instead. (For more information on binning see Sect. 3.3.7.)

The last essential function to be discussed in this section is sorting. Sorting is inherently a complex topic and is the focus of many computer science courses and textbooks. Index sorting has previously been demonstrated and the primary tool to accomplish most of the sorting applied to a dataframe or series is the optimized `sort_values` function.

A key initial point to note about `sort_values` is that it may not work as expected. Consider the following code:

```
Partial code: full program in sort_values_example1.py:

df = df['birth_year'].sort_values()
```

What does this function produce? The variable `df` is now assigned a series of the column `birth_year`, **not** the dataframe sorted based on the column `birth_year`. Consequently, it's important to remember to include the `by` parameter when using the *sort_values* function. The following examples show various ways to use the *sort_values* function:

```
Partial code: full program in sort_values_example2.py:

df.sort_values(inplace=True, by=['birth_year'])  # sorts birth_year numerically (smallest first)

df.sort_values(inplace=True, by=['title'])  # sorts title alphabetically (A-Z)

# sorts numerically in descending order:
df.sort_values(inplace=True, by=['birth_year'], ascending=False)

# sorts title then birth_year:
df.sort_values(inplace=True, by=['title', 'birth_year'])
```

Specifying the `key` parameter in the `sort_values` function allows you to sort columns in a customized way. You either pass a built-in or lambda function as the `key` parameter to define how the values are compared for sorting purposes. The `key` parameter is mentioned here as an introduction. Detailed explanations about customized sorting are beyond the scope of this book and we suggest referencing a comprehensive data structures book from computer science if you are interested in researching this topic.

3.3.11   Different Encodings for Categorical Data

Another important consideration when wrangling data is changing the representation or encoding of categorical features. The main types of encoding that we will cover are mapping categories to numerical representations and transforming categories from a single column to multiple binary columns.

In many cases certain algorithms, especially machine learning algorithms and statistical models, require that categorical information be represented in numeric form. This conversion is sometimes called normalizing categories to numbers. For example, consider mapping the different types of royal titles into numbers.

There are many ways to accomplish this transformation. However, one of the easiest approaches is to use the *LabelEncoder* class that is part of the *sklearn* package. The following is an example:

```
File: labelencoder_example1.py:
import pandas as pd
from sklearn import preprocessing

df = pd.read_csv('royal_line.csv', index_col='ID')

le = preprocessing.LabelEncoder()  # create the object
le.fit(df['title'])  # fit or learn the data
print(le.classes_)  # not necessary, but to show the unique titles

df['title_int'] = le.transform(df['title'])  # create an int representation
print(df[['title', 'title_int']].head(10))
```

The above code initially creates a *LabelEncoder* object, which is then utilized to fit, or learn, the data. In other words, *LabelEncoder* scans the unique titles (called classes) and creates a number representation for each class. The statement `print(le.classes_)` displays a list of all the unique titles. The conversion operation that assigns numerical values to unique categories is invoked through the `transform` function. Instead of doing this in two lines (calling the `fit` function then later calling the `transform` function) as demonstrated in the previous code, you can simply call the `fit_transform` function to accomplish both tasks in one statement as follows:

```
Partial code: full program in labelencoder_example2.py:

# fit (learn) and transform (create) an int representation in one step:
df['title_int'] = le.fit_transform(df['title'])
```

A convenient and useful function provided by the *LabelEncoder* class is *inverse_transform*, which will return the class name (in this case the royal title) associated with a given numerical mapping. The following code calls the *inverse_transform* function to decode 285 back to the title "Queen of England:"

```
Partial code: full program in labelencoder_example3.py:

# given a list of numbers, what are the titles?
print(le.inverse_transform([285]))  # prints 'Queen of England'
```

Another common conversion scenario required by machine learning algorithms is called one-hot encoding or dummies. *One-hot encoding* or *dummies* is the result of transforming a single column of categories such as our list of titles to a set of multiple columns, where each new column corresponds to an individual category label and indicates through a binary value if an observation either belongs (1) or does not belong (0) to that category.

More formally, one-hot encoding converts a single feature with N categories to N separate features, and dummies encoding converts a single feature with N categories to N-1 separate features.

As demonstrated in the following examples, the difference between one-hot encoding and dummies is that one-hot encoding returns a one-to-one relationship between the number of unique classes in the data and the resulting number of columns. That is, N classes produces N columns.

Three classes listed in the following example (cat, dog, and turtle) will produce three columns.

On the other hand, dummies generate a more compact result of N-1 columns for N classes.

Dummies is an older statistical method based on linear dependencies that are not as relevant in modern machine learning applications. Unless you need dummies for a specific reason, the default of one-hot encoding is recommended.

To demonstrate one-hot encoding, assume a single column or series containing three unique animals:

```
File: one_hot_encoding1.py:
import pandas as pd
s = pd.Series(['dog', 'dog', 'cat', 'turtle', 'turtle', 'turtle'])
print(s)
```

Which produces the following output:

```
0        dog
1        dog
2        cat
3     turtle
4     turtle
5     turtle
dtype: object
```

One-hot encoding will convert these categorical variables to a new multi-column-based format that generally allows machine learning algorithms to be more effective at prediction tasks.

```
Partial code: full program in one_hot_encoding2.py:

results = pd.get_dummies(s)
print(results)
```

Which produces the following output:

```
   cat  dog  turtle
0   0    1     0
1   0    1     0
2   1    0     0
3   0    0     1
4   0    0     1
5   0    0     1
```

Even though the Pandas function is called `get_dummies` it produces a one-hot encoding by default because that is what most algorithms expect. If you truly want to get dummies then you need to specify the `drop_first` option as shown here:

```
Partial code: full program in one_hot_encoding3.py:

results = pd.get_dummies(s, drop_first=True)
print(results)
```

Which produces the following output:

```
   dog  turtle
0   1     0
1   1     0
2   0     0
3   0     1
4   0     1
5   0     1
```

One-hot encoding can also be accomplished using the *OneHotEncoder* class included in the *sklearn* package. There are a number of options in the *sklearn.preprocessing* package to create numerical mapping from a single column of categorical data, including the *LabelEncoder*, *OrdinalEncoder*, *LabelBinarizer*, *OneHotEncoder*, and *MultiLabelBinarizer* classes.

An example of the *MultiLabelBinarizer* class is illustrated in Sect. 6.1.7 where we work with genres from movies. In that particular example, one-hot encoding would not be appropriate because that approach prohibits more than one category per row. This requirement would limit an individual movie to only one genre, such as comedy or romance. However, the *MultiLabelBinarizer* class allows a single movie in the dataset to have multiple genre labels permitting a movie to be categorized as both comedy **and** romance.

Exercises

1. What is data wrangling?
2. What causes data to become information?
3. What causes information to become knowledge?
4. How can you help other people gain knowledge from a dataset?
5. CSV files are often preferred over XLS or XLXS files. Why? Do you agree or disagree with the reasoning?
6. When would you use a Pandas dataframe?
7. When would you use a Pandas series?
8. When do you use loc and when do you use iloc?
9. How would an organization (e.g., a business) use the binning capabilities of cut and qcut?
10. Explain different situations when you would use apply, map, and applymap.
11. Explain three scenarios when the value_counts function could be useful.
12. Provide two scenarios when you might need to transpose a dataframe.
13. What is the difference between the unique and nunique functions?

Chapter 4
Crash Course on Descriptive Statistics

What are "statistics?" For our purposes, how are "statistics" employed in the service of data science?

Even if you enrolled in and completed a statistics course you may not have realized how useful these analytical techniques would be in the long term. A good portion of the most important insights into any dataset would not be uncovered without statistics.

Also, for some context it is important to be acquainted with a common phrase regarding statistics: "Statistics is not second-rate mathematics, but first-rate science." Without statistics we would not have achieved some of our most substantive scientific insights.

The purpose of this chapter is not to displace any formal statistics course, but instead to augment what we have found many students and professionals interested in data science might be lacking. Even if you have successfully completed a statistics course and are comfortable with all the concepts in this chapter you may not know how to effectively apply many of the analyses and visualizations in Python also included in this chapter. Feel free to scan through this chapter and gain the knowledge that best applies to your current data science explorations.

Also, please realize that this chapter should be regarded only as an ***introduction*** to many statistical concepts. It is our firm belief that a data scientist or analyst can never learn enough about statistics. Like most professions, data scientists should be constantly seeking to improve and expand their knowledge base, especially in areas involving both traditional and contemporary numerical analysis.

Statistics is the discipline of developing and using different methods for collecting, analyzing, interpreting, and presenting numerical data. Probabilities, distributions, uncertainty, randomness, experimental design, and evaluating results are all major topics in statistics. Given the broad spectrum of topics encompassed by statistics, it is important to realize that statistics is its own field of study with many different variations and techniques.

In this book we will be focused on the Exploratory Data Analysis (EDA) aspect of statistics. *Exploratory Data Analysis (EDA)* defines a particular way of analyzing numerical and categorical values that focuses on maximizing insight into these data using statistics. As suggested by its name, EDA *explores* the data and attempts to decipher it by looking specifically at what is typical (e.g. its distribution and structure), what is not typical (e.g. outliers), and what are the most important insights. EDA is both a microscope, by allowing us to drill down into critical and informative details, and a telescope, by helping us to articulate "big picture" observations and inductions about the data set in question.

We will begin our practical investigation into EDA by using the Iris dataset. The Iris dataset, available since the 1930's, has become one of the de facto training datasets to teach basic statistics and machine learning. The dataset includes four variables (also known as features or attributes) that describe the dataset and contains 150 observations with 50 records each for three type of Iris flowers: Iris setosa, Iris virginica and Iris versicolor. The four variables are width and length of the flower's petals, and width and length of the flower's sepals.

An *observation* in statistics is a discernible and measurable fact. For instance, an observation might be a record of how many times a flipped coin reveals "heads" or an observation might be the record of how tall you were on your twelfth birthday.

© The Author(s), under exclusive license to Springer Nature Switzerland AG 2022
R. Ball, B. Rague, *The Beginner's Guide to Data Science*, https://doi.org/10.1007/978-3-031-07865-1_4

To understand the Iris dataset we need to first review petals and sepals as botanical parts of a flower. For illustration purposes, Fig. 4.1 shows an example Gaillardia flower. A "petal" is the colorful part of a flower, the red and yellow part of the flower in Fig. 4.1, and a "sepal" is the (usually) green covering that protects the flower as it grows from bud to full bloom. The sepal often looks like a green leaf under the petals.

Fig. 4.1 Evolution from bud to flower for the Gaillardia flower (also known as Blanket Flower), Gaillardia pulchella, with sepals and petals labeled. (Image created by Robert Ball)

There are many different perspectives someone may adopt about the Iris dataset depending on their interest. This tendency for individuals to notice particular aspects of an object or dataset based on their prior experience or current goals is called *observer bias*. For our purposes let us suppose the person who collected the data is a botanist (a plant scientist) and seeks our help analyzing their data. The botanist might be interested in trying to identify and categorize the different Iris plants. In this case, the domain expert would be the botanist and we (the data scientist) would be responsible for helping them understand specific features and trends related to the three Iris species.

As with all data science projects it is vital to the success of our endeavor that we understand **what we are doing** and **why we are doing it**. In this example we are being asked to assist a botanist with the identification of plants (the three Iris species). From a "big picture" perspective of botany, plant taxonomy is the classification of plants into their respective hierarchical structure, i.e., their correct kingdom, phylum, class, order, family, genus, and species. For our Iris flowers, the first six categories, kingdom through genus, are already known. We are focused only on helping the botanist with the identification of the species.

In other words, if a group of botanists were to visit a large meadow filled with hundreds or thousands of Iris flowers, what insight into the identification of the different species could we offer the botanist to help them efficiently assign a selected flower to one of the three Iris species (Setosa, Versicolor, and Virginica)?

With that in mind, before we begin to examine the observations from the Iris dataset it is important for us to ask a series of questions that will help us elucidate our task. The following are sample questions to ask about this particular collection of data:

- What is similar about all three species?
- What is different about all three species?
- How do the different variables (i.e., the petal length and width, and sepal length and width) relate to one another for each different species?
- Are there examples that qualify as outliers (i.e., extreme values) which might confuse our identification strategy or make classification more difficult?
- What is the cheapest or most economical way of identifying the different species?

Now that we have clarified the main questions we wish to explore, let us look at the data.

The Pandas library in Python offers many of the most commonly used statistical functions. The following code example demonstrates one of the easiest methods to load the Iris dataset into a dataframe structure. See Sect. 3.3 for more information on dataframes.

```
File: iris.py:
import pandas as pd

iris_df = pd.read_csv('iris.csv')
print(iris_df)
```

Which produces the following output:

	sepal_length	sepal_width	petal_length	petal_width	class
0	5.1	3.5	1.4	0.2	Iris-setosa
1	4.9	3.0	1.4	0.2	Iris-setosa
2	4.7	3.2	1.3	0.2	Iris-setosa
3	4.6	3.1	1.5	0.2	Iris-setosa
4	5.0	3.6	1.4	0.2	Iris-setosa
..	...	...	...	...	...
145	6.7	3.0	5.2	2.3	Iris-virginica
146	6.3	2.5	5.0	1.9	Iris-virginica
147	6.5	3.0	5.2	2.0	Iris-virginica
148	6.2	3.4	5.4	2.3	Iris-virginica
149	5.9	3.0	5.1	1.8	Iris-virginica

```
[150 rows x 5 columns]
```

The 150 observations in the dataset are considered a "sample" of the overall "population" of all the Iris flowers in the entire world. In statistics, it is important to differentiate between when you have a *sample*, which is only part of all the possible data, or the *population*, which includes **all** the data. We must also be especially careful to ensure that our sample is sufficiently representative of the population, through both size and diversity, before asserting general statements about the broader population based on analysis of the sample observations.

For the Iris flowers example, the population would be all of the Iris flowers everywhere on the globe. Taking the measurement of every Iris flower in every field and meadow all over the world is impractical, so botanists only take samples to ascertain some useful knowledge about the worldwide population of Iris flowers.

In statistics many of the formulas are different when dealing with a sample compared to the entire population. However, a vast majority of the data you will likely analyze will be sample data that is representative of the overall population. For our purposes, all the equations in this chapter will presume that we are working with sample data and not population data.

One of the most basic categories of statistical measurement is called "descriptive statistics" – the subject of this chapter. The application of *descriptive statistics* attempts to summarize the data into a few concise but meaningful measurements.

Before we begin inspecting the data, we need to understand what the term "univariate" means. When used with statistics, *univariate* analysis means investigating a single variable. In the case of the Iris dataset, we will be looking at each variable (petal length, petal width, sepal length, and sepal width) in isolation. In other words, univariate analysis assesses the individual variables separately and independently with no regard to how they might interact.

Dependent and independent are both important concepts in statistics. A *dependent variable (DV)* is a value that depends on another variable. For example, the final grade that you receive in a class depends on the amount of time that you devote to studying. In this case, the quantity of study time is considered the *independent variable (IV)*.

In other words, the resulting outcome of a dependent variable is directly related to the value of the independent variable. To further illustrate this relationship, consider the following examples:

- The amount you pay on taxes (DV) is based on how much income you earn (IV).
- The height of a plant (DV) depends on the quantity of water it receives (IV).
- The chances of being diagnosed with lung cancer (DV) depends on exposure to smoke and pollution (IV).
- A person's weight (DV) depends on how much food the person consumes (IV).
- How strong you are (DV) depends on how much you exercise (IV).

4.1 Min and Max

Two of the most basic univariate functions (functions which analyze only one variable) that provide insight about the data are the minimum and maximum. The definition of *minimum* is obvious and is defined as the "lowest" value (or equally lowest – for multiple occurrences of the same lowest value). The definition of *maximum* is also obvious and is defined as the "highest" (or equally high) value. We can use the respective built-in min and max functions to perform this calculation.

```
Partial code: full program in min_and_max_example.py:

print(iris_df['sepal_length'].min())
print(iris_df['sepal_length'].max())
```

Which produces the following output:

```
4.3
7.9
```

4.2 Count

Knowing the total number of observations in the dataset is also important. In our example, there are 50 observations for each of three Iris species, resulting in a total of 150 observations. We can use the built-in `count` function to determine this value.

```
Partial code: full program in count_example.py:

print(iris_df['sepal_length'].count())
```

Which produces the following output:

```
150
```

4.3 Mean

The mean value of any particular variable is also helpful information. (In mainstream society the term "average" is often used synonymously with the term "mean.")

However, there are many types of mean calculations available. The most familiar mean is called the "Arithmetic Mean." The general approach to compute the arithmetic mean is to sum all the values of interest then divide by the total number of observations. The "x-bar" $\bar{x}$ notation often represents the sample mean. We can utilize the built-in `mean` function to calculate the arithmetic mean. The following is the mathematical notation for determining the arithmetic mean followed by associated Python code:

$$\bar{x} = \left(\frac{1}{n}\right)\sum_{i=1}^{n}(x_i)$$

```
Partial code: full program in mean_not_nice_example.py:

print(iris_df['sepal_length'].mean())
```

Which produces the following output:

```
5.843333333333334
```

4.4 Standard Deviation

Another important procedure to achieve insights about a given dataset is calculating the sample *standard deviation*, or the spread of the data. The lower-case letter s often represents sample standard deviation. Given the sample arithmetic mean $(\bar{x})$, the sample standard deviation is a summary measure of the distance between each data point and the sample mean. For example, if there is a relatively expansive spread, in other words, many values situated at a wide range of distances from the mean, then the standard deviation will be large. However, if most of the numbers are grouped closely around the mean, even if there are a few outliers (eccentric values at a significant distance from the mean), then the standard deviation will be small.

Variance and standard deviation are tightly coupled calculations. *Variance* is by definition an average of the sum of the squared differences from the mean and is often represented as s^2. Standard deviation is the square root of variance. The following is the mathematical notation for the unbiased (note division by $n - 1$) sample standard deviation:

$$s = \sqrt{\frac{\sum_{i=1}^{n}\left(x_i - \bar{x}\right)^2}{n-1}}$$

The familiar example of student grade distribution effectively illustrates the power of descriptive statistics. If we were informed that the mean score for an exam was 75% we would only know one aspect of overall student performance. If the standard deviation for that exam was also reported as 2.5, then we gain a more meaningful insight about student grades, namely, most students scored close to 75%. Specifically, we would know that most of the class scored from 72.5% to 77.5% (75% ± 2.5). On the other hand, if the standard distribution were higher, say 10, then the range of scores for most of the students would be from 65% to 85% (75% ± 10).

Figure 4.2 displays the normal distribution with 0 representing the mean (the location of our 75% score in the exam example). One standard deviation away from the mean (from $-1s$ to $1s$) encompasses 68.2% of the data (34.1% + 34.1%) – the light blue area. When two standard deviations are included (from $-2s$ to $2s$) then 95.4% of the scores would fall into that range – the light blue and light green areas together. Adding the light red area (from $-3s$ to $3s$) comprises approximately 99.7%, or close to 100% of the scores.

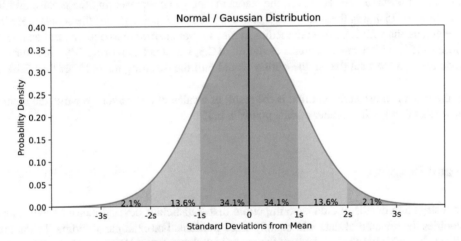

Fig. 4.2 The normal or Gaussian distribution (also known as the bell curve) with zero (0) representing the mean. The light blue area depicts one standard deviation away from the center, the light green area depicts two standard deviations away from the center, and the light red area depicts three standard deviations away from the center. (Image created from normal_distribution.py)

The following code shows how to derive the sample standard deviation:

```
Partial code: full program in standard_deviation_example.py:


print(iris_df['sepal_length'].std())
```

Which produces the following output:

```
0.828066127977863
```

4.5   "Bell Curve" or Normal Distribution or Gaussian Distribution

The concept of *spread* (also known as a *distribution*) is very important in grasping the information content inherent in a dataset. Whenever you try to extract insights from a dataset it is essential to understand where most of the data resides. In many cases the data are randomly generated, the situation for most natural phenomena, and like most exam results, the distribution will appear as shown in Fig. 4.2. The bell-like curve is often called exactly that: the "bell curve." However, it is more formally labeled the "normal distribution" or the "Gaussian distribution" (named after Carl Gauss – a famous mathematician).

In normal distributions, the three standard deviations from the mean are collectively expressed as the *empirical rule* or the *68-95-99.7 rule* to reflect the data density intervals described above.

4.6   Median

The distribution of your data can initially shed light on the path toward a proper interpretation of that data. One of the important measures derived from distributions is the *median*. In statistics, the "middle" value of a distribution is generally described as the *central tendency*. The most common measures of central tendency include the mean (Sect. 4.3), the mode (Sect. 4.10), and the median. Revisiting Fig. 4.2, we observe that half the data falls to the left of the middle "zero" line and the other half falls to the right of the middle "zero" line. The value which divides the data distribution into two equal halves is the median.

What is the difference between the median and mean? The mean sums all the numbers in the dataset and divides that result by n – the number of observations. However, the median is the value representing the precise middle of the distribution. For our exam example, 75% was the mean, but it could also be the median *if* half the scores were less than 75% and half the scores were greater than 75%. Unlike the median, a mean is very sensitive to outliers. For example, assuming most students scored close to 75% with a small standard deviation of 2.5, if instead of scoring 74% one student did very poorly and received a grade close to 0% then that single outlier would pull the resulting mean lower than 75% without affecting the median.

In other words, the mean is a measurement that is the result of a calculation of average while the median is the result of determining the value that divides the number of data points in half.

4.7   Quantile and Boxplots

Understanding the distribution details of data is so important that statisticians devised measurement intervals called quantiles. A *quantile* codifies the division of data into equally sized portions. For example, dividing the data into two quantiles generates the median value, which is the point where the quantile division occurs. Dividing the data into four quantiles (also known as *quartiles*) produces the values where the data is divided among four equally sized sections which forms the basis of the *boxplot visualization*.

Figure 4.3 depicts a boxplot visualization in the top part with the associated normal distribution below. A boxplot has two main features: the box and the whiskers. The box is shown as purple in the image and represents the *interquartile range (IQR)*. The IQR is the median value plus the two adjacent quartiles, which comprise 50% of the data. The whiskers, the horizontal lines extending outward on either side of the box, indicate either one of two measures. They either terminate at the minimum and maximum of the data or, if there are outliers present, they span from each side of the box plus 1.5 * IQR with outliers represented as circles. Unfilled circles indicate suspected outliers while filled circles signify known outliers. By definition, known outliers are more than 3 * IQR above Q3 or below Q1, and suspected outliers are more than 1.5 * IQR but less than 3 * IQR above Q3 or below Q1. For illustration purposes only, Fig. 4.3 depicts suspected outliers to the far left and known outliers to the far right.

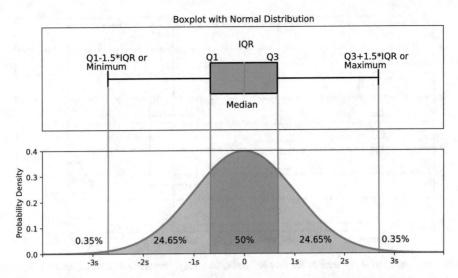

Fig. 4.3 Example of how the boxplot visualization maps onto the normal distribution. (Image created from boxplot_with_normal_distribution.py then modified with annotated text)

As an example, consider the different quantiles generated by the following Python code with dataset values of 1, 2, 3, 4, 5, 6, 7, 8, 9, and 10.

```
File: quantiles_example.py:
import pandas as pd
quantile_df = pd.DataFrame([1,2,3,4,5,6,7,8,9,10], columns=['a'])

print(f"50% quantile = {quantile_df['a'].quantile(.5)}")
print(f"median = {quantile_df['a'].median()}")
print(f"25%, 50%, and 75% quantiles are: \n{quantile_df['a'].quantile([.25,.5,.75])}")
```

Which produces the following output:

```
50% quantile = 5.5
median = 5.5
25%, 50%, and 75% quantiles are:
0.25    3.25
0.50    5.50
0.75    7.75
```

As we can see from the above program output, the median and the 50% quantile are the same. Also, the 25% and 75% quantile values are displayed using the *linear interpolation method*.

The following code creates a boxplot visualization for the four variables in the Iris data with results shown in Fig. 4.4:

```
Partial code: full program in boxplot_example.py:

import matplotlib.pyplot as plt
iris_df.boxplot()
plt.show()
```

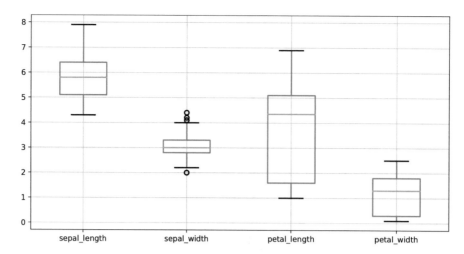

Fig. 4.4 Example of boxplot visualizations for each of the four Iris variables. (Image created from boxplot_example.py)

Note in Fig. 4.4 that none of the four features of the Iris dataset exhibit the canonical symmetric boxplot shape associated with the normal distribution shown in Fig. 4.3. The locations of the boxes, which encompass 50% of the sample data in each of these plots, are generally biased toward the lower end of the range defined by the whiskers. The sepal width boxplot shows suspected outliers which are plotted as separate unfilled circles.

4.8 Pandas "Describe" Function

Finding the count, arithmetic mean, sample standard deviation, min, max, and three quantiles (25%, 50%, 75%) is so common that Pandas created the "describe" routine to compute each of these queries using one function call.

This example shows the statistics generated by the "describe" function for each of the features in the Iris dataset:

```
Partial code: full program in describe_example.py:

print(iris_df.describe())
```

Which produces the following output:

```
         sepal_length  sepal_width  petal_length  petal_width
count    150.000000    150.000000   150.000000    150.000000
mean       5.843333      3.054000     3.758667      1.198667
std        0.828066      0.433594     1.764420      0.763161
min        4.300000      2.000000     1.000000      0.100000
25%        5.100000      2.800000     1.600000      0.300000
50%        5.800000      3.000000     4.350000      1.300000
75%        6.400000      3.300000     5.100000      1.800000
max        7.900000      4.400000     6.900000      2.500000
```

4.9 Z-Score

A measure similar in spirit to standard deviation is the "z-score." The *z-score*, also called the *standard score*, indicates the distance of a data point from the mean in standard deviation units. The *z-score* formula as shown here is the data point minus the sample mean divided by the sample standard deviation:

$$z = \frac{(x - \bar{x})}{s}$$

A negative z-score indicates the data point is below the mean. Similarly, a positive z-score indicates the data point is above the mean. A z-score of approximately zero would signify the data point is near the mean. For example, if a student received a 98% on an exam and had a z-score of 1.3 then the student would be 1.3 standard deviations above the mean. If another student received a 55% and a z-score of −1.8 then the student would be 1.8 standard deviations below the mean.

The following code calculates the z-score for each of the scores from an exam:

```
File: z_score_example.py:
from scipy import stats
import pandas as pd

scores = pd.Series([55, 60, 72, 78, 80, 81, 82, 83, 90, 98, 100])
print(scores.describe())
print('\n',stats.zscore(scores))
```

Which produces the following:

```
count    11.000000
mean     79.909091
std      13.895715
min      55.000000
25%      75.000000
50%      81.000000
75%      86.500000
max     100.000000
dtype: float64


0    -1.880067
1    -1.502681
2    -0.596956
3    -0.144093
4     0.006862
5     0.082339
6     0.157816
7     0.233293
8     0.761633
9     1.365450
10    1.516404
dtype: float64
```

The above results display the z-score of each exam. Note that 55 (the first value in the series) has a z-score of −1.88, and the value 60 has a z-score of −1.5. The value with the smallest z-score, which is almost 0, is 80, which is practically equivalent to the calculated mean score of 79.91.

4.10 Mode

The last univariate statistic we will discuss is the *mode*, which is simply the value that appears most frequently in a dataset. As mentioned above, the mode is considered a measure of central tendency. Either one or several modes are possible for a given variable. Only one mode is reported for the case in which there is only one observation more common than any of the others. If there are multiple values that each occur at the same frequency then all are reported. The following code initially displays the mode for the sepal length from the Iris dataset followed by several other representative examples.

```
File: mode_example.py:
import pandas as pd
iris_df = pd.read_csv('iris.csv')
print(f'mode of the Iris sepal length is \n{iris_df["sepal_length"].mode()}')

mode_df = pd.DataFrame([1, 2, 3], columns=['a'])
print(f'\nmode of [1, 2, 3] is \n{mode_df["a"].mode()}')

mode_df = pd.DataFrame([1, 2, 2, 3], columns=['a'])
print(f'\nmode of [1, 2, 2, 3] is \n{mode_df["a"].mode()}')

mode_df = pd.DataFrame([1, 2, 2, 3, 3, 5], columns=['a'])
print(f'\nmode of [1, 2, 2, 3, 3, 5] is \n{mode_df["a"].mode()}')
```

Which produces the following output:

```
mode of the Iris sepal length is
0    5.0
dtype: float64

mode of [1, 2, 3] is
0    1
1    2
2    3
dtype: int64

mode of [1, 2, 2, 3] is
0    2
dtype: int64

mode of [1, 2, 2, 3, 3, 5] is
0    2
1    3
dtype: int64
```

Note for Iris sepal length a single mode is calculated, but the second example with dataframe [1,2,3] calculates three modes since each value occurs only once. Although mode is useful, it is not used nearly as often as the other statistical functions we have covered.

However, consider which descriptive univariate measure would allow for easier interpretation when assessing data obtained from asking a group of individuals the following customer satisfaction question: "How was your experience – 1. Excellent 2. Good 3. Fair?" Unless the qualitative "distance" between 'Excellent' and 'Good' can be proven to be identical to the qualitative "distance" between 'Good' and 'Fair', a mean value of say 2.6 would be difficult to decipher because the intervals between values cannot be presumed equal. In this case, the mode would be much more useful in that it would return the most common response(s).

4.11   Data Visualization Using Distributions

As we have already seen, one of the most important concepts in descriptive statistics is the distribution of data. A distribution explains how the data points are organized relative to one another. In other words, a distribution shows us what to expect from our data in terms of key descriptive statistics such as the mean, median, variance, and standard deviation.

For our Iris dataset, we can first visualize the petal length distribution by plotting it in a histogram. A *histogram* is a specialized bar chart that bins (or aggregates) the data.

For example, binning data in the Iris dataset might entail taking all real number values and aggregating them into whole number value groupings. Namely, if a petal length were 1.1 and another were 1.4 then they would both be placed in the "1" bin. Binning allows researchers to obtain a better overview of data without being bogged down with small value fluctuations. Binning does not necessarily have to aggregate precisely into whole numbers. For example, we could bin the Iris data in *half* integer values: 1.1 and 1.4 would be placed into the "1.0" bin, while 1.5 and 1.6 would be assigned the "1.5" bin.

The following code produces a histogram that reveals a coarse-grained view of the petal length distribution for the Iris dataset:

```
File: histogram1.py:
import pandas as pd
import matplotlib.pyplot as plt  # this is needed to show visual charts

iris_df = pd.read_csv('iris.csv')
ax = iris_df['petal_length'].plot.hist()
ax.set_title('Distribution of Petal Length')
plt.show()
```

Figure 4.5 shows the histogram result. The histogram aggregates all the petal length data into 10 separate bins based on total number of occurrences. (For more information about binning see Sect. 3.3.7.) For example, a quick inspection of Fig. 4.5 reveals the most common petal length is in the 1 cm range with a count slightly greater than 35.

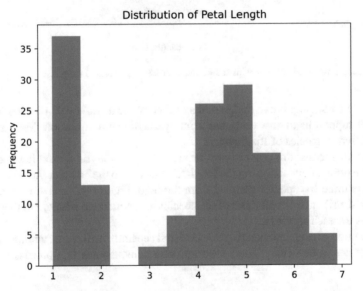

Fig. 4.5 Distribution of Petal Length in the Iris dataset. (Image created from histogram1.py)

Although the histogram furnishes us with an approximation of the overall spread of the data for petal length, it is still quite rough. Using the `distplot` function from the Seaborn library adds a smoothing curve to the plot that enhances our understanding of the overall shape and trend of the distribution.

Smoothing provides an estimated trend that is easier to visualize and understand. Smoothing is useful in a variety of applications, improving the clarity for both distributions and time series, a topic covered in Sect. 10.5.

The following code produces a slightly different binned histogram from the previous program along with a smoothing curve. The resulting plot is displayed in Fig. 4.6.

```
File: histogram2.py:
import matplotlib.pyplot as plt
import pandas as pd
import seaborn as sns
```

```
iris_df = pd.read_csv('iris.csv')

ax = sns.distplot(iris_df['petal_length'])
ax.set_title('Distribution of Petal Length')
plt.show()
```

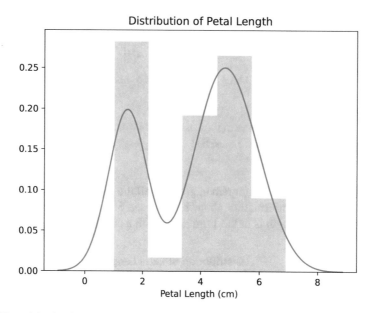

Fig. 4.6 Distribution of Petal Length in the Iris dataset with smoothing curve. (Image created from histogram2.py)

Figure 4.6 demonstrates a smoothing curve that exhibits two peaks or a "bimodal" distribution, but it really is not very helpful because it does not support us in answering our primary classification question. More specifically, how does petal length relate to the three different species of Iris flowers?

What we really need to remember at this point is why we are looking at the Iris dataset in the first place. At the beginning of this chapter recall that the point of our analyzing the Iris dataset is to help the botanist (i.e., the business domain specialist) more efficiently identify the three Iris species sampled from the field. Figures 4.5 and 4.6 are both helpful when trying to understand the petal length distribution for all three of the species as a collective **whole**, but they do not help us understand the **individual** distributions for each species.

To investigate the distribution of each individual species we first separate the Iris data among the three species then visualize these distributions together in one chart. The following code demonstrates this process with the result displayed in Fig. 4.7.

```
File: histogram3.py:
import matplotlib.pyplot as plt
import pandas as pd
import seaborn as sns

df = pd.read_csv('iris.csv')

setosa = df[df['class'] == 'Iris-setosa']
versicolor = df[df['class'] == 'Iris-versicolor']
virginica = df[df['class'] == 'Iris-virginica']

ax = sns.distplot(setosa['petal_length'], label='Iris Setosa')
ax = sns.distplot(versicolor['petal_length'], label='Iris Versicolor')
ax = sns.distplot(virginica['petal_length'], label='Iris Virginica')
ax.set_title('Distribution of Petal Length')
plt.legend()
plt.show()
```

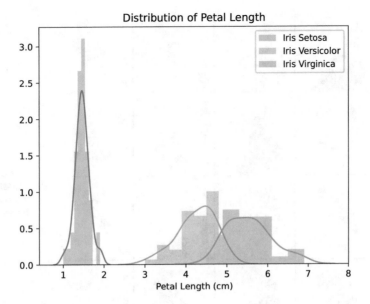

Fig. 4.7 Distribution of Petal Length for each species in the Iris dataset. (Image created from histogram3.py)

Figure 4.7 distinguishes the distribution of each of the three species using three separate colors. We can see the Setosa species rendered in light blue has the smallest spread of the three species. However, even though the mean and the standard deviation of each of the three species is different, we observe the petal length measurements for each approximate the normal distribution.

We note the central tendency for each of the three species is different with roughly 1.5 cm for Setosa, about 4.5 cm for Versicolor, and 5.5 cm for Virginica. This is highly informative for supporting the botanist's effort to more easily identify the three Iris species.

More precisely, we can observe the spread, or amount of variance, for the Setosa species is smaller compared to the other two species. We can also notice for the Setosa species there are very few outliers located a significant distance away from the central tendency of 1.5 cm. In addition, we also can easily see that Versicolor and Virginica overlap in petal length values whereas the Setosa distribution remains distinct from the other two.

So, for a practical example, if we were to go to a meadow populated with the three Iris species and measure petals with a length between 1 to 2 centimeters then we could conclude with high confidence that those Iris flowers belonged to the Setosa species.

Conversely, if we were to measure Iris flowers that possessed petal lengths between 4.5 to 5.5 centimeters then we could only state with some certainty that they are probably *not* in the Setosa species, but at the same time we could *not* with confidence assign these flowers to either the Versicolor or Virginica species because the distributions of those two species overlap.

Consequently, the next step in our Exploratory Data Analysis would be to examine the trends and distributions of other variables (e.g. petal width, sepal length, and sepal width) to help us better identify which particular flower belongs to which Iris species. More information will increase the likelihood of an accurate classification.

Although we could generate many distribution plots at this point for each of the other three variables, the Seaborn library has a `pairplot` function that will allow us to get an overview of all the distributions in one visualization. Please note in the following code the use of the `hue` keyword in the `pairplot` function to indicate to the function that we specifically would like to show the three different species in different colors (the results of the code are in Fig. 4.8):

```
File: pairplot_example.py:
import matplotlib.pyplot as plt
import pandas as pd
import seaborn as sns

df = pd.read_csv('iris.csv')
sns.pairplot(df, hue='target')
plt.show()
```

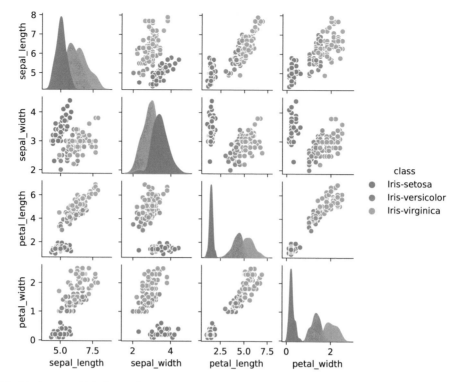

Fig. 4.8 Pairplot of the Iris dataset, which shows a four-by-four matrix of charts. The diagonal from top-left to bottom-right displays the univariate distributions for each of the four variables. For example, the third chart from the top-left to bottom-right shows the distribution of petal length and is the same graph we created in Fig. 4.7. (Image created from pairplot_example.py)

The plots along the diagonal from top-left to bottom-right display the univariate distributions for each of the four Iris variables. The other charts provide *bivariate visualizations* called scatterplots. The "bi" in "bivariate" means "two" just like the "uni" in "univariate" means one. So, a bivariate statistic or visualization is a statistic or visualization that looks at the relationship between two variables. A *scatterplot* is a bivariate visualization that illustrates how two variables relate to each other with one variable along the X-axis and the other along the Y-axis. The following summarizes the differences between univariate vs bivariate vs multivariate analysis:

- Univariate analysis is the analysis of one ("uni") variable.
- Bivariate analysis is the analysis of exactly two ("bi") variables.
- Multivariate analysis is the analysis of more than two ("multi") variables.

Due to the nature of the plot matrix in Fig. 4.8, six out of the sixteen charts are redundant. For instance, the second chart along the top row is the same scatterplot as the second chart down the first column with the X and Y axes reversed. Both of these plots show the relation of sepal width and sepal length values for the three species.

By visually analyzing the different distributions and scatterplots we can derive various insights. Although other conclusions might emerge regarding the identification of different species, the following list of insights can be readily uncovered by our simple analysis up to this point:

- Setosa is easier to identify due to its shorter petal length and shorter petal width.
- In terms of petal length and petal width, although Versicolor and Virginica overlap, Versicolor tends to have smaller values.
- Relative to petal length and petal width, sepal length and sepal width considered independently are less effective identifiers because of the significant overlap between the three species.
- The correlated measurements of sepal length and sepal width provide improved classification primarily for Setosa.

4.12 Basic Distribution Concepts

Knowing the distribution your data approximates is essential in the quest to understand your data. However, even if your data closely mimics a theoretical distribution the contours of the distribution curve may still appear different when compared with other distribution plots of the same category. For example, Fig. 4.9 shows three different normal distributions, but their location on the X-axis is based on their respective means and their relative height and width depend on their individual measure of standard deviation. This same concept of varying distributions is exhibited in Fig. 4.7 which displays the merging of three normal distributions for the petal lengths of each species in the Iris dataset.

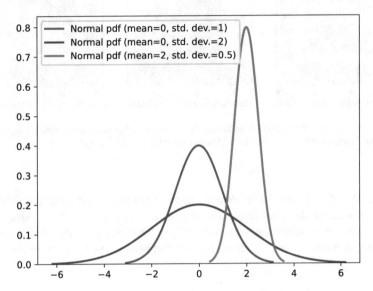

Fig. 4.9 Examples of three theoretical normal distributions with different means and standard deviations. (Image created from normal_distribution_different_means.py)

In addition, real data seldom aligns exactly with a theoretical distribution. In fact, a theoretical distribution, is just that – *theoretical*. Theoretical distributions such as those shown in Fig. 4.9 help us to make sense out of finite collections of data samples. If the mean is 0 and standard deviation is 1 then we call that theoretical distribution a *standard normal distribution*.

For example, Fig. 4.10 demonstrates how each of the 50 observations of the sepal length variable for the three species maps to a theoretical normal distribution matching the data's descriptive statistics. The orange histogram provides the actual values of the binned observations while the blue line shows what the theoretical distribution should look like given the mean and standard deviation of the dataset. The blue dots on the line correspond to selected sepal length values in the data and show where their frequency values would be if the data exactly followed the theoretical normal distribution.

You might notice that even though the actual bin values of the histogram **roughly** follow the normal distribution, there remains many inconsistencies. For example, the median for the Iris-Setosa species (middle graph) is greater than its theoretical counterpart (approximately 5.00) would predict. A normal distribution is symmetric and therefore the mean and median values are the same.

So, how could this small subset of sepal length measurements be a normal distribution? Empirical studies using the ideas of limits from calculus have shown that as the number of observations goes from zero to infinity then histograms as shown in Fig. 4.10 increasingly resemble the actual theoretical distribution. In other words, if we had thousands of observations instead of only 50 then the orange histogram and the blue line would look more similar. Extending this argument even further, if we had millions of observations then the orange histogram and the blue line would be almost identical.

Histogram of sepal length with normal distribution plotted

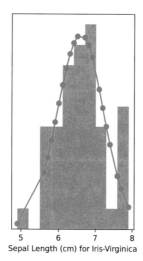  

Fig. 4.10 Detailed view of the individual distributions of the sepal length similar to the cumulative histogram plot in Fig. 4.5.The dotted curve provides theoretical distributions based on descriptive statistics. (Image created from sepal_length_normal.py)

This concept of limits such that as the number of samples grows from one observation to infinity the distribution increasingly resembles the theoretical normal distribution is called the *central limit theorem*. The central limit theorem allows us to use the assumptions of normality (*normality* means that a variable is distributed according to the normal distribution) and lets us leverage normal distribution functions and reasoning on our data even when our data do not appear similar to the ideal theoretical.

4.13   Probability

One essential numerical description that distributions help communicate and clarify is the concept of probability. Let us review basic probabilities then address how probabilities apply to distributions.

At its core, a probability is the frequency of an event based on the sample size. For example, given a fair coin, what is the probability that a single flip will result in heads or tails? Since there are two possibilities and only one possible outcome for a single flip then the probability is one out of two or 50% probability that it will land on one particular side. Alternatively, suppose there are three marbles in a jar, one red and two blue. What is the probability that you will select a red marble given an equal chance of choosing any of the three marbles? Since there is one occurrence of a red marble and there are three possibilities, then you would have a one out of three, or 33.33% probability.

This kind of basic, precise probability is fine with datasets that are small, but how do probabilities help us understand sets with much larger numbers of data points, such as the Iris dataset?

With the Iris dataset there are many possible values for every measurement. For example, with the Iris-Virginica, what is the probability that a flower's sepal length is 4.9 centimeters? (The value of 4.9 centimeters is the lowest value recorded in the Iris dataset for the Virginica species, as shown in the leftmost plot in Fig. 4.10.)

There are two ways to look at this question:

- What is the probability of a 4.9 centimeter sepal length measurement for the Viginica species for **only** the Iris dataset measurements we have recorded? In that case, since there are 50 observations per species and only one of them has the value of 4.9 centimeters, the answer is one out of 50 or 2%.

- What is the probability of a 4.9 centimeter sepal length measurement for the Viginica species for **all** Viginica flowers everywhere? This question is much more robust and useful. In this case, the 50 observations that were collected are only samples and are used to help us understand the distribution of values for the entire population of all the Iris flowers across the globe.

4.14   Percentile

To answer the second question above we need an appreciation of percentiles and percentile rank. For example, a percentile rank is how the doctor compares your height to the rest of humanity. Simply recording an isolated measurement of how tall you are does not assist the doctor in evaluating your growth.

Instead, your doctor contrasts your height to that of all other people of similar sex and age to determine a percentile rank. For example, according to the CDC,[1] a boy of 7.5 years and 135 cm is ranked in the 95th percentile for his sex and height. In other words, if 100 other boys 7.5 years old were gathered together then the boy being measured by the doctor would be as tall or taller than 95 other boys in that group. On the other hand, according to the CDC, a 11.5-year-old boy with the same 135 cm height would be in the fifth percentile for his sex and height. In his case, out of 100 other boys there would likely only be five boys his age and height that were exactly his height or shorter.

This concept of percentile rank can be applied in many data analysis scenarios. For example, when enrolled in a course a student listed in the top 25th percentile in performance indicates that 75% of the class is most likely going to earn a better grade. This percentile rank measure is commonly applied in sports, health, finance, and any other field where there is a continuous range of possible values found in a distribution.

4.15   Cumulative Distribution Function (CDF) and Probability Density Function (PDF)

In statistics, the tool employed to map a value to its percentile is the *Cumulative Distribution Function (CDF)*. In the example of the boy's growth, mapping 135 cm to the 95th-percentile would be represented as follows: CDF(135 cm) = 0.95.

The *Probability Density Function (PDF)* is defined as the derivative of the CDF. Where the CDF returns the percentile of a value compared to the whole, the PDF returns a measure of the relative likelihood of that value based on a given distribution.

For example, returning to the example of boys' growth patterns, if a 7.5-year-old boy were 135 cm tall then he would be in the 95th-percentile and the datapoint for that boy would appear on the far right of the CDF graph. The CDF is important because it reports how tall the boy is in comparison to all the other boys of his age.

However, a calculation of the area under the PDF for a small range of values around 135 cm would return a probability of about 2% revealing that only about 2% of all the boys are in that height range. Another boy that is 7.5 years old and 125 cm would be in the 50th percentile based on the CDF. It turns out that 125 cm is very common and an area calculation under the PDF for a suitable range of values would return a probability of about 40% indicating that about 40% of all boys that age are approximately 125 cm tall.

Figure 4.11 plots the theoretical CDF and PDF of the distribution of sepal length for Iris-Virginica. The graph on the left displays the CDF of the sepal length and the graph on the right plots the associated PDF. Both the CDF and PDF operate under the assumption that all probabilities for a given distribution add up to 100% or 1.0 using the properties of calculus. In other words, the sum of all the probabilities or area under the curve for the PDF is equal to 1.0, whereas the value of the CDF at sepal length x would be the area under the PDF from minus infinity to x.

---

[1] The Centers for Disease Control and Prevention (CDC) is a national public health institute in the United States.

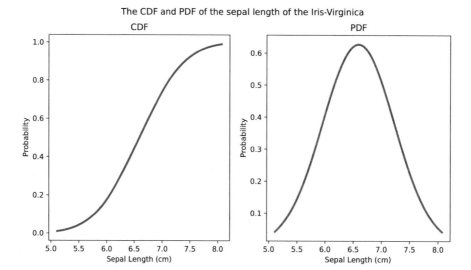

Fig. 4.11 Example visualizations of the CDF and PDF functions for the Iris-Virginica species. (Image created from iris_cdf_and_pdf.py)

Detailed inspection of the Iris-Virginica data shown in Fig. 4.11 indicates that a sepal length of 4.9 cm is both small in comparison to other values and not very common and thus registers both a low CDF and low relative PDF. In other words, the CDF reveals that 4.9 is roughly the minimum measurement available in the distribution and the PDF confirms that this particular measurement is very rare across all possible values of sepal length.

Conversely, from the CDF a sepal length of 6.5 cm returns approximately 50% indicating that 6.5 cm is roughly the median value of the dataset and the PDF returns the highest relative value of 60% indicating that most of the measurements in the distribution are grouped near that value.

4.16 Percent Point Function (PPF)

A *Percent Point Function (PPF)* is the inverse of a CDF. In other words, if a CDF returns the percentile based on a value, then the PPF returns the associated value for a given percentile. For example, using the boy's height example, the CDF of 135 cm for a 7.5 year old would return 95th-percentile (CDF(135 cm) = 0.95). Conversely, passing 0.95 to the PPF for the same distribution would return 135, as in 135 cm height, (PPF(0.95) = 135).

Using the CDF in conjunction with the PPF can be very useful. For example, knowing his percentile of height, the boy could approximately estimate how tall he will be at later stages in life. For instance, if he were in the 95th percentile at 135 cm when 7.5 years old, how tall would he be if he were still in the 95th-percentile when 20 years old? To determine the value of his height at 20 years old we would use the PPF, the inverse of the CDF. More precisely, given the CDC distribution of 20 year old men, the boy's expected height at the 95th percentile would be PPF(0.95) = 189 cm (approximately 6 foot 4.5 inches).

This prediction strategy could be utilized in many areas of investigation. For example, an 18-year-old weightlifter who bench presses 53 kg (116 lbs.) would discover that she is in the 50th percentile for her body weight, sex, and age for this bench press amount: CDF(53 kg) = 0.5. Assuming she would belong to the same percentile 10 years later, she could estimate she will be able to press 63 kg (138 lbs.) – an additional 10 kg – when she is 28 years old by using the 28-year-old woman's distribution for the bench press: PPF(0.5) = 63.[2]

4.17 Skewness

When referring to distributions, *skewness* is a measure of symmetry. Figure 4.12 displays three normal distributions. Both the distribution on the left and the one on the right exhibit skewness in opposite directions. The distribution on the left is considered to have a *negative skew* while the distribution on the right is considered to have *positive skew*.

---

[2] These weightlifting values are based on real analyzed data from powerlifting competitions during the years 2012–2016 as reported by the USA Powerlifting federation (www.usapowerlifting.com).

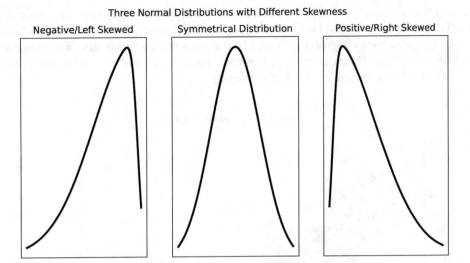

Fig. 4.12 Examples of skewness. The left figure shows a left-skewed normal distribution. The middle figure shows no skewness. The right figure shows a right-skewed normal distribution. (Image created from skewed_distribution.py)

Another way to express the same idea of a negative skew is to call it *left tailed* or *left-skewed*. The "tail" is the skew. Likewise, a positive skew is often referred to as *right tailed* or *right-skewed*. For a normal distribution, a negative skewed distribution will have the mean to the left of the median value, whereas a positive skewed distribution will have the mean to the right of the median value.

4.18 Exponential Distribution

The exponential distribution is a common distribution that is often connected with many datasets. The exponential distribution is frequently associated with datasets that include a time measurement. For example, consider the time required for a person to shop for a particular item and successfully purchase it. Once a decision is made to buy something, most people will buy that item within a certain time. However, there are always stragglers that take a long time to decide, but do complete the purchase eventually, which is behavior associated with the long tail of the exponential distribution. An example visualization of the PDF of theoretical exponential distributions can be seen in Fig. 4.13.

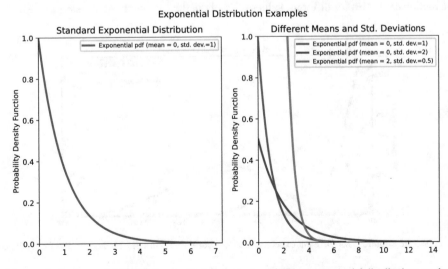

Fig. 4.13 Example of the standard exponential distribution on the left and various different exponential distributions on the right for comparison. Compare to Fig. 4.9 to see something similar for the normal distribution. (Image crated by exponential_distributions.py)

Exponential distributions are also appropriate models for problems involving horsepower (i.e., power to pull heavy loads) and fuel efficiency (i.e., the distance covered by the vehicle per gallon or electric charge). More vehicle fuel efficiency is typically associated with less horsepower. Figure 4.14 indicates that some vehicles have little horsepower and high fuel efficiency (shown as MPG), and some vehicles have high horsepower and low fuel efficiency, but most vehicles are clustered together exhibiting both moderate horsepower and fuel efficiency.

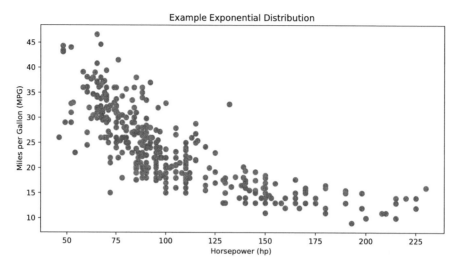

Fig. 4.14 Example of a real exponential distribution comparing fuel efficiency (MPG) and horsepower. (Image created from correlation_auto.py)

4.19   Poisson Distribution

As with the exponential distribution, the Poisson distribution is typically related to a time component, more specifically the rate at which a particular event occurs. For a fixed-length time interval, the Poisson distribution provides the probability that a certain number of events will take place during that interval. Since we count occurrences using whole numbers (an event either does or does not happen), the Poisson distribution is defined as a discrete probability distribution, in which probabilities are assigned to specific integer values 0,1,2,3, … The events being counted are assumed to be independent of each other.

One example is a traffic mitigation project that monitors the number of cars arriving at an intersection per minute. The average rate, or mean number of cars per minute, directly influences the shape of the distribution as shown in Fig. 4.15, which graphs Poisson distribution examples for different values of lambda, the expected rate of occurrences.

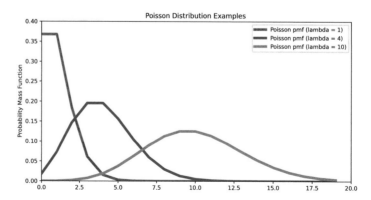

Fig. 4.15 Example of Poisson distributions for different expected rate of occurrences (lambda). (Image created from poisson_distributions.py)

Since the Poisson distribution models discrete events, Fig. 4.15 plots the probability mass function (PMF) defined only at integer values. The discrete points in the graph are connected for easier visualization of the overall PMF shape. Since the events are independent of each other, the Poisson distribution would not apply to exceptional situations such as a Presidential motorcade traveling through the observed intersection.

Let's assume the intersection being measured is in a remote, rural area with very light traffic patterns. The average rate of arrivals is 3.6/minute. What is the probability that 5 cars arrive at the intersection for any randomly selected 1 minute interval? The Poisson distribution would determine this probability as 13.77%, reflecting a somewhat low chance of slightly more frequent car arrivals above the average 3.6/minute.

4.20 Additional Distributions and Reading

There are many other common distributions that data scientists should explore. The following distributions are some of the most common:

- Uniform Continuous
- Uniform Discrete
- Binomial
- Lognormal
- Chi-squared
- Bernoulli

Figure 4.16 depicts the first two distributions from the above list: Uniform Continuous and Uniform Discrete. The uniform continuous distribution, shown on the left in Fig. 4.16, is a distribution that is uniform in nature, where every value between two points has the same probability. In other words, the occurrence of a value among the range of all possible values is equally likely. The uniform discrete distribution, the second distribution in Fig. 4.16, is similar to the uniform continuous distribution, except probability values are tied to discrete whole numbers.

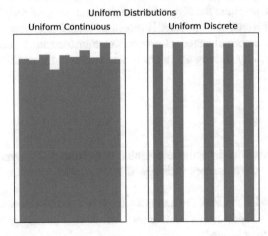

Fig. 4.16 Examples of two uniform distributions. The one on the left is continuous and the one on the right is discrete. (Image created from uniform_distributions.py)

There are many distributions that have been created to accurately model actual data. Knowing even the basics about distributions will help you better understand and explain your data.

4.21 Transformations

Transformations are important when your data exhibits a specific distribution and you need to change or transform these data to conform to another distribution. Normality (assuming the sample mean distribution is normal) is an important assumption for many parametric statistical techniques. In other words, there are many times when you might wish to perform certain statistical methods or functions that require your data to adhere to a normal distribution. If your dataset does not satisfy the properties associated with a normal distribution then it will need to be transformed.

Parametric, implying "parameters," refers to a set of fixed parameters that determine a probability model. The two defining parameters for a normal distribution are the mean and the standard deviation (see Fig. 4.9 for examples of normal distributions with different parameters). In other words, parametric methods typically apply to sample mean distributions that are approximately normal.

Nonparametric techniques are statistical techniques for which there are no assumptions about parameters for the population we are studying. Specifically, no assumptions about the normality of data are posited. Table 4.1 lists a small sample of different parametric and nonparametric tests available based on the normality of the data.

Table 4.1 Small sample of parametric vs nonparametric tests to use based on the data's normality

Parametric test	Nonparametric test
T-test	Wilcoxon test
Analysis of variance (ANOVA)	Kruskal-Wallis test
Pearson coefficient of correlation	Spearman's rank correlation

This discussion naturally brings up the question of why we would ever employ parametric tests that require normality? In other words, why go through all the trouble to conform to normal distributions and parametric tests when we could simply ignore the distribution and always use nonparametric tests?

The first reason is that nonparametric tests generally have a greater likelihood for producing false negative results, i.e. indicating no association between two variables when one actually exists (see Table 5.1 for more information about false negatives). This means that a nonparametric test typically requires a larger sample size to achieve the same predictive power as the corresponding parametric test.

Another drawback associated with nonparametric tests is their results are often more difficult to interpret compared to the results of parametric tests. Many nonparametric tests use data value rankings rather than leveraging the actual original data in the computation of descriptive statistics such as the mean or standard deviation. If nonparametric tests used data measurements like mean and standard deviation they would be parametric tests by definition.

In the end, parametric tests, which presume the distribution of the data, are generally preferable. As a result, further reading into transformations from one distribution to another is recommended. For example, the *Box Cox transformation* maps non-normal dependent variables into a normal shape. However, this transformation does have a requirement that all values be non-negative, which can be realized by shifting all values in the dataset in the positive direction.

4.22   Correlation

One of the best ways to describe data is to investigate potential correlations. A *correlation* is a concise, often numerical description of the relationship between two or more variables. If two variables are not related, then there is said to be no correlation between them.

First, correlation can be understood abstractly. For example, there is a positive correlation between ground moisture and a rainstorm. The straightforward explanation is that when it rains the ground usually gets wet. Another statement of correlation is "the taller a person, the more they weigh."

What is important to understand is that statistical correlation is all about numerical results and it ultimately falls on data analysts to determine the meaning behind those numbers. Specifically, statistical correlation is a function that receives a set of numbers and returns a result called an "R-value." This *R-value* is either positive or negative and ranges from −1.0 to 1.0.

A *positive correlation,* reflected by an R-value greater than zero, means that the relationship between the two sets of numbers is positively correlated. In other words, if one set of numbers generally increases the other set of numbers also increases. Analysts may refer to this as a direct relationship between variables.

For example, if the first set of numbers are values of a given city's precipitation during a month period and the other set of numbers is ground moisture during this same time then usually the two numbers are related.

Conversely, a negative R-value indicates a *negative correlation:* as one set of numbers increases the other set of numbers decreases. Analysts may refer to this type of correlation as an inverse relationship between variables. For example, a person's weight and their life expectancy may exhibit a negative correlation. Similarly, the higher the number of absences for a student, the lower the grade.

At this juncture, it is important to distinguish between causation and correlation. Although these can coexist, one cannot immediately deduce causation from correlation. *Causation* means that one thing **causes** another. For example, throwing a ball through a window **causes** the window to break.

However, the relationship between the quantity of hotdogs sold and the number of home runs recorded at a baseball stadium is a straight correlation. Selling more hotdogs did not *cause* more home runs nor did more home runs cause the purchase of more hotdogs. The likely relationship between hotdog sales and the number of home runs can be described as an indirect chain reaction: the greater success of a baseball team may be reflected in an increasing amount of home runs, leading to more excited fans, and thus more people attend the games resulting in more people buying hotdogs at those games.

From our initial correlation examples above, while we may be fairly confident through intuitive reasoning that a rainstorm in a desert climate with a low water table (controlling for all other variables) is the cause of wet terrain, we cannot with the same level of certainty say that an increase in a person's height is the cause of increased weight gain. Further experimentation is necessary to establish causality.

Things are not always straightforward with correlation measures. Generally, when it rains the ground gets wet. However, the ground can possess high water content for many reasons. For example, from sprinklers, from water balloons, and from incoming stream sources.

There are many examples of positive correlations between two variables where one measurement does not directly cause the other. For example, during a recession people tend to buy fewer cars and prefer to eat at restaurants less often. Buying fewer cars does not *cause* people to eat at restaurants less frequently even though there is a positive correlation between the two during a recession. The real root cause of both sets of data is that people tend to spend less money during a recession.

Most importantly, remember that correlation only indicates if sets of data are related in some capacity, not if one set of data causes the other.

Let us now explore an example from the Iris dataset. We know in humans there is generally a direct correlation between a person's height and weight. What about Iris petal measurements? Specifically, if the petal is longer, is the petal also generally wider too? The code listed here shows how to find the correlation between the petal length and the petal width:

```
File: correlation_example1.py:
import pandas as pd
df = pd.read_csv('iris.csv')
#Spearman, Pearson, or Kendall can be specified. Pearson is the default.
correlation = df['petal_length'].corr(df['petal_width'])
print(f'The resulting R-value of petal length to petal length: {correlation}.')
```

Which produces the following output:

```
The resulting R-value of petal length to petal length: 0.9627570970509663.
```

How do we interpret this result? Let's look at two specific characteristics:

- The direction: positive or negative slope.
- The strength: the R-value's proximity to -1 or 1.

Regarding direction, given an R-value of 0.96 we know the direction is positive. In other words, when the petal length increases then the petal width also increases. Concerning strength, 0.96 is **very** close to 1.0. So, we can with high confidence say that the Iris dataset reveals a **strong** positive correlation between petal length and width.

We often wish to inspect correlations among all the possible pairs in the dataset. Luckily, we can perform a pairwise correlation function on all four Iris variables (i.e. petal width, petal length, sepal width, and sepal length). The following code performs a pairwise correlation and renders the results in a heatmap visualization shown in Fig. 4.17:

```
Partial code: full program in correlation_example2.py:

correlation_table = df.corr(method='pearson')
ax = sns.heatmap(correlation_table, vmin=-1, vmax=1, annot=True)
ax.set_title("Correlation Table for Iris Data")
```

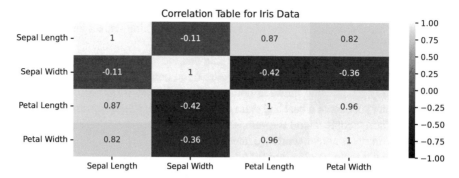

Fig. 4.17 Pairwise correlation table showing heatmap colors. Lighter (hotter) corresponds to higher R-values while darker (cooler) corresponds to lower R-values. (Image created from correlation_example2.py)

How do we interpret the results of the pairwise correlation table in Fig. 4.17? First, we recognize an R-value of 1.0, or perfect correlation, along the diagonals. This makes sense because there is always a perfect correlation when comparing a set of numbers to itself.

Second, we can see again that there is an R-value of 0.96 between petal width and petal length. In general, this correlation strength is rarely seen in data outside of selected textbook examples. An R-value of 0.96 signals a **very** strong correlation, not far from a perfect correlation. We can see other strong positive correlations between sepal length and petal length as well as sepal length and petal width with R-values of 0.87 and 0.82 respectively.

Interestingly, there is a moderate negative correlation between sepal width and petal length as well as sepal width and petal width with R-values of −0.42 and −0.36. respectively. The negative direction indicates that as sepal width increases petal length and petal width decrease and vice versa. However, the moderate values of −0.42 and −0.36 suggest that this trend may not be evident when comparing new sets of Iris flower data.

Finally, there is a **slight** negative correlation of sepal width and sepal length with an R-value of −0.11. Little to nothing can be interpreted from an R-value that is so close to 0.0, which represents a very weak correlation. Although the direction is negative, the strength of the correlation is minimal and any interpretations about the relation between sepal width and sepal length would need to be supported by additional data gathering and research.

There are many different types of correlation algorithms, each with their own assumptions and applications. The two main types of correlations that we will discuss are the following: Pearson and Spearman. Figure 4.18 shows different scatterplots that visualize how the Pearson and Spearman correlation values differ with the same data. For example, in the upper-left corner of Fig. 4.18 the data along the X-axis increases at exactly the same rate as the data along the Y-axis. In this case both the Pearson and Spearman correlation functions return an R-value of 1.0 indicating a perfect correlation.

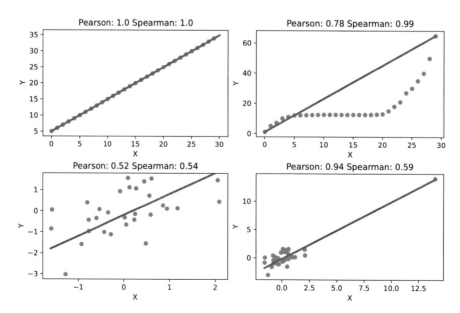

Fig. 4.18 Four different charts showing various data patterns and their resulting R-values for the Pearson and Spearman correlation functions. The red line represents the ideal line for finding linear correlation for the Pearson function. (Image created from correlation_pearson_vs_spearman.py)

The Spearman correlation function measures the monotonic relationship between data. *Monotonic* is a mathematical concept indicating that a function's slope either persists in one direction (positive or negative) or remains level but never trends toward the other direction. The chart in the upper-right corner of Fig. 4.18 demonstrates how the two correlation functions behave differently with monotonic data. For instance, in this case, the data along the Y-axis either increases or stays the same as the data along the X-axis increases. However, there is no instance in which the values on the Y-axis ever decrease as the values on the X-axis increase. The data in this chart describe a monotonically increasing function.

In this case, we can see the Spearman correlation function assigns the monotonic data a much higher value (0.99) than the Pearson correlation function (0.78).

The Pearson correlation function measures the linear relationship between two sets of data. The linear relationship is shown as a red line in each of the four graphs in Fig. 4.18. The closer the data is to the red line representing the ideal linear correlation, the higher the resulting R-value is. This can be seen most notably in the bottom two charts in Fig. 4.18. The cluster of data points around the ideal linear regression line in the bottom right chart leads to a much higher Pearson R-value when compared with wider scatter range of data in the bottom left chart. In contrast, the Spearman measurement between the two charts is very similar.

It is important to strategically decide when to use the Pearson correlation function and when to use the Spearman correlation function. However, there are times when neither of these correlation calculations report a strong relationship between two variables even when one exists. For example, consider Fig. 4.19, which shows a very strong relationship between two datasets visualized as a scatterplot. However, since the relationship is neither linear (Pearson) nor monotonic (Spearman) the resulting R-values indicate no clear relationship where one exists. In these situations, the use of nonlinear correlation functions would be appropriate.

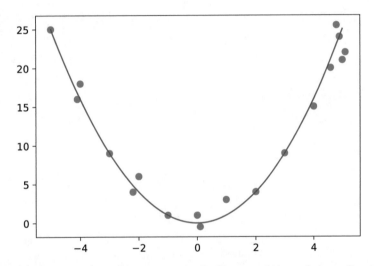

Fig. 4.19 An example where a correlation exists but cannot be detected using linear correlation techniques. (Image created from curved_line.py)

Exercises
1. What is the objective of descriptive statistics?
2. Provide three examples of when you might use EDA.
3. Explain the significance of the 68-95-99.7 rule.
4. Explain when to use mean vs median vs mode.
5. Under what circumstances would you use nonparametric tests? When could you use parametric tests on the same data?
6. What is the importance of a distribution?
7. What is the PDF? When would you use it?
8. What is the CDF? When would you use it?
9. What is the PPF? When would you use it?
10. What is a correlation?
11. What is the difference between the Pearson and the Spearman correlations?

Chapter 5
Inferential Statistics

***

Whereas the descriptive statistics covered in the previous chapter describe or summarize data, these calculations do not help us derive general observations about data. Inferential statistics is a branch of statistical analysis that allows us to infer (or generalize) trends about the larger population based only on the sample values we have collected and measured.

Whereas a correlation provides a numerical value of how different variables are related, inferential statistics investigate how one or more variables may affect or influence another variable.

For this section, we will delve into the data of the *RMS Titanic*. The RMS stands for "Royal Mail Steamer" indicating the ship was contracted with the British government to carry postal mail.

The *RMS Titanic*, popularly referred to as *The Titanic*, was a ship that was believed to be unsinkable. On its maiden voyage the steamer collided with an iceberg and sank in the early morning hours of April 15, 1912 and has long been a topic of fascination for many people. Although the exact number of people that were on board is unknown and most likely will never be established with certainty, there are many websites dedicated to the ship, its passengers and crew, and its ultimate demise. For example, Wikipedia.org has webpages with comprehensive lists of all the known people on the ship (e.g., datasets).

The *Titanic* dataset is interesting for various reasons including the following:

- It is real. Unlike the Iris dataset, the *Titanic* dataset is neither fabricated, fully populated (no NAs), nor compact.
- Because the dataset is real, information is not known and is therefore missing. For instance, the age of every person is not known. The occupation of many passengers is not known. Also, it is believed that there were possibly stowaways (passengers that did not pay) that are unaccounted for.
- It is complex. For example, there are individuals listed in both the dataset of the crew complement (e.g. the people paid to run the ship: officers, cooks, etc.) and the passenger complement (e.g. the people that paid for passage). An illustration of this double counting involves the musicians who played in the ship orchestra. These performers were considered crew because they were paid to work on the ship and were also considered passengers because they were provided rooms in the second-class section.
- It is ambiguous. Another example of people difficult to classify are servants who accompanied passengers. These people were maids, butlers, and support staff who traveled with their employers on the ship. They were not categorized as crew because they were not paid to help run the ship, yet are they really "passengers" if they did not pay for their own passage and were effectively workers on the voyage?

Considering all the above nuances and eccentricities, the *Titanic* datasets offer us a rich resource to explore the purpose and application of inferential statistics.

However, before we examine the *Titanic* datasets, as with any good analysis we should establish the question we wish to ask. The main question that we will try to answer through our analysis is the following: **What variables contribute the most to determining a person's likelihood of surviving the sinking of the *Titanic*?**

© The Author(s), under exclusive license to Springer Nature Switzerland AG 2022
R. Ball, B. Rague, *The Beginner's Guide to Data Science*, https://doi.org/10.1007/978-3-031-07865-1_5

5.1 Independent and Dependent Variables

To effectively apply and evaluate the inferential statistics necessary to help answer our question, we define the dependent variable. For our analysis, the dependent variable indicates whether the person did or did not survive. A *dependent variable* is a variable that depends or relies on other variables. What does a dependent variable depend on? The independent variable(s). The *independent variable(s)* are isolated and do not necessarily affect or interact with each other.

To illustrate, for the *Titanic* passengers we can identify two independent variables, age and gender, that strongly affect the survival dependent variable. A well-known phrase from that era in the context of a sinking ship or other life-threatening crisis was "Women and children first!" So, to be precise, survival is a dependent variable that is contingent on the independent variables of passenger age and gender. Given the crew's priority of saving women and children first it is not surprising that more females and people younger than sixteen survived when compared to males and older people.

Inferential statistics are synonymous with *scientific testing*. As such, employing inferential analysis allows you to find statistical significance and either accept or reject what is called the null hypothesis. Scientific testing either accepts or rejects hypotheses; the tests do not fundamentally explain the objectives of the analysis. Aside from inferential statistics and scientific testing, the imagination and creativity of the person doing the testing are the most important aspects for drawing conclusions and discovering insights.

The *null hypothesis* states that one or more sets of independent variables have no quantifiable or measurable effect on the outcome (dependent variable). For our example dataset, the null hypothesis to determine the influence of the gender variable would state there is no difference between the survivorship of males compared to females. More precisely, if the dependent variable is survivorship, then the null hypothesis states that males and females survived at approximately the same rate and if there is any difference between the number of males versus females who survived, then this can only be due to randomized chance and variance.

On the other hand, the *alternative hypothesis* states that one or more sets of independent variables do exert a quantifiable or measurable effect on the outcome (dependent variable). In our case, the alternative hypothesis states there *is* a difference between the survivorship of males to females and that the difference is *not* due to randomized chance and variance.

We know from history that many more females than males survived the sinking of the *Titanic*. However, is the difference between these survival rates significant?

Statistical significance is an especially tricky topic though fundamentally it is asking the question: Based on the results of the data analysis, should we reject or accept the null hypothesis? It is logically and implicitly understood that if we reject the null hypothesis then we may subsequently assert the alternative hypothesis.

In addition, when addressing the veracity of the null hypothesis the concepts related to probability are necessarily involved. Specifically, when making inferences based on statistical hypothesis testing, we must allow for variance in the data. More to the point, after establishing statistical significance in an initial test, what if similar data were generated and the analysis repeated. What is the likelihood we now detect no statistical significance? The overarching goal from our inferential analysis is to establish a probability number such that if 100 tests were conducted on other ships similar to the *Titanic* in the early 1910's that sank with the same complement of women, men, and children then the survivability results would be the same a significant majority of the time. In effect, we declare a level of confidence that our conclusions are not simply an artifact of random chance.

The "standard" probability of error that is generally accepted as a reasonable confidence threshold is a probability – called the *p-value* (probability value) – of 5%. In other words, most people that conduct statistical hypothesis testing often use a p-value of 5% (i.e., 0.05). Selecting this probability of error means that analysts allow for the fact that if there were hypothetically 100 *Titanic*-like ships in the early 1910's with a similar complement of women, men, and children then about 95% of the time the statistical tests would find that more women and children would be saved, thus establishing statistical significance and rejecting the null hypothesis. Conversely, 5% of the time the data analysis would **not** find a difference in the survivorship of women and children to men.

Another way to conceptually frame scientific testing is in terms of positive or negative results. For example, if a doctor were testing a tissue sample for cancer, then the results would either be a positive result indicating the patient has cancer or a negative result indicating the patient does not have cancer.

However, for every outcome there is the possibility that the doctor's diagnosis is incorrect. In this case, if the patient had cancer and the doctor's tests returned positive then this result would be considered a *true positive*. Similarly, if the patient did not have cancer and the doctor's tests returned negative then this result would be considered a *true negative*. In an ideal world, doctors' tests would only ever return positive when cancer truly is there (true positives) and negative when cancer is not present (true negatives).

However, given the ambiguity inherent in the real world mistakes are sometimes made. So, in the case where the patient does **not** have cancer, but the doctor's test returns positive, then the result is categorized as a *false positive*. In this case the patient might undergo surgery to remove healthy tissue that is thought to be cancerous. Another name for a false positive is *Type I Error*.

The other mishap, in which a patient **does** have cancer, but the doctor's tests return negative is considered a *false negative* result, which is unfortunately also possible. In this case the patient is believed to be cancer free, but in reality has contracted cancer. Situations like this can be very dangerous because the patient may die from undetected cancer. Another name for a false negative is *Type II Error*.

Table 5.1 displays these four possible outcomes in tabular form. In a perfect world Type I errors and Type II errors would never happen. However, we do not live in a perfect world and thus errors do occur. With scientific testing, the p-value indicates the percentage of time that type I or type II errors arise. This knowledge will help limit the number of unnecessary statistical tests. For example, if we were to run 100 statistical tests on the *Titanic* dataset with a p-value of 5%, or p = 0.05, then approximately five out of the one-hundred tests would be either type I or type II errors, thus returning the wrong conclusion.

Table 5.1 Type I and type II errors

	Null Hypothesis is correct	Null Hypothesis is not correct
Reject null hypothesis	Type I error (false positive)	Correct conclusion (true positive)
Fail to reject null hypothesis	Correct conclusion (true negative)	Type II error (false negative)

Similarly, a smaller p-value, say 1% (i.e., 0.01) would result in fewer mistakes. Theoretically, a p-value of 0.01 would result in 99 correct tests and only one type I or type II error.

Our p-values should be determined *a priori*. *A priori* comes from Latin and means "first" or "beforehand." In other words, you should decide what your level of comfort is with Type I and Type II errors and choose the p-value **before** you run your statistical tests. As a rule of thumb, if you are not sure what your p-value should be then stay with the accepted standard of p = 0.05.

The forthcoming sections show how to perform different kinds of statistical tests on the *Titanic* data, depending on the type of data being analyzed. The primary classes of data are continuous data and categorical data. Continuous data is a measurement like age, weight and height that is continuous in nature. For instance, if a person's age ranges continuously from 0 years old to 125 years old then every fractional value, like 5.1 years, would still be allowed and has meaning. Categorical data, on the other hand, uses more abstract discrete groupings like "toddler," "teenager," and "young adult."

In addition, continuous data can typically be binned or collected together to form categories. Binning is essential when visualizing continuous data using histograms. For instance, an analyst might leverage continuous age data to determine general weight patterns for all ages. However, the analyst might find it helpful to separate age into predefined groups to better understand how the "toddler" category may have a different weight pattern than "older adults." So, the analyst might bin (i.e., create groups) based on age: 0–2 years for "toddler," 3–8 for "preadolescent," 9–12 for "tween," and so on. See Sect. 3.3.7 for more information on binning.

5.2 Chi-Squared Analysis

We will now address a simple question: Scientifically speaking, during the sinking of the *Titanic* did more women and children survive compared to men?

To select the appropriate statistical test to perform we have to understand the datatype of our **dependent** variable. What type of variable is gender?

Although gender **could** be represented as a numerical value (e.g., 1 for female and 2 for male) the numbers would not possess any intuitive meaning. With numerical (e.g., continuous) variables, each value is systematically and consistently ordered when represented on a classic number line. For example, the value 5.3 is less than 6.3 and is exactly 1.0 less. Also, 7.4 is 1.1 more than 6.3. However, if gender were given numerical values, then you could neither subtract them nor do any other useful arithmetic to them to generate a meaningful result. For example, what does male multiplied by female mean?

Some categorical values do not have any intrinsic ordinal meaning. *Ordinal data* is data that can be ordered, like "wealthy," "middle class," and "poor." In this example, the data is categorical and can also be ordered, making it ordinal data.

You cannot order gender, so it is not ordinal data. As another example of a categorical variable that is not ordinal, consider eye color whose values imply no natural order. Categorical data that cannot be ordered is called *nominal data*. Any attempt at ordering eye color categories would be entirely arbitrary. For example, are hazel eyes less than blue? Are green eyes less than brown? You could order the colors based on the color order in the rainbow or other color theory concepts, but it would not be based on a widely accepted, formalized numerical order.

Given that we have now established that gender is categorical or nominal data, we need to employ the correct scientific algorithm to measure if more females than males survived the sinking of the *Titanic*.

The *Chi-squared test of independence,* or *Chi-squared test* for short, evaluates how likely it is that an observed difference in the data collected about two or more groups is due to chance. In other words, it seeks to provide a confidence measure (inferential statistic) of the relationship among two or more categorical variables under investigation assuming all observations of the sample population are independent of each other.

In our case, the Chi-squared test seeks to determine if the people on the *Titanic* survived or died independently of gender or if gender indeed influenced whether a particular category of person survived. If gender did not play a part in the survivorship of the passengers, then we would expect that the passengers died in a randomly distributed fashion and that even though one gender may have survived more than another this difference could be attributed purely to chance.

For another example in which a Chi-squared test can apply, suppose we find a fair coin and flip it four times. Since a perfectly balanced coin has a 50% chance of landing on either side (heads or tails) then one might expect it to land 2 times on heads and 2 times on tails. However, with such a small sample size it is not uncommon for the outcome to be four heads in a row. In fact, there is a 1/16 (6.25%) chance that four heads would be the result. If a class of 50 students all flipped a coin four times, then there would likely be several students that would report four heads in a row.

However, assume we now flip the perfectly balanced coin 1000 times. If it landed on heads at a frequency of 95% then what would you think? The chances that a "perfectly balanced" coin would land on heads 950 out of 1000 times is astronomically small. Most rationally intelligent people would then intuitively presume that the description of the coin as "perfectly balanced" is not accurate at all. In this case, the actual results do not align with our expectations of the distribution of the number of heads and tails. Another way to frame this experimental outcome is with distributions, more specifically when flipping a fair coin there is a very small tail probability that the occurrence of 950 out of 1000 heads can be attributed to chance.

This same intuition underscores the idea of the Chi-squared test which is also often called a *goodness of fit* statistic because it tells us how well the data matches what we would presume if the observations in the test were completely independent of each other.

5.3   Chi-Squared Example: *Titanic* Gender Example

Let us remember the big question: **Did women and children survive more than men during the sinking of the *Titanic*?** First, let us examine gender.

Before performing the test, we inspect the data. Running the following Python program produces the contents of Table 5.2, which provides numerical evidence from the *Titanic* dataset that many more males died than females.

```
File: contingency_table.py:
import pandas as pd


df = pd.read_csv('Titanic_Passengers.csv')
#'margins=True' shows the summary information
contingency_table = pd.crosstab(df['Gender'], df['Lived_Died'], margins=True)
print(contingency_table)
```

Table 5.2 is a good start, but it is often very useful to visualize the data in some instructive format. Figure 5.1 diagrams the data from Table 5.2 as a stacked bar chart.

Table 5.2 Contingency table with summary information comparing gender to survival. Contents based on the output of contingency_table.py

Lived_Died/Gender	Died	Survived	All
Female	143	339	482
Male	709	161	870
All	852	500	1352

From Table 5.2 and Fig. 5.1 it **appears** that more females survived than males. However, is it just by chance, like our coin coming up with four heads in a row, or is there a much greater likelihood that a real difference exists in the data?

To find out, we run the Chi-squared test using the following Python statements:

```
Partial code: full program in chi_squared1.py:

contingency_table = pd.crosstab(df['Gender'], df['Lived_Died'], margins=False)
print(stats.chi2_contingency(contingency_table)[0:3])
```

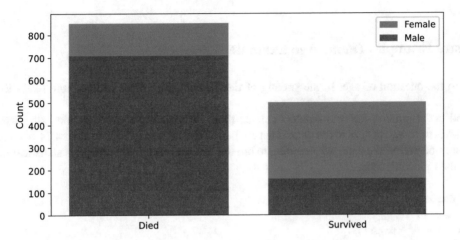

Fig. 5.1 Stacked bar chart showing the number of passengers on the Titanic that survived versus died. One can see there is a gender imbalance and more females survived compared to males. Created from stacked_bar_chart.py

Which produces the results:

```
(355.24355436145913, 3.0570688722545846e-79, 1)
```

The output is represented by three numbers. The first is called the Chi-squared statistic, our inferential result. This number incorporates the Chi-squared distribution and provides a measure of the confidence level of the difference between two categorical variables (female and male in this example). The higher the Chi-squared statistic, the less likelihood that the observed values were the result of pure chance.

Intuitively, the Chi-squared statistic is used in conjunction with the third number, the degrees of freedom, to ultimately produce the middle value, the p-value. The *degrees of freedom* for the Chi-square test are determined by observing the number of rows r and number of columns c in our contingency table (Table 5.2). The degrees of freedom are calculated as (r-1)(c-1), which in our case of two rows and two columns would result in $1 \times 1 = 1$. The more degrees of freedom in a computation (i.e., the more categories) the greater the Chi-square statistic must be to achieve statistical significance.

Since we have already set our target p-value to 0.05 (or 5%) we must compare 3.0570688722545846e-79 to 0.05. Since 3.0570688722545846e-79 is clearly less than 0.05, we can reject the null hypothesis claiming there is no difference in survivorship between males and females and accept with some level of confidence that there is a difference.

In other words, since 3.0570688722545846e-79 is less 0.05, we can conclude scientifically that more males died during the sinking of the *Titanic* and that their deaths were most likely **not** due to chance. Or can we? We need to be very careful in trying to answer this question. The reason is because the Chi-squared statistic relies on ratios, not on raw numbers.

Let us revisit the raw numbers comparing males to females again in Table 5.2. There are two important things to notice. First, there were almost twice as many males aboard the *Titanic* as females. Second, the percentage of females that died was about 30% (143/482) and the percentage of males that died was about 81% (709/870). Note that the Chi-squared test was testing whether the survival rates of the two genders would be about the same, **regardless of the raw numbers**. In other words, if no statistical significance was found then we would expect to have approximately the same percentage of females as males that died, independent of the fact that based on sheer numbers alone there were more males aboard.

So, **yes**, we can definitely conclude that more females lived than males (both in the confirmation of raw numbers and percentage) than what would be expected if gender and the likelihood of survival were truly independent of each other.

A word of **warning**: Many people, including students, teachers, and even professional scientists will use adjectives to describe their results.

For example, many people reporting their results might use phrases like "very" significant, "slightly" significant and various other qualifying adjectives associated with significance. However, statistical significance is binary: the difference either **is** or **is not** significant.

For example, if the resulting p-value is 0.056 and our *a priori* p-value is 0.05 then the result is no statistical significance. Many scientists fall prey to the idea that their results are "almost" significant, but that is more wishful thinking than science.

So be careful to not succumb to the temptation to add inaccurate adjectives that purport relative degrees of significance. Remember, your results are either statistically significant or they are not, nothing more.

5.4 Chi-Square Example: *Titanic* Age Example

Let us now turn to the question of age. In the sinking of the *Titanic*, were older people more likely to die compared to younger people?

Visualizing and understanding the data would be made more difficult if we were examine all possible ages simultaneously. The reason is the range of ages in our dataset is very large – from a few months to almost 74 years old. Consequently, in an effort to better understand the data, we are going to bin the data and determine if there is statistical difference between age groups using the following code:

```
File: bin_data.py:
import pandas as pd
from scipy import stats

df = pd.read_csv('Titanic_Passengers.csv')
df['age_bins'] = pd.cut(x=df['Age'], bins=[0,10,20,30,40,50,60,70,80])

contingency_table = pd.crosstab(df['age_bins'], df['Lived_Died'], margins=True)
print(contingency_table)

print('\n', stats.chi2_contingency(pd.crosstab(df['age_bins'], df['Lived_Died']))[0:3])
```

This code generates the following results which are plotted by another program to produce Fig. 5.2:

```
Lived_Died  died  lived   All
age_bins
(0, 10]       46     54   100
(10, 20]     156     71   227
(20, 30]     286    166   452
(30, 40]     158    107   265
(40, 50]     111     64   175
(50, 60]      40     30    70
(60, 70]      27      7    34
(70, 80]       5      0     5
All          829    499  1328
```

```
(24.465546982751874, 0.0009433264132943263, 7)
```

Both the above results and Fig. 5.2 disclose that the age counts roughly follow a normal distribution that is positively skewed (or right-tailed). Also, the results from the Chi-squared test show that with 7 degrees of freedom ((8 age categories − 1) × (2 survival categories − 1) = 7) the p-value is 0.0009433264132943263, which is less than our *a priori* p-value of 0.05 thereby indicating statistical significance.

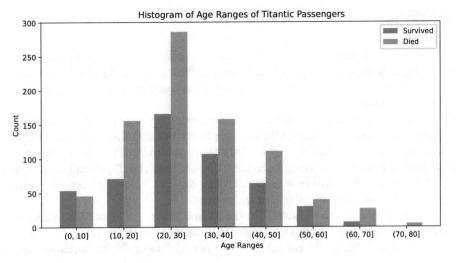

Fig. 5.2 Histogram of binned ages separated by survived or died. Image created from binned_ages_visualized.py

Armed with these results, we can conclude that age is statistically significant and accept the alternative hypothesis that the survival of more young people is **not** due to chance.

A great way to show how both age and gender interact is to visualize this relationship. Figure 5.3 renders a pivot table highlighted by a heatmap.

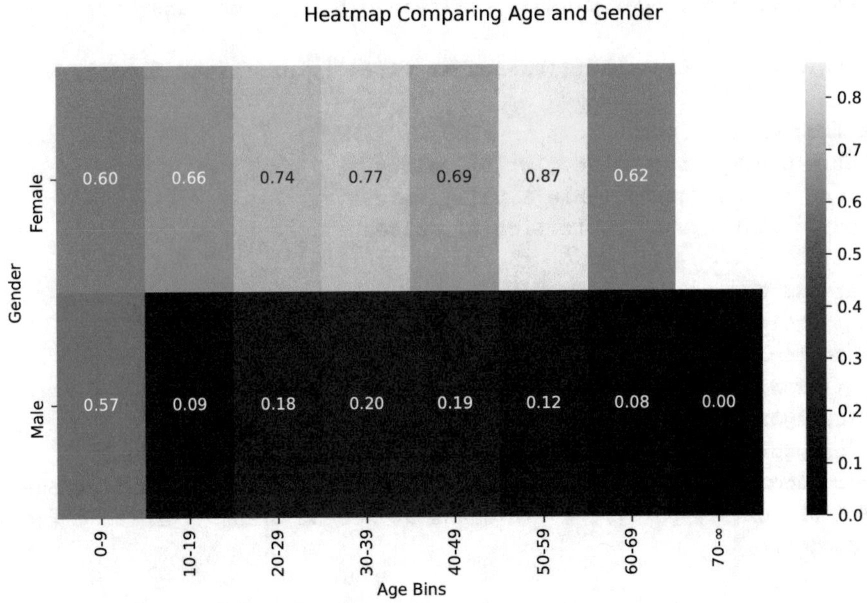

Fig. 5.3 Heatmap showing how age and gender are related. The decimal places (e.g., 0.57, 0.6, etc.) represent the percentage of survivors (e.g., 0.57 = 57% survived). Image created from titanic_age_heatmap.py

Figure 5.3 depicts the relationship between age **and** gender. The color legend on the right indicates the higher the survival percentage of a particular group, the brighter (or hotter) the color. Figure 5.3 reveals that for children less than 10 years old 60% of girls and 57% of boys survived. The figure also shows this general trend of higher percentages of women surviving continues for ages greater than or equal to 10 years old with the highest valued cell (women 50–59 years old) surviving at 87% and the lowest valued cell (women 60–69 old) surviving at 62%. (The last cell of 70 and older is blank because there were no women that were older than 69 listed in the *Titanic* dataset.)

In contrast, we note that the overall survival rate for men was relatively low. The survival percentage of boys 10 and older was particularly abysmal. For ages greater than or equal to 10 years old, we can observe that the highest valued cell (men 40–49) was only 19% while none of the five men older than 69 survived.

5.5   Chi-Square Example: *Titanic* Passenger Class Example

To continue our journey in understanding the profile of surviving *Titanic* passengers, let us examine the passengers not in terms of gender or age, but in terms of passenger class. Passenger classes aboard the *Titanic* were advertised as first class, second class, and third class, with first class being the most expensive and prestigious ranging to third class being the least expensive and the least prestigious.

In this case, the data are *ordinal* because we can order the categories from first to third class based on expense. However, these values are not intrinsically numerical because the numbers do not in themselves reflect meaningful mathematical relationships. For instance, you cannot add first class and second class to get third class. Thinking about these labels in another way, the name "first class" is arbitrary and could have been labeled "best class," "second class" could have been called "middle class," and "third class" could have been assigned "cheapest class." However, for this example, the data are still considered ordinal because they are based on cabin cost and prestige.

Let us now focus on a slightly different question: **Did the *Titanic* passenger class affect survivorship?** For example, were the travelers in first class more likely to survive compared to the travelers in third class?

To understand the data, let us visualize the entries shown in Table 5.3 using the histogram in Fig. 5.4 generated by the following code:

```
File: contingency_table2.py:
import pandas as pd
import matplotlib.pyplot as plt
import seaborn as sns


df = pd.read_csv('Titanic_Passengers.csv')
print(pd.crosstab(df['Passenger_Class'], df['Lived_Died'], margins=True))


contingency_table = pd.crosstab(df['Passenger_Class'], df['Lived_Died'], margins=False)


#Assigns the frequency values:
firstClassCount = contingency_table.iloc[0].values
secondClassCount = contingency_table.iloc[1].values
thirdClassCount = contingency_table.iloc[2].values


#Plots the bar chart
fig = plt.figure(figsize=(10, 5))
sns.set(font_scale=1.8)
categories = ['Died', 'Survived']
p1 = plt.bar(categories, firstClassCount, 0.55, color='#d62728')
p2 = plt.bar(categories, secondClassCount, 0.55, bottom=firstClassCount)
p3 = plt.bar(categories, thirdClassCount, 0.55, bottom=firstClassCount+secondClassCount)
plt.legend((p1[0], p2[0], p3[0]), ('1st Class', 'Second Class', 'Third Class'))
plt.ylabel('Count')
plt.show()
```

Table 5.3 Contingency table showing how many passengers survived or died based on their passenger class. Data from contingency_table2.py

Lived_Died/Passenger_Class	Died	Survived	All
First Class	149	201	350
Second Class	175	118	293
Third Class	528	181	709
All	852	500	1352

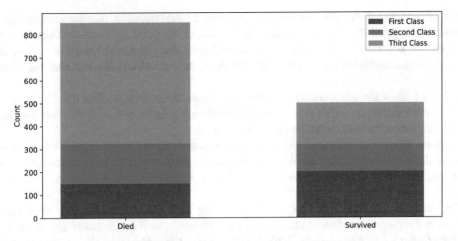

Fig. 5.4 Stack bar charts showing how many passengers survived or died based on their passenger class. Image created from contingency_table2.py

Table 5.3 and Fig. 5.4 suggest promising results that the alternative hypothesis claiming a difference in passenger class is true. Specifically, it appears that a higher percentage of people in first class survived compared to the other two classes. Now that we have reviewed the contingency table and stacked bar charts, we will run the Chi-squared test using the following code:

```
File: chi_squared2.py:
from scipy import stats
import pandas as pd
import numpy as np

df = pd.read_csv('Titanic_Passengers.csv')
contingency_table = pd.crosstab(df['Passenger_Class'], df['Lived_Died'], margins=False)

#Assigns the frequency values without getting the summaries
firstClassCount = contingency_table.iloc[0].values
secondClassCount = contingency_table.iloc[1].values
thirdClassCount = contingency_table.iloc[2].values

f_obs = np.array([firstClassCount, secondClassCount, thirdClassCount])
print(stats.chi2_contingency(f_obs)[0:3])
```

Which produces the results:

```
(104.05195288933228, 2.543346399916496e-23, 2)
```

The first thing to do is to verify the degrees of freedom (the last value). There are three passenger classes and two survivor classes, so the degrees of freedom should be 2 (2x1 = 2). Now we can proceed with the listed p-value: 2.543346399916496e-23. Since 2.543346399916496e-23 is less than 0.05 we can say there is statistical significance. But which specific passenger class is responsible for the significant difference in survival rates?

An essential point to emphasize at this juncture is the Chi-square test does not provide an answer as to whether first class survived more than second class or third class. The only insight the Chi-square results offer is that at least **one** of the passenger classes is different from what would be expected if the variables (survivorship and passenger class) were independent from each other.

Since we performed the Chi-squared test to answer our primary, general question, this particular method is considered an *a priori* test because we intended to perform an initial exploration of the data. As mentioned above, *a priori* comes from Latin and means "first" or "beforehand." In statistics, *a priori* indicates that you have planned something beforehand, like an analysis of data in which the p-value is pre-selected as 0.05. Also, an a priori test is a test that you planned to perform in advance.

However, now that we know there is statistical significance across the passenger class data, if we do further analysis to parse out exactly where the difference exists then we will be performing a *post hoc* analysis. *Post hoc* means "after this" or "after the fact" and describes additional unplanned tests in statistics.

For our *post hoc* tests, we want to compare every passenger class to every other passenger class to see if there is statistical significance between the two being contrasted. We know there is statistical significance somewhere within our passenger classes, but we are unclear where it actually resides in the data. Consequently, we will compare every passenger class to every other passenger class.

For this *post hoc* analysis we could apply a Chi-square test for each of the three separate comparisons. However, for smaller sample sizes in the range provided by our *Titanic* data, *Fisher's Exact test* is better suited for a 2X2 analysis. The Python code that runs *Fisher's Exact test* follows Table 5.4, which lists the individual 2X2 contingency tables for each of the three tests.

```
File: fishers.py:
from scipy import stats
import pandas as pd
import numpy as np

df = pd.read_csv('Titanic_Passengers.csv')
contingency_table = pd.crosstab(df['Passenger_Class'], df['Lived_Died'], margins=False)
firstClassCount = contingency_table.iloc[0].values
secondClassCount = contingency_table.iloc[1].values
thirdClassCount = contingency_table.iloc[2].values

print('Fisher\'s 1 and 2:')
oddsratio, pvalue = stats.fisher_exact(np.array([firstClassCount, secondClassCount]))
print(pvalue)

print('Fisher\'s 1 and 3:')
oddsratio, pvalue = stats.fisher_exact(np.array([firstClassCount, thirdClassCount]))
print(pvalue)

print('Fisher\'s 2 and 3:')
oddsratio, pvalue = stats.fisher_exact(np.array([secondClassCount, thirdClassCount]))
print(pvalue)
```

Table 5.4 All passenger classes compared to each other

Lived_Died/Passenger_Class	Died	Survived
First Class	149	201
Second Class	175	118

Lived_Died/Passenger_Class	Died	Survived
Second Class	175	118
Third Class	528	181

Lived_Died/Passenger_Class	Died	Survived
First Class	149	201
Third Class	528	181

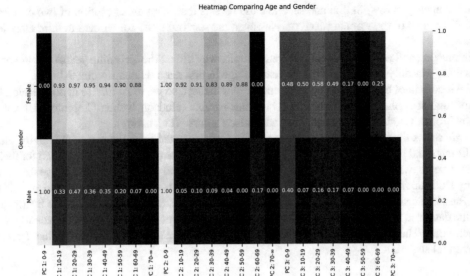

Fig. 5.5 Heatmap depicting survival percentage for groups organized by age, gender, and passenger class. "PC 1: 0-9" means passenger class 1 (or first class) ages 0–9. Similarly, PC 2 means second class and PC 3 means third class. The decimal places (e.g., 1, 0.9, etc.) represent the percentage of survivors (e.g., 1 = 100% survived; 0.9 = 90% survived). Image created from titanic_age_and_pclass_heatmap.py

The following is the generated output:

```
Fisher's 1 and 2:
1.8271492767038394e-05
Fisher's 1 and 3:
7.457205699028396e-24
Fisher's 2 and 3:
5.007621820660424e-06
```

Each of the Fisher's tests reveals statistical significance. Another way to conceptualize these results is, notwithstanding the different passenger class sizes (for example, there were 709 third class passengers compared to 350 first class passengers), there was a measurable significant difference in the percentage of the passengers who died, dependent only on the passenger class to which they belonged.

Table 5.5 confirms that each passenger class logged different percentages of people who died. Travelers in the most expensive and prestigious class had a higher survival percentage compared to the less expensive and prestigious classes.

Table 5.5 Table showing the percentage of each passenger class that died

Passenger_Class	Died
First Class	43%
Second Class	60%
Third Class	74%

Figure 5.5 displays a summary heatmap depicting the relationship among gender, age, and passenger class. At first glance, Fig. 5.5 may appear to be a jumble of rectangles with numbers and background colors. However, each rectangle represents the percent of people from that category that survived the sinking of the *Titanic*.

Focusing on the bottom-left cell, you can see that all boys in first class who were younger than 10 years of age survived (PC 1: 0–9). This is indicated by the lighter tan ("hotter") color and the number "1.00" which represents 100%. More detailed inspection of the dataset reveals that this particular group includes three boys below the age of 10.

Following along the bottom row of cells, a general pattern emerges that about one-third to one-half of males in first class survived from the ages of 10–49. However, for men in first class 50 or older the survival rate was much lower.

Crossing over from first class to second class in the bottom row an examination of the ninth cell from the left indicates that all boys under ten in second class survived. However, for males in second class 10 years or older, the survival percentages quickly decrease and remain low. The prospects for males in third class did not improve.

Looking at the women (top row) in first and second class reveals that, with the exception of two outliers (PC 1: 0–9 and PC 2: 60–69), 83% or more of the women survived. However, the survival trend for women in third class is closer to that of first-class males.

Outliers are interesting and worth investigating. For example, why did all the girls nine years or younger die in first class? The data reveals there was only one girl less than 9 years of age in first class: Helen Loraine Allison.

The reason she died cannot be explained by the values in our dataset, so we must turn to history. Helen Loraine Allison's story is both short and sad. She was on board with both her parents (Hudson Joshua Creighton Allison and Bess Waldo Allison), her baby brother (Hudson Trevor Allison) and Hudson's nurse (i.e., nanny).

When the *Titanic* struck the iceberg, Hudson and his nurse were separated from the rest of the family. History tells us that although Helen Loraine and her parents could have escaped on lifeboats, the parents instead searched for their baby and chose to stay behind. In a tragic twist, the reason why the Allison's could not find their baby was because the nurse had boarded a lifeboat with baby Hudson. In the end, Helen Lorain Allison and her parents died, but baby Hudson and his nurse survived.

In summary, Fig. 5.5 encapsulates the complex interactions of real life. Simply glancing at the heatmap will not provide full insight into the data. You must purposefully investigate how the different data compare independently as well as how they interact with each other. The next section will explore yet another interesting interaction with the *Titanic* passenger data using a new inferential statistic.

5.6   T-test Example: Fare and Gender

Whereas the Chi-Squared test and Fisher's Exact test focus on a categorical discrete dependent variable, the T-test and ANOVA tests target a continuous numerical dependent variable.

The big question we hope to answer in this section is the following: **How was fare (price of passage) on the *Titanic* affected by gender and passenger class?** To answer this question, we examine three separate but related questions in the following order:

1. Did one gender pay more fare than the other?
2. Is there a difference in fare between passenger classes?
3. Is there an interaction between gender and passenger class on the resulting fare?

The most common continuous numerical dependent variable tests are the T-test and the ANOVA tests. The *T-test* is used for comparing two populations (e.g., comparing male vs female) and the *ANOVA* is used when comparing three or more populations (e.g., comparing fertilizer one to fertilizer two to fertilizer three).

In this section we will focus on the T-test. Particularly, we will focus on the following question: **Did one gender pay more fare (price of passage) on the *Titanic* compared to the other?** Alternatively, to engage the services of a bad pun: was the fare fair?

Our first step is to visualize the data of interest with a boxplot. We use a boxplot because the fare is a continuous numerical value and the boxplot will allow us to rapidly view the mean, quartiles, and outliers. Recall the boundaries of the box figure in our boxplot define the "interquartile range" (IQR), which includes the median value plus the two adjacent quartiles, encompassing 50% of the data. Figure 5.6 displays boxplots for the fare passengers paid based on gender.

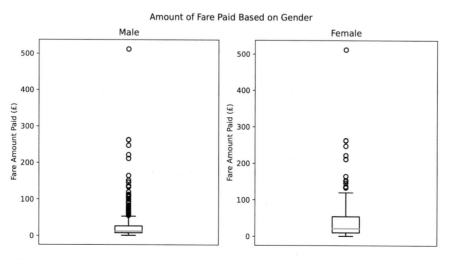

Fig. 5.6 Boxplots of the fare paid by the two respective genders. Image created from titanic_ttest.py

The following code employs the describe() function to provide the numerical values depicted in the boxplots.

```
Partial code: full program in titanic_ttest.py:

male = df[df['Gender']=='Male']          Name: Fare_decimal, dtype: float64
male['Fare_decimal'].describe()          female = df[df['Gender']=='Female']
                                         female['Fare_decimal'].describe()


count    839.000000                      count    482.000000
mean      26.226067                      mean      45.118299
std       42.514485                      std       63.149875
min        1.000000                      min        1.000000
25%        7.895000                      25%       10.460000
50%       12.350000                      50%       21.555000
75%       26.550000                      75%       54.525000
max      512.330000                      max      512.330000
                                         Name: Fare_decimal, dtype: float64
```

From both the summary descriptive information above and Fig. 5.6, we can deduce that females paid more for their fare in general when compared with their male counterparts. In other words, it appears that females were charged more and in effect (applying the pun introduced previously), the fare was **not** fair.

To determine if this assumption based on our visual inspection is true, we will run the T-test. The null hypothesis for the T-test can be formulated as follows: there is no difference between the fare for male and female populations. The alternative hypothesis is that one group paid more than the other. The following code executes the T-test:

```
Partial code: full program in titanic_ttest.py:

from scipy.stats import ttest_ind
statistic, pvalue = ttest_ind(male['Fare_decimal'], female['Fare_decimal'])
print(pvalue)


1.29767291117129e-10
```

Is the resulting 1.29767291117129e-10 less than our *a priori* p=0.05? Yes, which implies there is a statistical difference, and we can therefore reject the null hypothesis and assert the alternative hypothesis that one of the groups (i.e. male and female) paid more than the other. In particular, from the descriptive statistics we can see the fare paid by females has a mean of £45.12 and the fare paid by males has a mean of £26.23. So, we may conclude that a bias existed in which women were unfairly charged a higher rate for the same passage.

5.7 ANOVA Example: Price Differences Between Passenger Classes

Now that we know females paid more than males for the same trip, let us turn our attention to a similar question: **Is there a difference in fare among passenger classes?**

One way to contextualize this question is to consider the potential for price gouging on the *Titanic*. *Price gouging* is the practice of charging excessively high prices. Usually this occurs with products undergoing a sharp increase in demand. In our *Titanic* passenger case, if price gouging did exist (the null hypothesis) then we would likely see approximately the same price regardless of passenger class. In other words, if price gouging were present then we would not be able to distinguish the fare paid in first class from the fares paid in second and third class.

Instead of performing the T-test that specializes in comparing two populations as we did in the last section focused on gender, we will now turn to the ANOVA (Analysis of Variance) test. The ANOVA test allows us to evaluate the influence of three or more populations on a numerical continuous dependent variable. In this case, we focus on how the fare prices (continuous numerical dependent variable) are affected by the three passenger classes.

As already mentioned, the null hypothesis is there is *no* difference between the prices with respect to passenger class (e.g., price gouging) and the alternative hypothesis is there *is* a difference in price based on the passenger class (e.g., no price gouging).

The following code produces the boxplot visualized in Fig. 5.7 and computes the results for the ANOVA test:

```
Partial code: full program in anova1.py:

model = ols('Fare_decimal ~ C(Passenger_Class)', data=df).fit()

aov_table = sm.stats.anova_lm(model)  # the actual ANOVA test
print('ANOVA results:\n', aov_table)
```

```
ANOVA results:
                      df        sum_sq        mean_sq           F         PR(>F)
C(Passenger_Class)   2.0    1.201111e+06   600555.522821   338.116478   2.964452e-119
Residual          1318.0    2.341004e+06     1776.179397         NaN             NaN
```

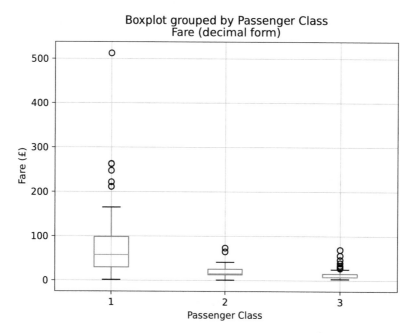

Fig. 5.7 Boxplot showing the different fares paid for the three passenger classes. Image created from anova1.py

You may not be familiar with the notational use of the "~" character that appears in the code example. Following R-style formulas, the string 'Fare decimal ~ C(Passenger Class)' translates to "Fare decimal variable is affected by the categorical variable (thus the `C') Passenger Class." When creating a model, the "~" character separates the left-hand side of the model from the right-hand side. The "+" character adds new separate columns to the design matrix. There is also the ":" (e.g., varA:varB) character which adds a new interaction column to the model matrix represented as the product of two variables. The "*" character (e.g., varA * varB) signifies that both the individual columns (e.g., varA and varB) and the combined interaction column varA:varB will be included in the analysis.

Functions can also be used in the model. For example, if an independent variable follows an exponential distribution, you may want to normalize this data by calculating the log of the column using the numpy library (typically imported as `import numpy as np`) as shown in the statement `np.log(variableName)`. Models can then be defined as shown here: `dependentVariable ~ np.log(IndependentVariable)`.

Given the resulting p-value of 2.964452e-119 generated by the above ANOVA is less than our *a priori* p-value of 0.05, we may conclude there is some difference in the prices relative to passenger class. Consequently, we should reject the null hypothesis stating there is no difference among the fares for the different passenger classes. What does this result imply for our question about price gouging? At this point, all we know is that there are at least two passenger classes that exhibit different prices. We will have to perform a *post hoc* analysis to determine which pairs of passenger classes had different fare prices and which pairs did not.

Analogous to how the Chi-squared test has the associated Fisher's Exact test to analyze 2X2 contingency tables, there is a post hoc procedure called the Tukey HSD test to compare each individual population from a statistically significant ANOVA. The Tukey HSD test is akin to applying a T-test on every combination of passenger classes. Specifically, we will run the following code to test how the fare paid by travelers compared for each pair of passenger class levels:

```
Partial code: full program in tukey_hsd1.py:

from statsmodels.stats.multicomp import MultiComparison
mc = MultiComparison(df['Fare_decimal'], df['Pclass'])
result = mc.tukeyhsd()
print(result)
```

Which produces the following output:

```
Multiple Comparison of Means - Tukey HSD, FWER=0.05
========================================================================
group1    group2    meandiff    p-adj      lower      upper      reject
------------------------------------------------------------------------
   1         2       -63.2576    0.001     -71.2564   -55.2588     True
   1         3       -70.9464    0.001     -77.4902   -64.4025     True
   2         3        -7.6888    0.0271    -14.6845    -0.6931     True
------------------------------------------------------------------------
```

The above results compare each of the "groups." Each "group" is a passenger class labeled with "1" for first class, "2" for second class, and "3" for third class. If no statistical significance is found, then the reject (last) column will indicate "False" – that is, the null hypothesis cannot be rejected. Conversely, if the reject column indicates "True" then we can reject the null hypothesis.

The boxplot in Fig. 5.7 displays several outliers (the circles above each of the whiskers), most notably for third class. There are only a few outliers for first and second class. However, the Tukey HSD test shows a "True" in the reject column for all three comparisons indicating that there is a price difference for every passenger class compared to every other passenger class, suggesting that, overall, each passenger class was set at a distinctly different price.

Both the box and whiskers in Fig. 5.7 reveal the biggest price variance within a passenger class occurred in first class. Phrased differently, there was a large amount of price variance within first class. However, the mean for each passenger class was distinct enough that we can safely rule out price gouging as a general practice although there are clearly individual cases where **some** price gouging took place as indicated by the outliers.

5.8 Two-Way ANOVA Example: How Gender and Passenger Class Together Affect Fare Price

One of the most intriguing aspects about data analysis is discovering how different variables interact with each other. We know that females paid more than males. We also know that each passenger class fare was different from that paid by the other passenger classes. What we want to know now is the following: **How was fare affected by both gender and passenger class?**

A two-way ANOVA provides the mechanism to analyze that exact relationship. Specifically, a two-way ANOVA allows us to determine the existence of (1) a main effect (the same as a one-way ANOVA described in the last section) and (2) an interaction between two or more variables.

Strictly speaking, if a two-way ANOVA was planned *a priori* then the previous T-test and ANOVA test would not have been necessary because these one-way results will be computed as main effects in the outcomes generated by the two-way ANOVA. Remember that 'C' designates a column containing categorical data. The following code produces a model that examines Passenger Class, Gender, and an interaction (denoted by the ":" character) between Passenger Class and Gender:

```
Partial code: full program in anova2.py:

from statsmodels.formula.api import ols
model = ols('Fare_decimal ~ C(Passenger_Class) + C(Gender) + C(Passenger_Class):C(Gender)',
data=fare_df).fit()
aov_table = sm.stats.anova_lm(model)
print(aov_table)
```

Which produces the following output:

	df	sum_sq	mean_sq	F	PR(>F)
C(Passenger_Class)	2.0	1.201111e+06	600555.522821	354.561594	6.964570e-124
C(Gender)	1.0	4.291327e+04	42913.269165	25.335538	5.485682e-07
C(Passenger_Class):C(Gender)	2.0	7.074753e+04	35373.767324	20.884296	1.177962e-09
Residual	1315.0	2.227344e+06	1693.797446	NaN	NaN

As we have previously discovered, there is a main effect (i.e., statistical significance alone) for both passenger class and for gender. There is also an interaction between both passenger class and gender. Note the degrees of freedom (df) are listed as 2 for the passenger class main effect (3 passenger classes −1), and 1 for the gender main effect (2 genders − 1). The degrees of freedom for the interaction effect are calculated as the product of the degrees of freedom for each of the main effects, or in this case 2 × 1 = 2. The interaction is best understood when visualized as shown in Fig. 5.8.

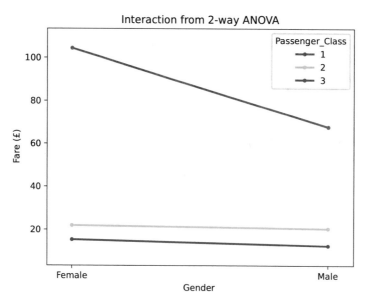

Fig. 5.8 Line graphs showing the interaction between passenger class and gender and how they affect fare (passage price). Image created from anova2.py

Figure 5.8 represents various conclusions of interest. Before examining the visualized data, note the continuous dependent variable (fare or passage price) is plotted along the dependent y-axis whereas the categorical values of gender and passenger class are denoted as distinct points along the x-axis and as separate colors, respectively.

Our first observation of the interaction graph centers on the main effect of passenger class and how the different colored lines reveal that at least one of the classes is priced much differently than the others. We can also see a confirmation of the results of the Tukey HSD test where **all** three of the passenger classes have distinctly different means from each other.

Second, we immediately observe the main effect of gender where males paid less in fare than females. The slopes of the second- and third-class linear plots exhibit a gentle downward trend, but the first-class linear plot displays a noticeable difference in the fares paid by men and women.

Lastly, we can see the interaction effect of gender and passenger class. For interaction plots, parallel lines signify no interaction effect between the two categorical values. Imagine if the first-class line demonstrated the same gentle negative slope as the second- and third-class lines, then we could say that men pay a little less than women to travel on the *Titanic*, but the small relative difference in pay does not depend on the passenger class. Conversely, let's assume the second class and third-class lines each exhibited the same sharp decrease as the first-class line. Again, we would say that women are paying a significantly higher fare, but this large difference between the passage price for men and women still does not depend on passenger class. Our particular plot reveals a visually recognizable difference among the slopes for the passenger classes, suggesting an interaction effect. Main effects should always be carefully interpreted in the context of measurable interaction effects.

Specifically, we can see the fare is not noticeably different between genders for second and third class, but there is a clear and noteworthy interaction effect between first class and gender. The gender main effect which measures the difference between female and male fares is likely primarily influenced by the first-class fare. Another way to frame this analysis would be our response to a woman asking if she will be paying significantly more than a man to board the *Titanic*. From our privileged view of the interaction plot, we would answer "It depends on the passenger class." Without digging deeper into the data and history we do not know why females paid more than their male counterparts in first class, but we now have a viable launching point from which to investigate.

Exercises

1. What is the difference between descriptive statistics and inferential statistics?
2. Explain the relationship between independent and dependent variables.
3. How does the p-value support scientific testing?
4. What is a Type I error? What is a Type II error? What specific issues do each of these errors introduce?
5. What is the difference between nominal data and ordinal data?
6. When deciding between tests such as Chi-squared and ANOVA, do you consider the dependent variable, the independent variable, or both? What variable characteristics are important?
7. For the Chi-squared test, how are the degrees of freedom determined?
8. Consider a scientist that did not pick a p-value *a priori*. Why might that be a problem?
9. When and why would you perform a *post hoc* test?
10. Under what conditions would it be better to use Fisher's exact test versus the Chi-squared test?
11. What is the purpose of the Tukey HSD test?
12. When do you use a one-way (traditional) ANOVA vs a two-way ANOVA?

Chapter 6
Metrics

This chapter is about metrics. Quite frankly, metrics are boring when studied in isolation. A metric is fundamentally a function that measures some quantity, like distance, similarity, or error, between one or more entities.

For example, what is the difference in temperature between 30 degrees and 32 degrees? 2 degrees. This is solved by something called Euclidean distance, an admittedly unexciting result.

Wait! Do not skip this chapter yet! The primary benefit of metrics is they help to clarify and **solve problems**! Without a problem to decipher, computing metrics can be considered a useless exercise.

What is a metric? A *metric* is a way to objectively judge or analyze the quality of something we are investigating.

For example, how do you decide what recipe to cook? By simply looking at a particular recipe, can you assert the bold claim that it is a good one? How can you tell?

The following are possible metrics related to the perceived quality of a recipe:

- What percentage of guest plates are completely empty by the end of the meal? Do the children eat the prepared food? Do people ask for seconds?
- How many calories per serving? What is the overall nutritional value with regards to minerals, vitamins, fat content and everything else?
- What is the cost per calorie? What is the overall cost?
- How must time does the recipe require to prepare?
- What are the total number of steps to prepare?
- A hybrid measure: take cost *and* time into consideration.
- Does it look appetizing?

Regardless of the metric you choose, a good metric should be deemed *objective* (unbiased), *universal* (can be applied to all recipes), and *concise* (provides a single measurable value). By returning a measurable value, you can then apply the metric to a large recipe dataset to rank the recipes and thereby find the top recipes for you or someone else.

This chapter is divided into three easily digestible examples demonstrating:

1. A way to obtain similar responses, e.g., retrieving movies with related content.
2. A way to measure performance, e.g., how well am I doing on my diet.
3. A way to measure prediction, e.g., how well a machine learning algorithm performs.

© The Author(s), under exclusive license to Springer Nature Switzerland AG 2022
R. Ball, B. Rague, *The Beginner's Guide to Data Science*, https://doi.org/10.1007/978-3-031-07865-1_6

6.1 Distance Metrics: Movies Example

Given the popularity of streaming entertainment services and cinematic productions, we can safely assume that most people in the world have viewed thousands of hours of movies, television shows, plays, and other visual media. In this section we will explain one of the more familiar and straightforward kinds of recommendation engines: one which offers movie selections based on prior viewing habits. A recommendation engine is simply software that uses metrics to predict or suggest a product or service, in this case a movie. You can read more about recommendation engines in Chap. 7.

The machine learning algorithm that we will use for our simple movie recommendation engine is known as *K-Nearest Neighbor (KNN)*. The KNN procedure is quite simple and comprises the following steps:

1. Pick K. For instance, let's arbitrarily select K = 10. In other words, we want 10 movie recommendations.
2. Pick a base case. In our example, select a base movie, say *Back to the Future* released in 1985.
3. Compare all other movies to *Back to the Future* using one or more metrics.
4. Sort the results of the comparison and display the top *K* (10) movies.

That's all there is to it. Let's go through this process in more detail.

We first obtain a list of movies. In the modern world this is not a very difficult task. In Chap. 2 we discuss various methods for gathering data. Applying the concepts from Chap. 2 we can locate a list of movies that either other people created or we create on our own. Although there are many lists of movies on the Internet, for our example we will create our own list using the IMDB database (imdb.com). Although we could web scrape the website, it will be faster and easier to use the IMDB Python package, which can be installed from the command line using the following code:

```
pip install IMDbPY
```

To begin, we need to decide what movies to target. If we start with the first movie in the IMDB database and extract the initial 20,000 movies, then we will collect movies in the time period from the 1890's to the 1920's. Although this collection may be of historical interest, these selections would clearly not be considered a modern list of movies. Since our base case *Back to the Future* was released in 1985, we should acquire other movies around that time period. According to the IMDB database, the id value for *Back to the Future* is 88,763. How many movies do we want? To have a robust and representative list of movies of modest size, we arbitrarily decided on 50,000 movies for our self-generated list.

We also need to determine the attributes of interest for the movies on our list. Once you have created an IMDB object in Python (`imdb_obj = imdb.IMDb()`) then you can obtain the movie instance that contains the relevant information about the movie (`movie = imdb_obj.get_movie(id)`). All the attributes available for each movie object may be displayed using the command `print(movie.__dict__)`. The attributes that we choose for our movie recommendation engine are the following:

- the IMDB id
- the title of the movie
- the year the movie was released
- the weighted mean of user review ratings, ranging from 1–10 stars
- the MPA rating (e.g., G, PG, PG-13)
- the genres
- the plot.

The following is Python code to gather 50,000 movies from the IMDB database (imdb.com) starting with *Back to the Future* (88763):

```
File: movie_list_creator.py:
import imdb

imdb_obj = imdb.IMDb()
with open("movies.csv", "w") as f:
    start_id = 88763  # start with 'Back to the Future'
    total_movies = 50000  # the number of movies you want
```

```
count = 0  # how many movies so far
id = start_id  # the current IMDB id
print("IMDB_id,title,year,stars,rating,genres,plot")
f.write("IMDB_id,title,year,stars,rating,genres,plot\n")
while count < total_movies:
    try:
        movie = imdb_obj.get_movie(id)
        # convert the list into a semicolon-separated list
        genres_string = ';'.join(movie['genres'])
        # get the first plot returned. Sometimes there are many.
        # Also, relace all double quotes with a single quote.
        first_plot = movie['plot'][0].replace('"', ''')
        certificates = movie['certificates']
        certificate_rating = ''
        for certificate in certificates:
            if "United States" in certificate and "TV" not in certificate:
                certificate_rating = certificate.split(':')[1]
        # this skips any movie that does not have a rating in the United States (e.g.,
        # G,PG,PG-13, etc.) It also skips TV ratings.
        if certificate_rating == '':
            continue
        # this line is so that you can follow the progress.
        print(f'{id},"{movie["title"]}",{movie["year"]},{movie["rating"]},
        {certificate_rating},{genres_string},"{first_plot}"')
        f.write(f'{id},"{movie["title"]}",{movie["year"]},{movie["rating"]}
        ,{certificate_rating},{genres_string},"{first_plot}"\n')
        count += 1  # only increment count if all parts of the movie that we want are found
    except:
        continue  # if there is any error then skip this movie
    finally:
        id += 1  # move to the next movie no matter what
```

Much of the above code handles exceptional conditions, including missing plots, genres, or any other random error. The exceptions help us to manage the quality control for our dataset. Most movies that either do not have a plot or MPA rating (e.g., G, PG) are not generally well-known movies and, therefore, are not included in the dataset. In addition, by excluding empty fields our movie recommendation engine will avoid potential processing issues related to missing data. These steps represent an effort to wrangle, clean, and prepare the dataset.

The file `movies.csv` generated by running the above code can be found in the chapter files that accompany this book. To use `movies.csv` with our example be sure to place it in the same directory as the Python code.

You should feel free to change the code to modify the starting movie (*Back to the Future*), to adjust the total number of movies (50,000), or to add/remove other attributes for the movies in your list. However, be aware that as more movies and more information are acquired from the IMDB database, the resulting final file will be larger and will negatively affect the processing speed of your recommendation engine. Also, even though this technique has the advantage of extracting a wide variety of movie lists, it is rather slow. Securing 50,000 movies with the IMDB Python package will require many hours to complete, so we recommend that you plan accordingly.

From the 50,000 downloaded movies, we chose the base case movie *Back to the Future*. The assumption is that we like *Back to the Future* and we wish to view other similar movies. If you have seen *Back to the Future*, then you might be able to recommend several similar movies. However, what is this "similarity" measure? In other words, how do you reliably compare *Back to the Future* to other movies? Once we have a technique to consistently compare *Back to the Future* to other movies and generate an associated interpretable and reliable numerical measure, then we can sort the resulting list and show the top K similar movies.

The selection of K is arbitrary - if you want to determine the top two similar movies or the top 100 movies that is entirely up to you. However, in terms of user experience a list of two similar movies will likely be disappointing whereas a list of 100 similar movies can be overwhelming, so ten movies is a good starting place.

Now that we have (1) secured our data saved as `movies.csv`, (2) chosen K, and (3) selected our base case to be *Back to the Future*, we now leverage the following Python code to execute our KNN recommendation engine:

```
File: knn_stub.py:
import pandas as pd

def metric_stub(base_case_value, comparator_value):
    return 0

# read the list of movies into a dataframe:
df = pd.read_csv('movies.csv', index_col='IMDB_id', on_bad_lines='skip')

# setting our K to 10. In other words, we will get the K(10) closest matches
K = 10

# get the row that has 'Back to the Future' - our base case
base_case = df.loc[88763]  # 88763 is the IMDB id for 'Back to the Future'
print(f"Comparing all movies to our base case: {base_case['title']}.")

comparison_type = 'genres'  # change this to the type of metadata to be compared

# The following evaluates each movie with the given metric.
# 'x' is the value in each row in the 'comparison_type' column
df['metric'] = df[comparison_type].map(lambda x: metric_stub(base_case[comparison_type], x))

sorted_df = df.sort_values(by='metric')
sorted_df.drop(88763, inplace=True)  # drop the base case
print(sorted_df['title'].head(K))
```

Running the code above will result in a recommendation list containing the first ten movies in the movies.csv file because we do not actually compare anything in our function `metric_stub`, which simply returns 0 and does not perform any real functionality yet.

Before we start comparing metrics, we need to understand what we are comparing. In this case we are comparing movies. However, what does it actually mean to compare movies?

Our real objective is to compare movie metadata. *Metadata* is data about something. For example, some of the metadata about you is your name, your mailing address, your weight, and your height -- your metadata is all the data that describes you.

For movies, the metadata would be the IMDB id, the title, the release year, the number of review stars, the MPA rating, the genres, and the plot. Movie metadata could also include any other data you specified when you generated your `movies.csv` file such as the actors, the directors, or other production information.

The next subsections will introduce metrics and provide examples of how to utilize them in the context of comparing the movie data that we collected with the goal of designing a basic movie recommendation engine.

6.1.1 KNN with Euclidean Distance

How do we compare movie release years or the number of rating stars a movie received? Another way to ask the same question is, how do we effectively compare numbers?

The most common approach to comparing numbers is to use Euclidean distance. *Euclidean distance* is named after the mathematician Euclid and is simply a method for determining the distance between points in N-dimensional space. More concisely, the Euclidean distance is the straight-line distance between two points, or as the old expression attests: "…straight as the crow flies." The Euclidean distance represents the magnitude of the shortest distance between two points – it is a difference determination that only returns positive numbers.

The graph (in 2D space) in Fig. 6.1 provides the concept underlying the calculation of Euclidean distance. Given two points, how do you calculate the distance between those points. The answer is commonly recognized as "the distance formula," but is more precisely called Euclidean distance which is simply an application of the Pythagorean formula as diagrammed in Fig. 6.1.

Fig. 6.1 Example of two points p and q on a graph. The Euclidean distance is simply an application of the Pythagorean formula. (Image created from euclidean_image.py)

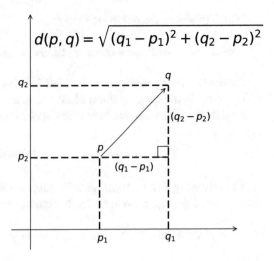

The following distance formula generalizes to any dimension (1D, 2D, 3D, etc.):

$$Distance(p,q) = \sqrt{(p_1 - q_1)^2 + (p_2 - q_2)^2 + \cdots + (p_n - q_n)^2}$$

The formula sums the squares of the difference for every dimension and then computes the square root of the summation. The following formula is a condensed version of the previous expression:

$$Distance(p,q) = \sqrt{\sum_{i=1}^{n} (p_i - q_i)^2}$$

To solidify our understanding, the following is example Python code that calculates the distance between two 2-dimensional points x and y:

```
File: two_d_distance.py:
import math
# Example points in 2-dimensional space...
x = (0, 0)
y = (4, 4)
distance = math.sqrt((x[0] - y[0])**2 + (x[1] - y[1])**2)
print(f"Euclidean distance from x to y: {distance}")
```

Which produces the following output:

```
Euclidean distance from x to y: 5.656854249492381
```

The following is an example in Python that calculates the distance between two 3-dimensional points x and y:

```
File: three_d_distance.py:
import math

# Example points in 3-dimensional space...
x = (5, 6, 7)
y = (8, 9, 9)
distance = math.sqrt(sum([(a - b) ** 2 for a, b in zip(x, y)]))
print(f'Euclidean distance from x to y: {distance}')
```

Which produces the following output:

```
Euclidean distance from x to y: 4.69041575982343
```

Euclidean distance between two points in one-dimensional space, such as comparing single value movie release years, will conform to the same equation. However, determining a one-dimensional distance may be simplified to the absolute value of the difference between the base case (*Back to the Future* release date 1985) and the movie of interest.

$$Distance(p,q) = \sqrt{(p-q)^2} = |p-q|$$

The following Python function will return the absolute value of the difference between the base case year (1985 in our example) and the release year of another movie:

```
Partial code: full program in knn_euclidean_year.py:

def euclidean_distance(base_case_year, comparator_year):
    return abs(int(base_case_year) - int(comparator_year))

...

comparison_type = "year"
df['euclidean']=df[comparison_type].map(lambda x: euclidean_distance(int(base_case[comparison_
type]), int(x)))
sorted_df = df.sort_values(by='euclidean')
```

Note these code revisions change the `comparison_type` variable to `year` and sort the data by the newly generated `'euclidean'` column. With these modifications in place, the previous KNN recommendation engine code now returns the top K (10 in our example) "most similar" movies to our base case. Since our base case is *Back to the Future* released in 1985 the Euclidean distance metric used by our K-Nearest Neighbor recommendation engine returns all 1985 movies because the distance between two movies released in the same year is 0, which means that any other movie introduced in 1985 is the most "similar" to *Back to the Future* based on our limited criterion.

The collection of movies in our dataset includes many 1985 movies! Since all other 1985 movies would have a Euclidean distance of 0 with respect to *Back to the Future* and there are more than 10 movies released that same year, the setting K = 10 presents to the user any subset of those 1985 movies.

However, if there were fewer than 10 movies released in 1985 then the next set of years that would be presented would be 1984 and 1986, which would register a Euclidean distance of 1 since they are both 1 year away from 1985. Similarly, movies in 1955 and 2015 are both 30 years "distance" from the year in which *Back to the Future* was released.

We could run the same code above but change the variable `comparison_type` to stars (`comparison_type = 'stars'`) at which point we would generate the top 10 movies that are closest to 8.6 stars out of 10. Using Euclidean distance on stars instead of years may produce a more satisfying, informative result. *Back to the Future* has 8.6 stars out of 10, which is highly rated. By using the comparison of stars instead of years then you would determine similarly highly rated movies instead of movies that were released in the same year.

Although identifying movies that have a release date similar to *Back to the Future* may not be the best metric to power our recommendation engine, the year of release could still be an important differentiating factor if other criteria were involved in the selection process. For example, if our base case is *Back to the Future* and we found movies that were similar using several metrics other than release year, a 1990 movie might be a better movie to suggest than a movie from 1912. In this situation, using year of release as a filter might be a better idea than employing it as a full-fledged distance metric.

A *filter* preprocesses the data before applying metrics. In our case we could apply a filter that would eliminate any movies released prior to 1980. Implementing a filter is completely arbitrary – in a real application the software would allow the user to set whatever filter they want. The following modification to the above code will filter out all movies not released in 1980 or later:

```
df = df[(df['year'] >= 1980)]
```

6.1.2 KNN with Jaccard Similarity Index

Another important distance metric is the Jaccard similarity index. We will demonstrate its use by comparing movie genres.

The *Jaccard similarity index* (sometimes called the *Jaccard similarity coefficient*) compares members of two individual sets to determine which members are shared and which are distinct. This index measures the similarity between the two sets of data, with a range from 0 to 1 inclusive, where 0 means no similarity and 1 indicates an exact match.

The nice thing about the Jaccard similarity index is that we can compare two sets of any items. The formula for the Jaccard similarity index in set notation is provided here:

$$J(X,Y) = \frac{|X \cap Y|}{|X \cup Y|}$$

Although the results of the index are easy to interpret, very small samples or data sets with missing observations may lead to erroneous results.

The following provides a simple example to determine the similarity between set X and set Y:

$$X = \{0,1,2,5,6\}$$

$$Y = \{0,2,3,4,5,7,9\}$$

$$J(X,Y) = \frac{|X \cap Y|}{|X \cup Y|} = \frac{|0,2,5|}{|0,1,2,3,4,5,6,7,9|} = \frac{3}{9} = 0.\overline{33}$$

In other words, set X and set Y are approximately one-third (or 33%) the same.

Let us return to our movie recommendations. When comparing genres, where each genre is a "member" of the set of genres, how does one movie stack up against another? According to IMDB, the genres assigned to *Back to the Future* are the following: Adventure, Comedy, and Sci-Fi. How does *Back to the Future* compare to another 1985 movie such as *Mad Max Beyond Thunderdome* whose IMDB assigned genres are Action, Adventure, and Sci-Fi?

$$X = Back\ to\ the\ Future : \{Adventure, Comedy, SciFi\}$$
$$Y = Mad\ \max\ Beyond\ Thunderdome : \{Action, Adventure, SciFi\}$$

$$J(X,Y) = \frac{|\{Adventure, SciFi\}|}{|\{Action, Adventure, Comedy, SciFi\}|} = \frac{2}{4} = 0.5$$

The interpretation of the above result demonstrates that when comparing *Back to the Future* and *Mad Max Beyond Thunderdome*, they are 50% similar in terms of genres.

To install genre comparisons as the metric in our KNN recommendation engine code example, we need to change the comparison_type variable to genres instead of years (comparison_type = "genres"). Additionally, we also modify how the results are sorted. With Euclidean distance we wanted the sort_values function to show the smaller numbers first (ascending order). However, with the Jaccard similarity function 0.0 means no similarity while 1.0 means an exact match. So, we add ascending=False to the sort_values function to display the higher numbers first. Our new result (including the optional filter of considering only movies that were released in 1980 or later) is the following:

```
Partial code: Full program in knn_jaccard_genres.py:

def jaccard_similarity(base_case_genres: str, compartor_genres: str):
    base_case_genres = set(base_case_genres.split(';'))   # cast list to a set
    compartor_genres = set(compartor_genres.split(';'))   # cast list to a set

    numerator = len(base_case_genres.intersection(compartor_genres))
    denomenator = len(base_case_genres.union(compartor_genres))
    return float(numerator) / float(denomenator)   # cast as float
```

```
comparison_type = "genres"

df = df[(df['year'] >= 1980)]  # filter out movies before 1980

df['jaccard'] = df[comparison_type].map(lambda x: jaccard_similarity(base_case[comparison_
type], x))

                sorted_df = df.sort_values(by='jaccard', ascending=False)
```

The results from the above code display ten movies similar to *Back to the Future* based on genre. The additional filter on release date further narrows down the movies listed in our results.

6.1.3 KNN with Weighted Jaccard Similarity Index

The traditional Jaccard similarity distance works quite well when performing one-to-one comparisons between **individual** movies. However, what happens when you want to accurately *combine* the features of more than one of your preferred movies to be the basis for future recommendations? In other words, we want to combine the genres of a group of preferred movies to compare with an individual movie at a later time to see if it is similar to our group.

For example, what happens when your favorites include *Mad Max Beyond Thunderdome* and *Back to the Future*? The assigned genres for *Mad Max Beyond Thunderdome* are action, adventure, and sci-fi and the assigned genres for *Back to the Future* are adventure, comedy, and sci-fi. Considered collectively as a set we obtain the following: {action, adventure, comedy, sci-fi}. Because sets only allow unique elements, the genre "adventure" only appears once in our combined set even though it is listed individually for both films.

Consider the following problem: We have built a preferred list of 100 movies that are all very similar with "adventure" and "action" appearing most of the time in the genre list. However, one of our selected movies has "horror" in its genre list. In general, we do not like horror movies and only watch them rarely, preferably on Halloween night. In general, we want our recommendation engine to present us with more movies that belong in the "adventure" and "action" genres and do not generally wish to watch any "horror" movies.

With the Jaccard similarity distance as previously defined, the fact that we watched "adventure" and "action" movies much more frequently than "horror" movies is not taken into account because the set definitions treat each genre equally. Instead, a comparison metric should be devised that includes the fact that we prefer "adventure" and "action" much more than "horror."

One solution is an adaptation of the *weighted Jaccard similarity index*. For our purposes, we first build a weighted dictionary for each genre of the movies in our preferred list. The following code illustrates this point:

```
Partial file: full program in weighted_dictionary_example.py:

# put our selections of 'Back to the Future'(88763) and
# 'Mad Max Beyond Thunderome'(89530) into a list:
selections = [df.loc[88763], df.loc[89530]]

# genres_weighted_dictionary is needed for the weighted Jaccard similarity index:
genres_weighted_dictionary = {'total': 0}
for movie in selections:
    for genre in movie['genres'].split(';'):  # the genres are separated by a semicolon
        if genre in genres_weighted_dictionary:
            genres_weighted_dictionary[genre] += 1
        else:
            genres_weighted_dictionary[genre] = 1
        genres_weighted_dictionary['total'] += 1

print(f'genres_weighted_dictionary = {genres_weighted_dictionary}')
```

Executing the code above results in the following weighted dictionary of the combined genres of the two movies *Mad Max Beyond Thunderdome* and *Back to the Future*:

```
genres_weighted_dictionary = {'total': 6, 'Adventure': 2, 'Comedy': 1, 'Sci-Fi': 2, 'Action': 1}
```

We then use that weighted dictionary in our comparison metric as illustrated in the following Python code:

```
Partial file: full program in weighted_dictionary_example.py:
```

```python
def weighted_jaccard_similarity(weighted_dictionary: dict, comparator_genres: str):
    # weighted_dictionary is based on all the selections that the user has made so far
    # comparator_genres is another movie's genres that is being compared
    numerator = 0
    denominator = weighted_dictionary['total']
    for genre in comparator_genres.split(';'):
        if genre in weighted_dictionary:
            numerator += weighted_dictionary[genre]

    return numerator / denominator
```

One question to consider regarding the traditional Jaccard similarity index versus the weighted Jaccard similarity index is how the results differ between the two metrics when the first argument `weighted_dictionary` reflects the genre preferences for only one movie. In other words, if only *Back to the Future* is currently preferred then how do the two metrics compare? This is an important question so we can confirm the consistency between the two metrics. Consider the following two inputs:

```
Partial code: full program in weighted_vs_normal_jaccard.py:
```

```python
comparator = 'Adventure;Comedy'
weighted_jaccard_similarity({'total': 4, 'Adventure': 1, 'Comedy': 1, 'Sci-Fi': 1, 'Action':
1}, comparator)
jaccard_similarity('Adventure;Comedy;Sci-Fi;Action', comparator)
```

Which produces the following output:

```
0.5
0.5
```

The above output reveals that when the weighted Jaccard similarity index represents only one movie in its dictionary then the results are equivalent to the traditional Jaccard similarity index.

Now that we have established the validity of the weighted Jaccard similarity index let's add two additional filters: one to only recommend movies with at least five stars, and one to only recommend family movies (G, PG, and PG-13). Putting all the pieces together we obtain the following Python code:

```
Partial file: full program in knn_weighted_jaccard.py:
```

```python
def weighted_jaccard_similarity(weighted_dictionary: dict, comparator_genres: str):
    # weighted_dictionary is based on all the selections that the user has made so far
    # comparator_genres is another movie's genres that is being compared
    numerator = 0
    denominator = weighted_dictionary['total']
    for genre in comparator_genres.split(';'):
        if genre in weighted_dictionary:
            numerator += weighted_dictionary[genre]

    return numerator / denominator
```

...

```
# put our selections of 'Back to the Future' (88763)
# and 'Mad Max Beyond Thunderome' (89530) into a list:
selections = [df.loc[88763], df.loc[89530]]

# genres_weighted_dictionary is needed for the weighted Jaccard similarity index:
genres_weighted_dictionary = {'total': 0}
for movie in selections:
    for genre in movie['genres'].split(';'):  # the genres are separated by a semicolon
        if genre in genres_weighted_dictionary:
            genres_weighted_dictionary[genre] += 1
        else:
            genres_weighted_dictionary[genre] = 1
        genres_weighted_dictionary['total'] += 1

print(f'genres_weighted_dictionary = {genres_weighted_dictionary}')

# three filters:
df = df[(df['year'] >= 1980)]  # filter out movies before 1980
df = df[(df['stars'] >= 5)]  # filter out movies less than 5 stars
# select only movies that are G, PG, or PG-13
df = df[(df['rating'] == 'G') | (df['rating'] == 'PG') | (df['rating'] == 'PG-13')]

comparison_type = "genres"

df['weighted_jaccard'] = df[comparison_type].map(lambda x: weighted_jaccard_similarity(genres_
weighted_dictionary, x))
```

The printed results display the weighted genres dictionary and the genres for each of the top ten movie recommendations so that you can assess how close the movie results are to the weighted genres dictionary.

6.1.4   KNN with Levenshtein Distance

Another well-known distance metric is the Levenshtein distance. The Levenshtein distance is the most common form of edit-based metric, which generally quantifies the work required to transform a string from an initial sequence to a target sequence. The Levenshtein distance is the metric to determine the difference between two sequences, typically a character sequences.

Informally, the Levenshtein distance between two words is the minimum number of single-character edits (insertions, deletions, or substitutions) required to transform one word into the other word. Different edit-based metrics have different penalties for the specific manipulations required to change one string to another. However, the Levenshtein distance has the same penalty score of 1 for an insertion, deletion, or substitution.

To install the Python package to use the Levenshtein distance, enter the following command:

```
pip install python-Levenshtein
```

The Levenshtein distance is often used for spelling mistakes, DNA sequence analytics, and other character-based analytics. Consider the following Python code:

```
File: levenshtein_example.py:
import Levenshtein

print(Levenshtein.distance('woodman', 'woodland'))
```

Output:

```
2
```

This common example demonstrates that inserting "d" at the end of "woodman" and substituting "l" for "m" converts the word "woodman" to "woodland." With only two modifications, one insertion and one substitution, the Levenshtein distance is 2.

For our example, the most obvious shortest distances between movie titles will be sequels. For our code modification we will use the Levenshtein package with the following Python code:

```
Partial code: Full program in knn_levenshtein_title:

comparison_type = "title"

...

df['levenshtein']=df[comparison_type].map(lambda x: Levenshtein.distance(base_case[comparison_
type], x))

sorted_df = df.sort_values(by='levenshtein')
```

When executing the above code to compare with our *Back to the Future* base case, we obtain a list of movies that often include the words or phrases "Back," "to the," and "Future." Although these results are interesting, we would probably generate more informative and meaningful recommendations if we were to include more words related to the movie content, such as the plot description. Unfortunately, the Levenshtein distance is not especially suited for a long list of string comparisons. However, the cosine similarity metric will resolve that exact problem.

6.1.5   KNN with Cosine Similarity

How can we compare movie plots represented as descriptive text? Comparing large blocks of narrative leads us to the rapidly expanding field of research and development known as Natural Language Processing (NLP), discussed in more detail in Chap. 9. However, at this point in our treatment we can present an effective distance metric for comparing large blocks of text.

How can a series of words be compared? We could include each word in the text as a unique element in a set and use the Jaccard similarity index. However, the problem with this approach is the length of the text would greatly affect the ultimate result for the Jaccard similarity index.

For example, if we were to compare all the words from the Bible, which contains tens of thousands of words, to the Gettysburg address by Abraham Lincoln, which has about 300 words, the difference in word count would be staggering and greatly affect the results produced by the Jaccard similarity index, which provides more reliable results when the sets being compared are balanced (i.e, approximately equivalent size).

To overcome this problem, we first put all the words into a vector. For our purposes a vector of words is essentially a line with a magnitude (or length). Different word vectors will have different magnitudes, where the vector that represents the Bible would have a much larger magnitude relative to the vector that represents the Gettysburg address. Vectors are not only quantified by a magnitude, but also a direction which can be translated as a specific angle from a set of axes or frame of reference. This angle attribute is where the cosine similarity metric can be especially helpful when comparing two different texts.

In reality, through normalization the cosine similarity distance metric does not actually measure **distance** but rather **angles**. The key concept behind cosine similarity is to determine vector proximity in terms of angles in a coordinate system. We care primarily about angle so that our metric is not biased toward vectors with large magnitudes, in other words, we do not want to rely on a metric biased toward text with more words than another text.

Specifically, the cosine similarity formula takes two vectors A and B and performs the dot product of the two divided by the product of their magnitudes. The formula returns values that range from −1 to positive 1. A result of −1 indicates that the vectors have opposite direction (highly dissimilar) and a positive 1 result indicates the vectors are essentially equivalent. The cosine similarity formula is shown here:

$$Cosine\ similarity\left(A,B\right)=\frac{A\cdot B}{\|\mathbf{A}\|\,\|\mathbf{B}\|}=\frac{\sum_{i=1}^{n}A_{i}B_{i}}{\sqrt{\sum_{i=1}^{n}A_{i}^{2}}\,\sqrt{\sum_{i=1}^{n}B_{i}^{2}}}$$

Since the above formula and our discussion about vectors and angles may not be particularly clear, let's refer to some trigonometry and an illustrative example.

The angle of what? Vectors of words? The key idea is that each word in the text under consideration is represented by its own axis or dimension. For example, if one vector is composed of the words "Hi, there" and the other one is composed of "Hello, there" then there would be a total of three axes: one for "Hi," one for "Hello," and one for "there." Figure 6.2 shows a visual example of the 3D vector space for the axes and the two vectors mapped to that space.

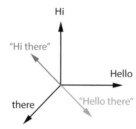

Fig. 6.2 Visual 3D visualization of comparing the vector "Hello there" to the vector "Hi there." Cosine similarity returns the angle between the blue vector and the green vector

Now that we have turned the expressions "Hi, there" and "Hello, there" into vectors we can compare their angles.

If each vector is basically a line that extends from the origin of a coordinate system, then what we are trying to figure out is essentially one angle of a triangle. Recall from trigonometry that if given two sides of a triangle, for our purposes the hypotenuse and the adjacent side, then the angle between the two sides can be determined using the cosine definition:

$$\cos\theta = adjacent\ /\ hypotenuse$$

Assume the length of the hypotenuse is 5 and the length of the adjacent side is 3. To determine the angle in degrees you use the inverse (arccosine) of the above formula: $\cos^{-1}(3/5) = 53.13$ degrees. Figure 6.3 illustrates this principle.

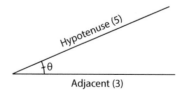

Fig. 6.3 The trigonometric principle of cosine. The angle θ is determined by $\cos^{-1}(3/5) = 53.13$ degrees

The cosine similarity metric is based on this relationship. The location of vectors relative to each other in a coordinate system can be used to determine their difference in angle. For our purposes, we can now compare the words in the plot description of *Back to the Future* to the words in the plot descriptions of all the other movies in our dataset.

Although we cover this approach in much more detail in Chap. 9, the first step to compute our cosine similarity is to convert a movie plot from a string to a vector of words. The following Python code accomplishes this task:

```
from sklearn.feature_extraction.text import TfidfVectorizer
tfidf_vectorizer = TfidfVectorizer()
word_vector = tfidf_vectorizer.fit_transform(input_string)
```

This metric is much slower to calculate than the other metrics employed thus far because more work is required to convert each plot description into a vector to compare it to other vectors.

As mentioned previously, the cosine similarity metric returns results between -1 and 1 with a result of -1 indicating that the vectors have opposite direction within the multidimensional frame of reference and a result of positive 1 indicating that the vectors are equivalent.

The following code modifies our current movie recommendation engine to leverage cosine similarity in the analysis of the text for plots:

```
Partial code: Full program in knn_cosine_similarity_plot:

def cosine_similarity_function(base_case_plot, comparator_plot):
    # convert the plots from strings to vectors in a single matrix:
    tfidf_matrix = tfidf_vectorizer.fit_transform((base_case_plot, comparator_plot))
    results = cosine_similarity(tfidf_matrix[0], tfidf_matrix[1])
    return results[0][0]


tfidf_vectorizer = TfidfVectorizer()

…


comparison_type = "plot"

df['cosine']=df[comparison_type].map(lambda x: cosine_similarity_function(base_case[comparison_type], x))

sorted_df = df.sort_values(by='cosine', ascending=False)
```

As with the Jaccard similarity measures, we sort cosine similarity results in *descending* order. The generated output list utilizing the cosine similarity index for the KNN movie recommendation engine depends significantly on the style of the author who composed the plot. However, not surprisingly, these results include sequels to *Back to the Future*.

Cosine similarity is sometimes called the **traditional** cosine similarity. The **soft** cosine similarity calculation offers a slight variation by grouping words that are semantically alike. Looking at the axes depicted in Fig. 6.2 we could argue that the words "Hi" and "Hello" are equivalent or similar in most situations.

The concept behind soft cosine similarity is the same as cosine similarity except that it includes the extra step of recognizing that some words have similar meanings. Advanced NLP (Natural Language Processing) typically incorporates the meaning of words and would use soft cosine similarity to determine near equivalences of meaning. For example, "Hi" and "Hello" are very similar as are "Bob" and "Robert."

The gensim Python package includes an implementation of the soft cosine similarity metric by utilizing a model of predefined words similar to one another. There are many reference models to choose from. For example, you can select words based on the Google news feed or Twitter data.

For our example expressions "Hi, there" and "Hello, there" we obtain the following results:

- Traditional cosine similarity: 0.34.
- Soft cosine similarity: 0.43.

However, we offer a warning. The current implementation of gensim's soft cosine similarity requires a large amount of memory and is very slow. If you were to include gensim's soft cosine similarity as part of a recommendation engine, then you should consider a preprocessing module in the design. *Preprocess* means to perform a preliminary operation on the data to prepare it for the core computational steps. For example, you might compare all movie description plots to each other using soft cosine similarity and store these results in a file or database before running your recommendation engine otherwise the user of your program might have to wait hours before receiving a recommendation.

6.1.6 Combining Metrics and Filters Together

We have so far discussed using one metric at a time and optionally using filters. How many metrics can you use simultaneously? How many filters?

Filters are easier to understand. Each additional filter you add effectively reduces the data available for analysis. For example, if you employ a single filter to weed out all movies before 1980 then you have reduced your dataset based on release date. If you add more filters, such as only considering movies awarded more than 5 stars, then you have reduced your data even further. You may add as many filters as necessary; our example in Sect. 6.1.3 used three filters.

There are two main concerns with filters. First, you can make filters too complicated and consequently produce inaccurate or erroneous responses. As with complicated SQL (Structured Query Language) queries (see Sect. 2.5 for more information on SQL), you can add numerous logical conditions like **and**, **or**, and **not** and end up confusing yourself. The more complicated the filter system the more likely you are to design one that performs incorrectly. Second, if you are too strict with your filtering criteria then you may end up without any results. For example, if you only seek movies released in 1985 with 9 stars or higher and with genres "Animation," "Horror," "Adventure," and "Comedy" then you are not likely to have any remaining movies to process and recommend.

Using multiple metrics can also be confusing but is strongly suggested to achieve robust recommendation system performance.

The main thing about combining metrics is to keep in mind the importance of generating only **one** result. Let us take an example of using cosine similarity results from the "plot" comparison type and the weighted Jaccard similarity for the "genres" comparison type.

To successfully combine the answers from the two metrics we must consider:

- Weight of each metric
- Normalization of the combined metric

Assume we want to combine the results from the cosine similarity metric (plot) and the weighted Jaccard similarity metric (genres). Our first task is to establish how much each metric will be emphasized individually in our resulting measure. Should they have equal weight?

For example, should both metrics contribute an equal amount (50%/50%) split, or should the genres be more important, with a division of 20% plot and 80% genres? Ideally, this weighting could be determined and adjusted by the user of our procedure.

After we have defined the weights for our individual metrics, we need to be careful about how we combine the results. The cosine similarity metric returns a range of output values from -1 to 1 whereas the weighted Jaccard similarity metric returns a range from 0 to 1. For each of these measures, the degree of similarity is directly proportional to the output. In other words, the greater the similarity the higher the associated value. Some metrics such as Euclidean distance exhibit scales that have inverse relationships with similarity – a result of 0 for Euclidean distance implies equivalence. Combining metrics that have different "directions" in their similarity measure will require an additional transformation to ensure that all metrics manifest the same relationship with similarity, either direct or inverse.

To effectively combine the results, we need to normalize the metrics. To *normalize* is to simply ensure that both results are within the same range and have the same similarity relationship before combining so that our weighting model returns a single value that is an accurate assessment of similarity.

Here is a concrete example where we seek a 20% plot/80% genres weighting from the cosine similarity metric and the weighted Jaccard similarity metric respectively.

Let us suppose that the result from the cosine similarity metric (plot) is 0.4 and the result from the weighted Jaccard similarity metric (genres) is 0.8.

There are various ways to normalize these results, or a mathematician might summarize this situation more accurately as "an **infinite** number of ways." So, we will simply choose one of the easier approaches: we map the range of the cosine similarity metric from the original -1.0 to 1.0 to 0.0 to 1.0. Since the magnitude of the range or Euclidean distance from -1.0 to 1.0 is 2 we first add 1.0 to the result as a bias to ensure that it is positive then divide that by 2.0 to scale the range to the target 0.0 to 1.0. Given our initial result of 0.4 for the cosine similarity metric, we calculate the following normalized result: $(0.4 + 1)/2.0 = 0.7$, which is an accurate mapping of the original metric 0.4 along a range of 0–1.

The next step is to multiply each normalized result by their assigned weight:

- Cosine similarity result: 0.4 normalized to 0.7 then multiplied by 20% (0.2) = 0.14
- Weighted Jaccard similarity result: 0.8 (already normalized) then multiplied by 80% (0.8) = 0.64

The individual results from each metric are added together to obtain the combined metric: 0.14 + 0.64 = 0.78.

The following code modifies our movie recommendation engine to utilize cosine similarity for the plot text and weighted Jaccard similarity for genres. Since we are using weighted Jaccard similarity we have assumed that the user has selected *Back to the Future* and *Mad Max beyond Thunderdome* as the base case movies. We also integrate the same three filters we used in the weighted Jaccard similarity example: release year, stars, and MPA rating.

```
Partial file: full program in knn_cosine_similarity_and_weighted_jaccard.py:

def cosine_and_weighted_jaccard(genres_weighted_dictionary_dict: dict, plots: str, compara-
tor_movie: pd.core.series.Series):
    # Perform the cosine similiarty and weighted Jaccard metrics:
    cs_result = cosine_similarity_function(plots, comparator_movie['plot'])
    wjs_result  =  weighted_jaccard_similarity(genres_weighted_dictionary_dict,  comparator_
    movie['genres'])

    # Normalization:
    # The weighted Jaccard similarity result has a range from 0.0 to 1.0.
    # The cosine similarity result has a range from -1.0 to 1.0. We need to change the range
    # for the cosine similarity result.
    # First, add 1 to the cosine similarity result so that it has a range from 0.0 to 2.0
    # Second, divide the result by 2.0 so that it has a range from 0.0 to 1.0:
    cs_result = (cs_result + 1) / 2.0

    # Weights:
    # Use a weight of 0.2 (20%) for the cosine similarity result:
    cs_result *= 0.2
    # Use a weight of 0.8 (80%) for the weighted Jaccard similarity result:
    wjs_result *= 0.8
    return wjs_result + cs_result

...

df['multiple_metrics'] = df.apply(lambda x: cosine_and_weighted_jaccard(genres_weighted_dic-
tionary, plots, x), axis='columns')
```

When calculating the final metric, since 80% weight was assigned to genres we obtain very similar results to the recommendations generated by applying *only* the weighted Jaccard similarity metric. However, there are noticeable updates to the resulting list that can be attributed to the cosine similarity index.

The above example demonstrates that any number of metrics may be combined as long as their individual results are appropriately normalized and the weights applied sum to 1.0.

6.1.7   *Mahalanobis Distance*

The Mahalanobis distance provides a measure from a vector to a distribution. The fundamental concept behind the Mahalanobis distance is to engage all the multivariate data – many variables, such as all the genres (28 genres!), stars, MPA rating, and year – to determine how closely they are related. This metric is especially good at finding outliers or anomalies given the multiple comparisons with different variables.

When investigating multiple variables simultaneously there most likely will be data that are correlated. For example, there is a direct correlation between the genre "short" and the early years of films because, in general, most of the movies made in the 1890's to 1920's were relatively short – 1 to 20 minutes in length. However, a movie made during that time **not** labeled "short" can be considered an outlier. On the other hand, movies made in the 1980's labeled "short" would be outliers because most movies in the 1980's were not short (relatively speaking), but usually about an hour and a half long. To find these outliers, we would consider both the year **and** genres instead of a single feature.

Other outlier examples might be movies with many genres. For example, if most movies in the 1990s only had two genres, such as "Comedy" and "Romance" then a movie that had three or four genres during that same time period might be considered an outlier. Alternatively, if most movies in the 1990s were labeled as either "'Adventure' and 'Action'" or "'Comedy' and 'Romance'" then movies that deviated from those paired genres could be considered outliers.

Real datasets with multiple variables are often correlated – in other words, specific variables in the dataset sometimes change together. If you study the movie data, you will find many correlations even among the specific genres. For example, do you think it is much more likely to find the genres "Animation" and "Comedy" or "Animation" and "Horror" together? Looking through the data, you would rarely discover the combination "Animation" and "Horror" but would likely encounter "Animation" and "Comedy". So, the paired genres "Animation" and "Horror" would be considered an outlier.

The Mahalanobis distance takes correlations among variables into account. In fact, the entire distribution of the data and the different variable interactions are all under consideration when determining the Mahalanobis distance.

The mathematics behind the Mahalanobis distance are some of the most misunderstood in the field of distance metrics. Like PCA (see Chap. 8), the Mahalanobis distance changes the axes of the data and applies a rescaling operation. Using the newly scaled axes the Euclidean distance from each point to the centroid of the distribution is determined. Eigenvalues and eigenvectors are leveraged (which derive from linear algebra and are also used in PCA), but, in general, the objective is to reshape the data onto axes that emerge from the newly scaled and newly uncorrelated data and then find the distance from the centroid to each point. The "magic" behind the Mahalanobis distance is that a data-driven approach is used to construct a new coordinate system instead of relying on the traditional X and Y axes learned in elementary school. Once the data has been scaled and a new coordinate system is created the centroid, the center of the data, is the new origin of the coordinate system.

There are two main applications to make use of the Mahalanobis distance:

- Compare two vectors to each other
- Compare a vector to the distribution

For our purposes we can use both methods with our movie recommendation engine. The first method will allow us to compare multiple attributes of *Back to the Future* to multiple attributes of all the other movies and then have our KNN movie recommendation engine suggest the top matches. This is useful when we seek to compare many attributes of a movie at the same time instead of using weighted metrics as described in the previous subsection. For example, how does the plot description, stars, genres, and release year compare between *Back to the Future* and *Mad Max Beyond Thunderdome*?

The second method provides an ability to find movie outliers or anomalies. For example, given a list of movies, which ones are noticeably different from the majority? More precisely, you might presume movies that boast of a high number of stars in their reviews are fundamentally different from the rest of the lot because they are rated higher, a deduction that leads you to sort the movies based on number of stars. However, this approach focuses on one and only one variable. As described above, the classification of a multivariate outlier is determined by the interplay of more than one variable. For instance, movies from 1980–1986 may generally receive an average of 6 stars, but most movies from 2004–2009 receive an average of 4 stars. The two potential outliers in this context might be a movie from 1983 that only received 4 stars and a movie from 2006 that received 6 stars. Both the release year and the star ratings need to be considered when determining an outlier case.

Compare Two Vectors to Each Other
Comparing two vectors to each other means comparing two movies with multiple variables while keeping the overall distribution of data in mind. For example, the Mahalanobis distance calculated between *Back to the Future* and *Mad Max Beyond Thunderdome* compares the variables such as the release years, stars, and MPA rating between the movies of interest while establishing the overall reference data distribution from all the movies in the dataset.

To use the Mahalanobis distance to compare *Back to the Future* to the other movies in the dataset for our KNN movie recommendation engine, we should conceptually understand where *Back to the Future* resides in terms of the distribution space.

In other words, we want to obtain the Mahalanobis distance of *Back to the Future* to the centroid of the distribution and then find and rank all the other movies based on their proximity to the distance assigned to *Back to the Future*. For instance, *Back to the Future*'s Mahalanobis distance is determined to be 26.50579, so we want to find all the other movies that have a distance close to 26.50579 to rank them appropriately.

An analogous scenario to comparing the Mahalanobis distance of two movie vectors is that of a teenager in high school who wishes to identify friends similar to herself. She might be interested in art, music, and ancient Rome. She does not really care about the interests of the average high school student at her school, but implicitly she hopes to find students who set themselves apart from the reference high school student population in a similar way. If the "average high school student" is the centroid of the distribution and her "distance" from the centroid were 26.50579 based on her interests, then she wants to be around other students with a similar "distance".

The following code defines the Mahalanobis function, reads the movies into a dataframe, generates a column for each of the 28 genres, computes the Mahalanobis distance on the year and genre data, sorts the dataframe based on the Mahalanobis distance, then prints out the 10 closest movies to the location of *Back to the Future* in the distribution. (The MultiLabelBinarizer class is explained in Sect. 3.3.11.)

```
Partial file: full program in knn_mahalanobis.py:


# create function to calculate Mahalanobis distance
def mahalanobis(data_input: pd.core.frame.DataFrame):
    data_mu = data_input - np.mean(data_input)
    cov = np.cov(data_input.values.T)
    inverse_cov_matrix = np.linalg.inv(cov)
    left = np.dot(data_mu, inverse_cov_matrix)
    mahal = np.dot(left, data_mu.T)
    return mahal.diagonal()

...


# calculate the Mahalanobis distance for each movie:
finalDf['mahalanobis'] = mahalanobis(finalDf)
```

The resulting ten movies are determined through a mixture of the genres and release year. For our *Back to the Future* base case, the results represent an accurate sample of similar genres in the 1980's.

Compare a Vector to the Distribution for Detecting Anomalies
Computing the Mahalanobis distance on all the movies would detect movies that deviate from the norm. This does not necessarily mean that you will identify movies you want to watch, but you will find movies that are distinctly unique from the rest of the dataset – the very definition of anomalies.

Let us proceed with an example where we seek anomalies for movies based on their genres and release years. As noted above, a sweeping but accurate observation is that the popular genres of movies change as time progresses. The following program is similar to the code above, except that instead of sorting the results post hoc, we employ the concepts of statistical significance for the Chi-Squared distribution (for more information see Chap. 5) which reduces the dataframe to only a few movies that satisfy the sensitivity requirement. The smaller the **sensitivity** variable in the following code the fewer the number of resulting movies.

```
Partial file: full program in mahalanobis_anamoly_detection.py


# create function to calculate Mahalanobis distance
def mahalanobis(data_input: pd.core.frame.DataFrame):
    data_mu = data_input - np.mean(data_input)
    cov = np.cov(data_input.values.T)
    inverse_cov_matrix = np.linalg.inv(cov)
    left = np.dot(data_mu, inverse_cov_matrix)
    mahal = np.dot(left, data_mu.T)
    return mahal.diagonal()

...


# calculate the Mahalanobis distance for each movie:
finalDf['mahalanobis'] = mahalanobis(finalDf)

# calculate p-value for each mahalanobis distance
finalDf['p'] = 1 - chi2.cdf(finalDf['mahalanobis'], len(finalDf.columns) - 1)
```

```
# only get the biggest outliers:
sensitivity = 0.000000001
finalDf['ss'] = finalDf['p'] < sensitivity
only_ss = finalDf[finalDf['ss'] == True]

print(only_ss[['title', 'year', 'genres']])
```

What do these results mean? The first result is *Billy the Kid and the Green Baize Vampire*, which was released in 1987. Aside from the unique title, the genres for that movie are the following: comedy, horror, musical, sport, and western. Any musical horror film would most likely be an anomaly, regardless of the release year, but a musical horror film that is comedic, set in the wild west and sports related is truly an anomaly!

In this case, the Mahalanobis distance calculation detects anomalous or unique movies based on release year and genres. This metric could be applied to any other variables as well. The Mahalanobis distance is often used in practice to detect anomalies in counterfeiting, physics, medicine, and many other disciplines. The mathematics are heavily based on Linear Algebra and are not easily understood, but that should not deter you from utilizing this powerful metric to identify useful and informative anomalies in a variety of datasets.

6.1.8   Additional Metrics

At this point in our discussion, several additional metrics can be examined. However, it is not our intention to fill up a book with the details of hundreds of potential metrics that could be applied in your data science research. Instead, we suggest that you become familiar with the more common metrics then branch out to others as time permits. We offer a few additional brief descriptions of metrics that you may find helpful in your current or future investigations.

The Manhattan distance, also known as the taxicab distance, measures distances in straight line, right-angle segments. If the Euclidean distance is the distance "as the crow flies", then the Manhattan distance is the distance that a cab drives in a downtown city block arrangement where only 90-degree turns are allowed, as in the borough of Manhattan. The Canberra distance is a weighted version of the Manhattan distance.

The Chebyshev distance, or chessboard distance, measures the absolute magnitude of the differences between points in a coordinate system. In chess terms, the Chebyshev distance measures the number of moves required for a King to travel from its origin square to a destination square.

The three distances are compared visually below in Fig. 6.4 with explicit distances calculated for different grid points:

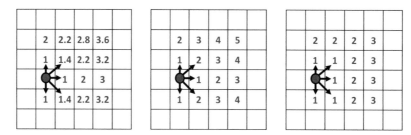

Fig. 6.4 Differences among Euclidean, Manhattan, and Chebyshev grid distances. (Image created by Brian Rague)

The Minkowski distance is a more complicated distance metric that encompasses the behavior of the above three metrics. This metric computes the distance based on a parameter p. When p = 1 we calculate the Manhattan distance; when p = 2 we obtain the Euclidean distance; and as p tends to infinity, we compute the Chebyshev distance. The Minkowski distance formula follows:

$$\left(\sum_{i=1}^{n} |x_i - y_i|^p \right)^{\frac{1}{p}}$$

There are various methods for measuring "distance" between digital string encodings. For example, the Hamming distance between two vectors is the sum of corresponding elements that differ. Since a string may be considered a vector of individual characters, the Hamming distance can be applied in this situation to determine the difference between two strings. The Hamming distance can also be used to find the number of different corresponding bits between two binary strings. For example, 7 is represented as the binary number 0111 and 5 is translated to 0101. The Hamming distance between 7 and 5 is 1, since the second rightmost (least significant) digit between the two numbers is different.

The Soundex metric compares words and returns how similar they are phonetically. Specifically, this metric encodes consonants and then detects *homophones* – words that sound the same. The following are examples of word pairs that are considered the same based on the Soundex metric: dam/damn, too/two, and color/colour. Other similar phonetic metrics include Daitch–Mokotoff Soundex, metaphone, double metaphone, and match rating approach.

In summary, each metric measures entities in different ways and is customized for specific purposes. Like all algorithms, you must provide good, viable input using preprocessing if necessary and invest the effort to understand the output. *GIGO* means "Garbage In, Garbage Out." If you provide a metric with meaningless input, then you will be the recipient of meaningless output. It is up to you to make sure that you understand how a metric works before utilizing it as part of your analysis regimen. Once you can identify the correct metric to utilize in a particular situation, then throughout your professional career you will be able to apply these tools in effective ways to resolve your unique problems.

6.2   Regression Metrics: Diet Example

The previous section described many varieties of distance metrics, which in each case constitutes an effort to determine a relevant quantitative value to express the similarity or difference among two or more entities. This section is about regression metrics – how well a statistical or machine learning model performs compared to the actual data. In other words, if a model is created, we want to ask the following question: How good is that model? In other words, does the model legitimately reflect the properties and behavior embodied by the data?

Currently, many individuals in the wealthier countries around the world are often trying to lose weight. Figure 6.5 shows actual body weight measurements of a person who weighed himself every morning during a diet spanning a two-month time period.

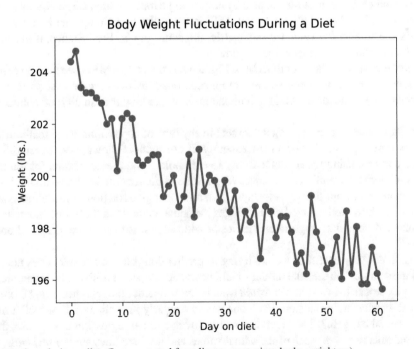

Fig. 6.5   Body weight fluctuations during a diet. (Image created from linear_regression_body_weight.py)

Since this graph represents data, we can analyze it! However, data science is useless without motivating questions. We will try to answer the following question with these plotted data: Given that we only have approximately 60 days of data recorded from this diet, if the dieter continues to lose weight at a consistent rate, how much will he weigh in X more days?

For example, on his first weight measurement, the dieter recorded 204.4 pounds and on the last day he weighed 195.6 pounds (see the data in the `linear_regression_body_weight.py` file), thus losing a total of 8.8 pounds in approximately 60 days. Presuming the dieter continues to lose weight in a manner suggested by the graph, how much more decrease in weight would he experience during an additional 60 days?

To answer this question, we create a model, in this case a regression line, that best "fits" the data. Specifically, we will create four different polynomial lines (see Fig. 6.6) then use a metric to see how well each line best fits or matches the data.

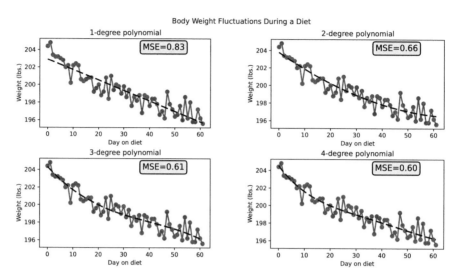

Fig. 6.6 Single individual real body weight fluctuations during a diet. The code for this image is in metrics_initial_regression_lines.py in the accompanying book files. (Image created from linear_regression_body_weight2.py)

We can measure how well each line best fits the data by employing a metric called *Mean Squared Error (MSE)*. The MSE values associated with each plot, 0.83, 0.66, 0.61, and 0.60, indicate the best fitting line is the bottom-right quadrant in Fig. 6.6. In other words, the smaller the value for this metric, the better the fit. For instance, if the MSE had a value of 0.0 then the dashed black line would exactly match the red line.

Even though the line we created in the bottom right of Fig. 6.6 has the lowest MSE, we need to consider the potential for *over-fitting* the data. In other words, just because the bottom right quadrant delivers the best score, it is possible the line is too good in its capacity to fit the data, while simultaneously degrading the quality of future weight measurement predictions.

How can a model be "too good" when tracking a data set? In any type of approximation of real-world information, we can sometimes be misled by outlying data that occur infrequently. If our model follows these eccentricities in the dataset too rigorously, then we are compromising the model's ability to extrapolate and predict results for the more normative cases, effectively decreasing the model's overall performance. We must be vigilant not to be drawn down this rabbit hole of tightly mapping to our observations, since our dataset is always just a representative fraction of the entire system.

Regression metrics tell us how well a model characterizes the actual data. As in the above example of determining which line best fits the continuous dieting data, regression metrics provide an assessment of the quality of our model.

Why should we care?

In brief, a bad model is not very useful. If we are trying to predict things and our model does not sufficiently reflect the behavior and trends of the sample data then the model should not be developed or utilized. For example, if we were modeling the stock market to identify stocks to invest in, we would want to know that our model is accurate before we invest our money.

We will cover several metrics in this section. Each one works similarly in evaluating how well a model compares to the actual values. For instance, take Fig. 6.7. The green line is the model, a linear regression in this case, that attempts to predict the actual values. The red dots represent each of the actual values and the blue lines are the *residuals*, the distance from the model to the actual values.

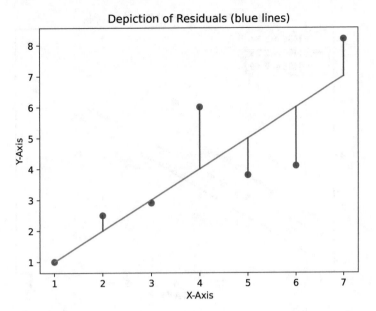

Fig. 6.7 Chart showing the residuals, differences between the model and actual values.. Each red dot represents the data. The green line is the model. The blue lines show the distance from the data to the model. (Image created from residuals.py)

In Fig. 6.7 the first data point is on the green line, so the residual at that location is 0. The second point is above the green line and has a residual of about 0.5, and the third is barely below the green line with a residual of 0.05. All subsequent points and residuals can be described in a similar way.

For each of the following metrics, $\hat{y}$ (pronounced "y hat") is the predicted value of the model. In Fig. 6.7 $\hat{y}$ represents the model value at a given x value. For example, if x = 2 then $\hat{y}$ =2 (the green line model) and y = 2.5 (the red dot observation).

6.2.1   Mean Squared Error (MSE)

Mean Squared Error (MSE) is one of the simplest regression metrics. MSE measures the mean (i.e., average) squared difference between the estimated values ($\hat{y}$) and the actual values (y). In other words, it squares each residual, adds them all up, and then computes the average.

Although MSE is easiest to understand, it probably is not the best metric primarily because of its simplicity. Note that MSE gives more weight to outlier differences, because it squares them as shown in the formula:

$$MSE = \frac{1}{n}\sum_{i=1}^{n}\left(y_i - \hat{y}_i\right)^2$$

Consider the example using the following data points from Fig. 6.7: (1,1), (2, 2.5), (3, 2.9), (4, 6), (5, 3.8), (6, 4.1), and (7, 8.2). Our model is the green straight line from (1, 1) to (7, 7). How well does the green line fit the actual data? Our model depicted in the figure predicts the points (1, 1), (2, 2), (3, 3), (4, 4), (5, 5), (6, 6), and (7, 7).

Applying MSE we obtain the following calculation:

$$MSE = \frac{\left(1-1\right)^2 + \left(2.5-2\right)^2 + \left(2.9-3\right)^2 + \left(6-4\right)^2 + \left(3.8-5\right)^2 + \left(4.1-6\right)^2 + \left(8.2-7\right)^2}{7}$$

$$MSE = 1.54$$

The MSE value of 1.54 by itself is useless. The MSE metric **only** has meaning when comparing two or more models. Consider the three models in Fig. 6.8, the green, blue, and orange lines. Which of the three lines most closely approximates the actual data (the red dots)?

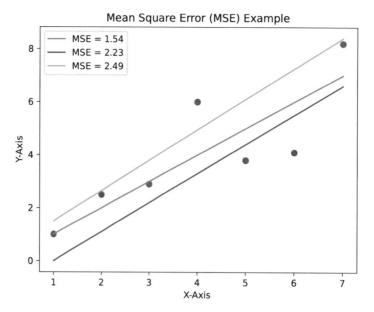

Fig. 6.8 Comparing three models shown as the green, blue, and orange lines using MSE. (Image created from mse.py)

The legend in Fig. 6.8 displays the MSE for each of the three models. Of the three values, 1.54, 2.23, and 2.49, which of the three is the closest to 0? The green line has the smallest value 1.54, which indicates it is the model that most closely approximates the actual data.

6.2.2 Root Mean Squared Error (RMSE)

The Root Mean Squared Error (RMSE) - also known as *Root Mean Squared Deviation (RMSD)* – calculates the standard deviation of the prediction errors. The square root operation is applied to ensure the units of the resulting error match the units of the target measurements. As with the MSE, this metric emphasizes large deviations produced by outliers, since large differences squared become larger and small (less than 1) differences squared become smaller.

$$RMSE = \sqrt{\frac{1}{n}\sum_{i=1}^{n}\left(y_i - \hat{y}_i\right)^2}$$

Applying the RMSE to the model in Fig. 6.7 (the green line), we derive the following:

$$RMSE = \sqrt{\frac{\left(1-1\right)^2 + \left(2.5-2\right)^2 + \left(2.9-3\right)^2 + \left(6-4\right)^2 + \left(3.8-5\right)^2 + \left(4.1-6\right)^2 + \left(8.2-7\right)^2}{7}}$$

$$RMSE = 1.24$$

6.2.3 Mean Absolute Error (MAE)

MAE measures the absolute magnitude of the prediction errors and produces a value in the same units as the target measurement units. This calculation allows for more easily interoperable errors – the error value is in the same reference frame as the measurements themselves – and also gives less weight to outliers. However, a model with a relatively good MAE can potentially have very high errors in specific cases since residuals with large magnitudes are not emphasized as much as the RMSE or MSE approach. The following is the formula for MAE:

$$MAE = \frac{1}{n}\sum_{i=1}^{n}\left|y_i - \hat{y}_i\right|$$

6.2.4 R² or R Squared: Coefficient of Determination

R^2 (pronounced *R Squared*) is the *coefficient of determination* and is the proportion of the variance in the dependent variable that is predictable from the independent variable(s). The coefficient of determination normally ranges between 0 and 1, and measures how well the regression model fits the data when compared to a model that always predicts the average of the observed data as the outcome.

The coefficient of determination provides a percentage measure indicating how many data points are within close proximity to the line defined by the model. The higher the coefficient, the better the model fits the data. For example, given an R^2 coefficient of 0.80, the sum of squares of the residuals is one fifth of the total sum of squares of the observed data with respect to the sample mean. A value of 0 indicates our model lazily predicts the average value of the observed data at each point, and a value of 1 indicates a perfect correlation between the model and the data.

If $\hat{y}$ represents the predicted values in the model, $\overline{y}$ (pronounced "y-bar") represents the mean of the sample data, and y represents each data point, then R^2 is calculated as shown here:

$$R^2 = 1 - \frac{\sum(y_i - \hat{y}_i)^2}{\sum(y_i - \overline{y}_i)^2}$$

6.2.5 Adjusted R-Squared ($\overline{R^2}$)

R^2 is quite useful in showing the correlation of the model and the actual data. However, as the number of independent variables in a model increases then the results of R^2 will improve significantly due to overfitting. However, this trend can be mitigated by using the *adjusted R^2 ($\overline{R^2}$)*. $\overline{R^2}$ takes each additional independent variable into account. If p is the number of independent variables or predictors in your model and n is the number of observations in your data, the formula for $\overline{R^2}$ follows:

$$\overline{R^2} = 1 - (1 - R^2)\left(\frac{n-1}{n-p-1}\right)$$

This formula reveals that as the number of observations n greatly exceeds the number of independent variables p, the value of $\overline{R^2}$ approaches the value of R^2. Figure 6.9 shows the same data as Fig. 6.8 with accompanying computed metrics given in the upper-left corner legend.

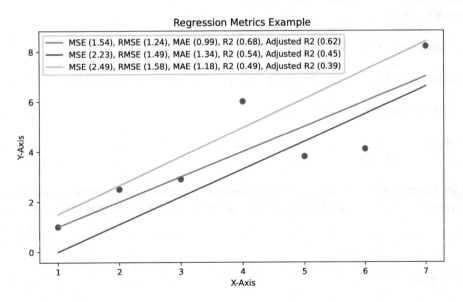

Fig. 6.9 Comparing three models shown as the green, blue, and orange lines for MSE, RMSE, MAE, R^2, and $\overline{R^2}$ (Adjusted R²). (Image created from many_residual_metrics.py)

6.3　Prediction Metrics

A prediction is simply a guess about what is going to transpire. For example, is it going to rain tomorrow? One prediction is "yes."

How do we measure the accuracy of this prediction? First, we can wait until tomorrow and observe the weather. Assuming we are making predictions about the weather for our specific local area, then based on that particular location we may be right or we may be wrong.

The most rudimentary kind of predictions are binary: yes or no. However, predictions may also involve more elaborate classifications. An example is a machine learning algorithm trained to differentiate between images of dogs, elephants, and monkeys. In this case there exists three possible outcomes instead of simply two as with the prediction about rain.

How do you measure the accuracy of a prediction?

This section will cover the prediction metrics associated with accuracy, the confusion matrix, and the classification report.

6.3.1　Accuracy

Accuracy is one of the more intuitive measures to assess the quality of predictions. The accuracy value answers the core question of how often we determined the correct outcome for each of our test samples.

For the weather example described above in which a single prediction is asserted about the occurrence of rain the next day, we would either be 100% accurate if our prediction were correct, or 0% accurate if our prediction were incorrect.

Alternatively, if we had predicted rain for 3 days in a row, but it only rained the first and third day then we would be accurate two-thirds of the time (66.67%). The following Python code illustrates two separate methods to determine accuracy: (1) a for-loop and (2) a call to the `accuracy_score` function from the *sklearn* package:

```
File: accuracy.py:
from sklearn.metrics import accuracy_score

prediction = [1, 1, 1]
actual_results = [1, 0, 1]

correct  = 0
for i in range(len(prediction)):
    if prediction[i] == actual_results[i]:
        correct  += 1

print(f'For loop accuracy = {(correct  / len(prediction))*100:.2f}%')

print(f'Sklearn accuracy = {(accuracy_score(actual_results, prediction))*100:.2f}%')
```

Which produces the following results:

```
For loop accuracy = 66.67%
Sklearn accuracy = 66.67%
```

Accuracy is a good metric because of the simplicity of the concept. However, the accuracy score alone does not provide details about the types of errors associated with our prediction, which leads us to the next subsection describing the confusion matrix.

6.3.2   *Confusion Matrix*

Let's return to our machine learning algorithm that predicts the images of dogs, elephants, and monkeys. Let's suppose that we have already trained our machine learning algorithm with a sufficient and balanced number of images. We then provide the machine learning algorithm new test images of dogs, elephants, and monkeys it has never seen before. How might we characterize the algorithm's performance on this specific recognition task?

Assuming an encoding of 0 = dog, 1 = elephant, and 2 = monkey (see Sect. 3.3.11 for more details on encodings), we can show images to our machine learning algorithm and interpret the results. For example, suppose that we presented our algorithm four dogs, followed by four elephants, and then four monkeys. This sequence of images could be represented in an encoded array as [0, 0, 0, 0, 1, 1, 1, 1, 2, 2, 2, 2].

We now inspect an array of the results of our algorithm which reveal a few mistakes: [2, 2, 0, 1, 1, 1, 1, 1, 2, 2, 0, 2]. In other words, our algorithm misinterpreted or confused the images of dogs and monkeys several times, but selected elephants correctly in each case. Invoking the `accuracy_score` function on these results would produce 66.67% accuracy, but this value does not provide any insight on the type of mistakes that occurred. To address this situation, the following code illustrates the calculation of a confusion matrix:

```
File: confusion_matrix.py:
from sklearn.metrics import accuracy_score, confusion_matrix

prediction = [2, 2, 0, 1, 1, 1, 1, 1, 2, 2, 0, 2]
actual_results = [0, 0, 0, 0, 1, 1, 1, 1, 2, 2, 2, 2]

print(f'Accuracy = {(accuracy_score(actual_results, prediction))*100:.2f}%')
print(f'Confusion Matrix:\n{confusion_matrix(actual_results, prediction)}')
```

Which produces the following results:

```
Accuracy = 66.67%
Confusion Matrix:
[[1 1 2]
 [0 4 0]
 [1 0 3]]
```

Specifically, the confusion matrix offers a more detailed view of where our algorithm was right and where our algorithm was wrong. The blue diagonal in Fig. 6.10 shows the correct predictions for our algorithm. Figure 6.10a depicts the ideal, perfect results, while Fig. 6.10b shows the results for our example. Red grid squares indicate an incorrect prediction.

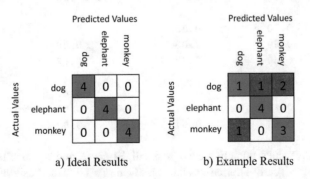

a) Ideal Results   b) Example Results

Fig. 6.10 (**a**) Shows the ideal results if the algorithm predicted correctly every time. The blue color indicates a correct prediction. (**b**) Shows the example predictions from the algorithm. The red color indicates an incorrect prediction. (Image created by Robert Ball and Brian Rague)

The rows (dog, elephant, and monkey) indicate the image that was presented to the algorithm while the columns list the algorithm's selection based on its training. For Fig. 6.10b, the top row ('dog') indicates that, when presented with four dog images, the algorithm correctly predicted 'dog' only once while also predicting 'elephant' once and 'monkey' twice. When presented with four elephant images the algorithm correctly predicted every time and when presented with four monkey images the algorithm correctly predicted three of them, but incorrectly predicted 'dog' once.

Overall, confusion matrices provide a more granulated view of the prediction errors. The entries provided in the confusion matrix may not be as intuitive as simply reporting the overall accuracy, but they express richer information. For example, after running this test and obtaining the results in Fig. 6.10b you might decide to give your machine learning algorithm more training data on dogs and monkeys, since these images caused the most prediction errors.

The last assessment tool we will investigate is the classification report, which offers another informative perspective of the prediction results.

6.3.3   Classification Report

If a confusion matrix provides more informative results than overall accuracy, a classification report represents even greater depth about the quality of the prediction task. The following code compares the overall accuracy, confusion matrix, and classification report:

```
File: classification_report.py:
from sklearn.metrics import accuracy_score, confusion_matrix, classification_report

prediction = [2, 2, 0, 1, 1, 1, 1, 1, 2, 2, 0, 2]
actual_results = [0, 0, 0, 0, 1, 1, 1, 1, 2, 2, 2, 2]

target_names = ['dog', 'elephant', 'monkey']

print(f'Accuracy = {(accuracy_score(actual_results, prediction))*100:.2f}%')
print(f'Confusion Matrix:\n{confusion_matrix(actual_results, prediction)}')
print(f'Classification Report:\n{classification_report(actual_results, prediction, target_names=
target_names)}')
```

Which produces the following output:

```
Accuracy = 66.67%
Confusion Matrix:
[[1 1 2]
 [0 4 0]
 [1 0 3]]
Classification Report:
                  precision       recall        f1-score        support

           dog       0.50         0.25          0.33              4
      elephant       0.80         1.00          0.89              4
        monkey       0.60         0.75          0.67              4

      accuracy                                  0.67             12
     macro avg       0.63         0.67          0.63             12
  weighted avg       0.63         0.67          0.63             12
```

Classification reports list many values, namely precision, recall, F1-score, and support. To fully understand the report, we need to examine Table 6.1, which visually summarizes the classifications TP (True Positive), FP (False Positive), FN (False Negative), and TN (True Negative).

Incorrect predictions correspond to type I and type II errors. The blue cells in Fig. 6.10b correspond to TP's (True Positives) for each category and red cells in Fig. 6.10b correspond to either FP's (False Positives) or FN's (False Negatives), depending on the reference category.

For example, the upper right cell entry in Fig. 6.10b indicates that two images of dogs were mistakenly identified as monkeys. From the perspective of the dog classification, these errors represent false negative predictions, but from the perspective of the monkey classification these errors represent false positive predictions.

Table 6.1 Similar to Table 5.1 showing type I and type II errors

		Actual results	
		True	False
Predicted	Positive	TP (true positive)(correct conclusion)	FP (false positive) (type I error)
	Negative	FN (false negative) (type II error)	TN (true negative)(correct conclusion)

However, this table describes the results in terms of predictions vs. actual results for better interpretation of prediction results

For each category listed in the classification report, these types of errors are folded into the calculations for precision and recall described below. For more information on type I and type II errors, see more details in Sect. 5.1.

Precision is defined by the following formula:

$$Precision = \frac{TP}{TP + FP}$$

What does precision mean in this context? Let's look at the precision for the elephant classification. For 'elephant' the classification report shows 80% precision. For precision we want to look down the elephant **column** in the confusion matrix (Fig. 6.10b) since any red cell errors in this column indicate false positive predictions of elephants. For elephants, we obtained 4 TP's and 1 FP, which generates the 80% result as shown in the following formula:

$$Elephant \; Precision = \frac{4}{4+1} = \frac{4}{5} = 80\%$$

Table 6.2 summarizes the numbers for each column related to precision and Fig. 6.11a provides a visualization of the entire classification grid with the middle column highlighted.

Table 6.2 Summary of precision for each category

	Dog	Elephant	Monkey
TP	1	4	3
TP + FP	2	5	5
Precision: TP/(TP + FP)	50%	80%	60%

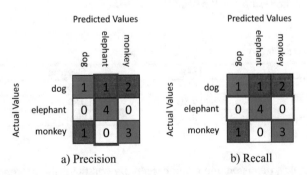

a) Precision   b) Recall

Fig. 6.11 Precision and recall for the elephant classification. (**a**) Purple box highlighting that precision is connected to the data in an individual column. (**b**) Purple box highlighting that recall is connected to the data in an individual row. (Image created by Robert Ball and Brian Rague)

Conceptually, *precision* is a measure of the positive predictive performance of the algorithm that answers the question of **how many of your positive predictions were correct?**

The overall precision is simply the average of the three categorial precisions: (50% + 80% + 60%)/3 = 63%.

Recall is defined by the following formula:

$$Recall = \frac{TP}{TP + FN}$$

As seen in Fig. 6.11b, the best way to think of recall is by **rows** since any red cell errors in a particular row indicate false negative predictions with respect to the category represented by the row. The following shows the elephant recall:

$$Elephant \; Recall = \frac{4}{4+0} = \frac{4}{4} = 100\%$$

The perfect recall result calculated above confirms the popular belief that "elephants never forget." Table 6.3 summarizes the numbers for each row related to recall and Fig. 6.11b provides a visualization of the entire classification grid with the middle row highlighted.

Table 6.3 Summary of precision for each category

	Dog	Elephant	Monkey
TP	1	4	3
TP + FN	4	4	4
Recall: TP/(TP + FN)	25%	100%	75%

Conceptually, *recall* (also known as *sensitivity*) is a measure of how well the classifier found all relevant samples and answers the question of **how many of the positive (actual) cases were predicted correctly?** Alternatively, what percentage of positive cases were correctly classified?

The *F1 score* calculation employs a harmonic mean as opposed to the traditional arithmetic mean. More precisely, the F1 score is the harmonic mean of precision and recall as shown here:

$$F1 \; Score = 2 * \frac{Recall * Precision}{Recall + Precision}$$

The F1 score for the elephant classification is the following:

$$Elephant \; F1 \; Score = 2 * \frac{1.0 * 0.8}{1.0 + 0.8} = 2 * \frac{0.8}{1.8} = 89\%$$

Table 6.4 summarizes the F1 score results for the dataset.

Table 6.4 Summary of F1 Score for each category

	Dog	Elephant	Monkey
Recall	0.25	1.00	0.75
Precision	0.50	0.80	0.60
F1 score: 2*(recall*precision)/(recall+precision)	33%	89%	67%

Since for each individual category, precision and recall essentially computes the percentage or rate of a different type of error (FP and FN) with respect to correct predictions (TP), the F1 score represents an alternative accuracy value that combines the precision and recall metrics. Note that the harmonic mean used to calculate the F1 score requires that both precision and recall contribute significantly to the overall accuracy. If either the precision or recall is 0, the F1 score will be 0. Conversely, both the precision and recall must be equal to 1 for the F1 score to attain its maximum value of 1.

The easiest of the four terms in the classification report to explain is support. *Support* is simply the number of instances of the given category that were input to the algorithm. In this case, the algorithm was provided four dog images, four elephant images, and four monkey images, resulting in a total of 12 samples. In practice, data rarely comprises a balanced equal amount from each category. Table 6.5 provides a summary description for each of the results generated by the classification report.

Table 6.5 Overview of classification report terminology

Precision:	How many of your positive predictions were right?
Recall:	How many of the positive (actual) cases were predicted correctly?
F1 score:	Accuracy measure combining precision and recall.
Support:	How many occurrences of each classification were there?

Exercises

1. What are the three requirements for any metric?
2. How is metadata useful for data science investigations?
3. When would you use Levenshtein distance over cosine similarity?
4. When would you use cosine similarity over Levenshtein distance?
5. What is the difference between traditional Jaccard similarity and weighted Jaccard similarity?
6. Explain one use case of the Chebyshev distance (or chessboard distance).
7. What is soft cosine similarity? When would you use it over traditional cosine similarity?
8. MSE and RMSE and MAE all appear to be roughly similar metrics. What criteria would determine which of these metrics to select?
9. What is the difference between R-Squared and adjusted R-Squared?
10. What is accuracy?
11. When would it be better to use a confusion matrix over accuracy?
12. What might be the purpose of a classification report?

Chapter 7
Recommendation Engines

Recommendation engines can influence human behavior and preferences. Most notably, these algorithms have altered how people shop, look up information, and consume news. For better or for worse, recommendation engines have significantly impacted how people conduct their daily lives.

Recommendation engines are software that filter data to a prescribed target subset based on different metrics. Section 6.1 demonstrates the importance of metrics and how they are used to create different modes of filtering for a movie recommendation engine.

Recommendation engines can range from very simple to highly elaborate depending on the complexity of the associated machine learning algorithms. However, regardless of the difficulty level of the recommendation engine, they all rely on an analysis of metadata linked to the primary items to be recommended.

For example, in Sect. 6.1 we described an example of a movie recommendation system. Any recommendation engine that suggests movies will necessarily be examining the metadata of those movies under consideration such as the title of the movie, the genre, reviews (e.g., 5 stars out of 10), the lead actors, and the director.

Most online stores use recommendation engines to help filter their data to precisely determine what the shopper will likely want to buy, typically based on preferences and previous purchasing habits. Although from a certain perspective recommendation engines can be described as manipulative, the main purpose of any recommendation engine is to predict what the customer wishes to buy, read, or view.

Compare a large physical store, also known as a brick and mortar store such as Walmart, to an online retail store like Amazon.com. One of the convenient features of the Walmart brick and mortar experience is the option to purchase a product immediately. However, have you ever been to large physical stores with a product purchase in mind, but neither you nor the store employees could locate the item? Recommendation engines are designed to filter and search to rapidly identify the precise product so the time to purchase is significantly reduced.

Another benefit of physical stores is to allow for serendipitous shopping. *Serendipitous* shopping or browsing from the consumer perspective is visiting a store seeking a particular item, but ultimately purchasing several other products which happened to capture attention because of the physical engagement in browsing the store selections. Another example of serendipitous browsing is visiting a library to search for one particular book but identifying several others that are also interesting but not especially related to the subject matter of the initial book.

Recreating a true serendipitous shopping or browsing experience with recommendation engines is a current area of research. This field of investigation presents a challenge because the main purpose of a recommendation engine is to narrowly predict, given certain information, what a person might want to buy or read. Serendipitous shopping or browsing is contrary to that laser-focused objective by its very nature.

As an example, consider news.google.com, a news aggregation recommendation engine. The site news.google.com is an effective recommendation engine as measured by its capacity to filter through tens of thousands of news articles and to predict which news articles the reader may want to read. The primary measure of success is when a user clicks on one of the suggested news articles.

R. Ball, B. Rague, *The Beginner's Guide to Data Science*, https://doi.org/10.1007/978-3-031-07865-1_7

The term describing this particular user action is called *click through rate (CTR)*. A 100% CTR means that the user has clicked on every suggested link which forwards them to a news article, product page, or other related site. A 0% CTR means the user was not interested in any of the recommendations presented.

A serendipitous experience as described above is often at odds with the metric of raising the CTR of an individual. In other words, what often happens in the recommendation engine environment such as news.google.com, is individuals are only exposed to the items of interest that the system has deemed appropriate based on the existing preferences of the user.

The problem with this preference-centric approach is that it can create the phenomenon of tribalism. As the CTR approaches 100% then the system only suggests links for which the user has already demonstrated prior interest, a self-perpetuating cycle. If the sole objective of the system is to maximize CTR then introducing novel, unexplored areas of potential interest would likely decrease the user's CTR.

Tribalism is when a person associates with only a select group of people who share a similar culture, religion, background, or other perspective while ignoring or shunning other people with opposing ideas or cultures. Similar terms are *echo chamber* and *preaching to the choir*. Tribalism provides a narrow view of life and does not allow other competing ideas and perspectives, such as unfamiliar religious or cultural views, to be seen or heard.

An example of the tribalism effect with regards to a news aggregator site like news.google.com is a user with certain political views will only receive suggestions and article links that align with those opinions. The more this person clicks on articles based on the same political perspective, the more the recommendation engine will suggest similarly biased articles. Consequently, tribalism emerges as a broad societal issue in that recommendation engines are effecting more political division primarily because the vendors of these recommendation engines are striving to maximize the CTR metric.

A related topic is confirmation bias. *Confirmation bias* is when a person seeks out evidence that confirms their hypothesis and ignores evidence that might refute their hypothesis. For example, consider a professor searching for validation that COBOL (a programming language from the late 1950's) should be currently taught as a university course. If this individual seeks and ultimately discovers a list of 100 facts about COBOL, even if there were 95 negative facts and only 5 supporting facts then a person driven to prove their point would likely ignore the 95 negatives and fixate on the 5 supporting facts as strong corroboration that COBOL should be taught as a university course.

This classic tradeoff problem of maximizing CTR while attempting to design a system that incorporates serendipity is analogous to balancing bias and variance in machine learning (see Chap. 8). Namely, as CTR increases then serendipity decreases and vice versa.

When discussing recommendation engines, people interacting with these algorithms may be categorized into different groups based on behavior. *White sheep* groups are those individuals that behave approximately the same way or have a high correlation with each other in tastes, opinions, and other measurable parameters. For example, there might be two groups of white sheep corresponding to the two main United States political parties. Each group is different from the other, but the white sheep in each political party have similar views and attitudes connected with each group. Establishing that a person belongs to one political party (e.g., Democrat or Republican) and is also a white sheep typically makes it easier to predict the person's likes and dislikes.

In contrast, *Black sheep* do not fit easily into conventional categories. People who are black sheep do not adhere to traditionally defined behavioral patterns. Taking the stance that the United States has only two political parties would necessarily assign as black sheep those individuals that belong to the dozens of minor political parties because these affiliations do not align with either the Democrats or the Republicans.

For recommendation engines, to combat the issue of numerous unpredictable black sheep classifications, typically more representative data is required. For example, if you have data on only 100 people and 95 of them are Democrats or Republicans then the 5 other people would be black sheep and by definition difficult to predict. However, if you were to increase the dataset so that more minor political parties were sufficiently represented then the result would be a smaller overall black sheep percentage because you would discover that Libertarians (another political party) tend to think alike, but are different from both Republicans and Democrats.

Simply obtaining more data may resolve the black sheep problem but the grey sheep problem remains. A *grey sheep* is someone who is so unusual that they are very difficult to correlate with any group. Acquiring exceptionally vast amounts of data may help with the grey sheep representation problem, but the reality is that there will always exist individuals who think and act conspicuously different or contrarian from everyone else. Given this realization, it is quite acceptable to admit that the behavior of these eclectic people is extremely hard to predict and that a certain amount of error is inevitable in almost any system attempting to determine the preferences of a large population of users.

To summarize, a recommendation engine is a set of algorithms used for predicting (recommending) a particular product, service, or other target of interest the user would likely wish to buy or consume. Common examples of these algorithms include search engines (e.g., Bing, Google), product recommendations (e.g. Amazon), and entertainment recommendations (e.g. Netflix).

Although the underlying algorithms and user experience of recommendation engines can range from very simple to excessively byzantine, there are only a few key categories of standard recommendation systems, which include the following:

- Knowledge-based
  - Constraint-based
  - Case-based

- Content-based
- Collaborative filtering

  - User-based filtering
  - Item-based filtering

Some types of recommendation engines are *stateful* and some are *stateless*. In other words, some recommendation engines such as knowledge-based recommendation engines are stateless, meaning they do not require a user profile. However, content-based and collaborative-filtering types are both stateful, meaning that a user profile in necessary to function.

For example, a content-based recommendation system like that demonstrated in Sect. 6.1, requires a user to select at least one movie that he or she prefers before similar suggested content can be generated. The more preferred content the user provides to the system, the better the system will perform.

The *cold start problem* is a design issue particular to stateful recommendation engines that rely on some form of user input before offering the user recommendations. For example, if a user is interested in gathering recommendations for movies from a stateful recommendation engine then they will need to submit at least one film preference. This implicit requirement may be annoying to a first-time user who seeks an initial movie recommendation without having to select from a long list of movies they either like or do not like.

The most common type of recommendation engine is a hybrid recommendation engine that is a combination of one or more of the types listed above.

To achieve a productive recommendation engine experience, it is necessary to store user profiles for both content-based and collaborative-filtering recommendation engines. For demonstration purposes, we have downloaded over 100,000 recipes from the Internet and millions of corresponding reviews associated with those recipes. We used the principles of web scraping (see Sect. 2.7) to gather all the relevant data. The recipes and accompanying reviews from our web scraping can be found in the chapter files `recipes.zip` and `recipe_reviews.zip`.

Each recipe file in `recipes.zip` contains the original URL source, the title, the ingredients, the directions, the submitter, a description that the submitter provided, optional categories, the number of people who rated each recipe, the corresponding number of rating stars, the preparation time, the "ready in" time (e.g. from start to finish the recipe is ready in 15 minutes), the number of servings, and nutritional facts.

The information in these files was not altered, which means that there are many misspelled words. For example, we have identified 20 ways that people misspelled the word "garlic" such as garlik, garlick, galic, galric, garclic, garic, garlic, and garli. To remedy this situation, we have created a file that has the most common mistakes and their correct mapping in the `mapping_ingredients_list.txt` file. This somewhat painstaking process reveals once again the time and effort data scientists must devote to data wrangling. We also have a `synonym_list.txt` file that lists various ingredients that may be similar.

The `recipe_reviews.zip` contains the reviews for every recipe in `recipe.zip` at the time the recipe was web scraped. Note that many of the recipes will have additional reviews posted after the date the recipe was web-scraped. Consequently, our data are considered *stale*, meaning that our files do not contain the most recent data because the recipes and reviews do not reflect the current information posted on the web sites.

7.1 Knowledge-Based Recommendation Engines

Knowledge-based recommendation engines represent some of the most basic implementations of recommendation engines. These engines simply remove (i.e., filter out) unwanted content or query a database to retrieve the desired content.

Knowledge-based recommendation engines are stateless and are typically utilized for situations in which having a user profile would not be particularly helpful. Consider an average family buying a house. Buying a house is an act that most people do seldomly, typically with years between purchases. For example, if a family bought a three-bedroom house 6 years ago and now wishes to buy another house more appropriate to their family size then it is likely they now have more or less children than 6 years ago and their needs have drastically changed during that time.

In addition, there is most definitely not enough data about past purchases from similar buyers to employ more sophisticated recommendation engines, such as a collaborative filtering system. Most people do not buy more than 5 houses in their lifetime in the United States, thus eliminating the collaborative filtering system as a potential recommendation engine solution.

Therefore, a recommendation engine for buying houses does not need know what people have purchased or browsed in the past; it simply needs a list of current requirements for the new purchase. Knowledge-based recommendation engines are often used for relatively infrequent events such as house buying and wedding ring purchases.

There are two main types of knowledge-based recommendation engines:

- Constraint-based
- Case-based

A *constraint-based knowledge-based recommendation engine* is simply a user interface to a database backend. Advanced queries can be leveraged in any combination to create the SQL (Structured Query Language) processed by the database to produce the associated result. Zillow.com and Realtor.com are examples of these constraint-based systems.

For our recipe example, the reviews would not be relevant for this kind of recommendation engine because we do not care if other users like or dislike the recipes. We are interested in retrieving recipes based only on the queries we provide.

When building a constraint-based knowledge-based recommendation engine queries may be submitted against any attribute or metadata about the recipe. We may consider the core information for any recipe to be its list of ingredients, so an example query might include "duck" and exclude "sugar."

In addition to ingredients, you could allow the user to query based on calories per serving, time to prepare the food, words in the title, or any other metadata found in the recipe.

A constraint-based knowledge-based recommendation engine could also be leveraged to overcome the cold start problem. For example, as people search for recipes based on metadata from the database they could subsequently tag any particular recipe as one they like. In this way, a constraint-based knowledge-based recommendation engine could be used as a gateway or bootstrap to more complex recommendation engines.

To implement a constraint-based knowledge-based recommendation engine the initial step would necessarily involve converting the web scraped information into a database. We have created various SQL files in the associated sql folder for this purpose – you will find them with the rest of the files for this chapter. The SQL files are intended to be used in MySQL.

After a suitable comprehensive database has been created then the next step is to create a user interface. Creating a good user interface for this data is beyond the scope of this book. There are many books on user interface design and development that can help you discover appropriate techniques that enhance and facilitate the user experience.

A *case-based knowledge-based recommendation engine* uses a *case* or a selected example as a launching point. For instance, you might have a preferred recipe in mind, but you wish to locate a similar one with fewer calories or with the inclusion or exclusion of a certain ingredient.

An initial case is required and to resolve the cold start problem you could have a case-based system that has a front-end add-on to a constraint-based system. For example, suppose you had implemented the database and created the user interface for the constraint-based system described above. When a user retrieves and identifies a recipe they like, the case-based mechanism can now be activated to find other similar recipes.

Compared to a constraint-based system, case-based systems implement metrics and machine reasoning to determine similar recipes. For example, consider the following approaches:

- Ingredients - Use Jaccard Similarity (see Sect. 6.1.2) to compare ingredients. A common alternative would be to use Cosine Similarity (see Sect. 6.1.5).
- The calories per serving (most people care about this) - Use Euclidean distance (see Sect. 6.1.1) to compare calories.
- Do not show a recipe twice (if the recipe was rejected the first time there is no purpose to redisplay the same recipe to the user.)
- There are many other methods outlined in Chap. 6 to compare one piece of data to another to discern similarities.

Case-based knowledge-based recommendation engines are customized to specific industry needs. For example, the most important parts of a recipe are ingredients, which is meaningless for a vacation recommendation engine.

An example scenario for a case-based system could be the following: Using the constraint-based knowledge-based recommendation engine as described above, a user might search for recipes that contain the ingredients chocolate, flour, and eggs. Upon finding a preferred cookie recipe, the user might press a button that finds similar recipes to the selected cookie recipe, effectively switching from a constraint-based to a case-based knowledge-based recommendation engine.

The user would then be offered options to modify the original case, such as including additional ingredients, excluding ingredients, finding a recipe with fewer calories, or identifying a recipe with more calories. Each time the user adds more specifications they exclude or filter out more and more recipes.

The biggest concern for any case-based knowledge-based recommendation engine is the user adds too many requirements such that there are no relevant recipes to be accessed and displayed.

Overall, knowledge-based recommendation systems are less commonly used. The implementations of these systems are relatively straightforward and can be employed effectively if the user has a clear definition of what they are seeking. In contrast, the content-based and collaborative-filtering systems are more complex and assistive, usually generating more appealing or relevant results.

Although knowledge-based recommendation engines are by definition recommendation engines, many people often do not consider them to be included in this category. In fact, many erroneously consider the content-based recommendation engines described next to be the most basic recommendation engines.

7.2 Content Based

Content-based recommendation engines recommend results based on the metadata of the item or product. They are stateful and so are impacted by the cold start problem. The example of a movie recommendation engine used in Sect. 6.1 is a content-based recommendation engine.

For example, suppose a user chooses a soup with the ingredients garlic and cheese. A content-based recommendation engine would recommend similar items that also have garlic and cheese. In other words, to generate recommendations a content-based system compares the contents or metadata of the selected items to the rest of the data. In this case, garlic cheese bread and other recipes with garlic and cheese would be returned to the user.

Metrics such as Jaccard similarity and weighted Jaccard similarity would be appropriate to compare ingredients of different recipes. Cosine similarity would also be a good option. The fundamental mechanism for content-based recommendation engines is to compare the metadata of one recipe to the metadata of all the other recipes. See Chap. 6 for more common metrics that provide comparisons of metadata.

For our example recipe data, there are also many other options for comparison such as the calories, keywords in the directions such as boil or bake, and preparation time to name a few.

If the user has only selected one item, one recipe in our example, then the case-based knowledge-based recommendation and the content-based recommendation techniques are virtually the same. However, with a content-based system, users can choose more than one item and the system will optimize the metadata matches based on the selected items.

One significant advantage of content-based systems is that meaningful recommendations can be produced with no detailed demographic profile information about the user, such as age, gender, culture, and location.

As mentioned above, content-based recommendation engines suffer from lack of exploration or serendipity. In other words, content-based recommendation engines only recommend items similar to previous selections highlighted by the user, thus increasing the potential and probability of tribalism.

For example, for a user who enjoys recipes with chocolate, the content-based recommendation engine would recommend additional recipes that also contain chocolate. However, the user might also like chicken, but a content-based recommendation engine would not explore that avenue if chicken were not previously selected by the user. For content-based systems to exhibit serendipitous recommendations, a randomization component that arbitrarily chooses and displays unrelated options would need to be included in the implementation of the engine. Displaying recipes that have no obvious connection to the user's stated preferences would likely prove distracting and be summarily dismissed as a viable use case.

For a content-based system to be effective, comparisons need to be consistently and repetitively applied. Specifically, assume a user chooses ten recipes from the over 100,000 recipes in our example dataset. How can the engine compare these ten recipes to all the others that are available?

Before initiating any comparisons the exact metadata to be compared should be identified. To keep our example fairly simple, only the ingredients will be compared and a data structure will be designed that holds all the ingredients. Finally, a metric is chosen to perform the comparison operation.

Let's choose weighted Jaccard similarity as our comparison method because it allows greater emphasis on higher preference ingredients as opposed to the implicit equal weights inherent with the traditional Jaccard similarity metric. For instance, if the user previously selected 8 recipes with cheese and only 2 with chocolate, then cheese would be deemed more important in the weighted Jaccard similarity calculation compared to chocolate.

However, identifying the metadata and the metric is not enough. We are also compelled to consider two other issues with our recipe data, spelling and semantics. Consider the occurrence of user entry errors when inputting the recipe which will likely involve several misspelled words. There are generally two ways to address the misspelled words event:

1. Autocorrect the words using a spell checker.
2. Create a mapping of possible misspelled words to the correct, intended word.

The first option of autocorrecting misspelled words may on the surface seem to possess the advantage of being easier and faster, but employing autocorrect may not be the best approach since it suffers from lack of context.

By implementing autocorrection the system may not recognize valid but rare words. For example, if a particular exotic ingredient is not listed in the autocorrect dictionary that ingredient will most likely undergo significant modification to another word. Also, there may be cultural ways to spell words that differ from those included in the available dictionary. Canadians may express exasperation with American dictionaries for "correcting" the word colour to color, because colour is the correct way to spell this word in Canada.

We provide a file of word mappings for some of the most commonly misspelled ingredients in the `mapping_ingredients_list.txt` file.

Semantically and contextually, the word recognition component of our recommendation engine can get even more complicated by considering the difficulty in determining how to accurately identify an ingredient. Consider the following text listed as a single ingredient: "½ cup of finely diced green onion." What is the actual ingredient in this item description?

Most people would consider "green onion" – also known as scallions – as the ingredient. If this is the case, then how do you remove each of the words "½," "cup," "of," "finely," and "diced" from the description while retaining "green onion?" Also, under which conditions should "green" be preserved? Should the ingredient be "diced green onion?" These are all nuances inherent in the ingredient item to be considered.

The automated generation of a list of compound words like green onion is called creating *collocations* and is part of natural language processing. More insight about this process can be found in Sect. 9.7.

The next significant hurdle when comparing ingredients is deciding similarity semantics. For example, if the recipe lists a non-dairy milk, like almond milk, should all non-dairy milks be considered?

Often soy milk is interchangeable with almond milk, but this is not always the case. What about coconut milk, oat milk, rice milk, cashew milk, macadamia milk, hemp milk, and quinoa milk? Are they all interchangeable with almond milk? If cow's milk or simply "milk" is listed, should all non-dairy milks also be considered? These are all difficult but important questions that need to be confronted.

Similarly, what about "wine" vs "red wine" vs "red wine vinegar?" Are all three interchangeable? Would pouring a glass of wine be the same as pouring a glass of red wine vinegar? Obviously not. These are just a few of the thousands of examples of similar, overlapping, and marginally related ingredients that make creating a recipe content-based recommendation system difficult.

There are no convenient solutions to these questions. However, to provide some level of support, we have an example list of synonyms that can be found in our `synonym_list.txt` file.

However you proceed, there are many decisions required to reasonably and successfully compare one data item to another. The concept of a content-based recommendation engine appears simple enough, but in practice an appreciable amount of work and domain knowledge is involved.

7.3 Collaborative Filtering

Collaborative-filtering recommendation engines use data from the community to suggest similar recommendations. Aside from hybrid recommendations, they are the most common and the most powerful recommendation engines available.

Collaborative-filtering systems are stateful and require access to user profiles. The more data collected and archived about people and items, the better the results. One noteworthy reason collaborative-filtering systems are considered powerful is because of their capacity to abstract latent features.

A *latent feature* is a feature that cannot be explained directly by the data, but instead is captured by the system. These emergent properties are sometimes referred to as hidden features that explain some meaningful relationship or correlation. For example, most people who enjoy *Star Wars: Episode IV - A New Hope* also like the movie *Monty Python and the Holy*

Grail. Aside from the fact that the two movies were released within 2 years of each other there is very little in common between the two films.

Among movie fans, the correlation between the two films is so high that it is rare to find someone that likes only one of these movies. This individual is so rare that they could be classified as a grey sheep.

If a user selected *Star Wars: Episode IV - A New Hope* using a content-based system then other science fiction-oriented films, movies about astronomy, movies about wars, and movies with the same actors would be suggested. However, *Monty Python and the Holy Grail* would likely not be suggested by a content-based system because there is not much similarity between the metadata of these two movies.

Collaborative-filtering systems by contrast will leverage latent features such a culture, age, and gender attributes that are not strictly present in the metadata, but still remain influential when determining recommendations. These latent features are discovered by examining the likes and dislikes of other people with existing user profiles.

At this juncture, we emphasize that the true nature of latent variables is ineffable – that is, incapable of being articulated. For example, we know that a correlation between *Star Wars: Episode IV - A New Hope* and *Monty Python and the Holy Grail* is present, but we do not ultimately know why the correlation exists. Movie (domain) experts might be able to explain the reasoning underlying the connection, but the collaborative-filtering recommendation engines do not need to know why, they simply identify and exploit the similarities. This lack of information regarding the emergence of latent features effectively defines collaborative-filtering systems as black boxes and thus these systems do not fall in the category of explainable AI (XAI).

The greatest weakness for collaborative-filtering systems is the cold start problem. For a user who only selects a few items of interest the results generated by the system are suspect. In fact, the cold start problem is such a consequential shortcoming for collaborative-filtering systems that many good recommender systems will invoke a content-based engine at startup until such time that the user has chosen a sufficient threshold number of items of interest.

There are two main types of collaborative-filtering systems:

- User-based filtering
- Item-based filtering

User-based collaborative-filtering recommendation engines are based on the principle that people are fundamentally similar to each other. User-based filtering leverages known information about you and compares it to known information about other people.

For example, suppose you like the movies *Aliens* and *Lion King*. Alice (another user) also likes those two movies. If Alice enjoys *Bambi*, then maybe you will like *Bambi* too.

The proposition for user-based collaborative-filtering is that similar people select similar items, regardless of how well we understand the psychological or sociological mechanisms for this circumstance. For example, if someone is from the South in the United States and likes chicken-fried steak they may also like biscuits with gravy, another popular food in the South, since both are specific to the local cuisine.

We could apply an extensive amount of effort to collect information and correlations on local cuisines and expend a huge amount of time and money in the process, but our conclusions would still be less accurate than those deduced by a collaborative-based recommendation engine that simply monitors what people do (preference selection), not what they say they do (subject interviews).

To illustrate this important point, consider a family in which the mother hails from California and the father was raised in the South. Their family will likely eat foods that are popular in both cultures, rarely committing to only one and consistently ignoring the other.

This case study would confound a narrow focus on local cuisine preferences as this family would be considered a black sheep since they are a minority group and would not fit neatly into either a Californian food group or a Southern food group. Instead, if we are supplied enough data on what a wide range of people like to eat then we will likely identify a sufficiently sized group of people who exhibit a preference for both Californian and Southern food and be able to recommend related foods from that group.

In practice, the use of reviews is essential to effectively expand the breadth of the recommendation space. We have web scraped millions of reviews for the corresponding recipes.

To clarify the previous discussion, consider the example recipe 6698: Mom's Zucchini Bread. (You can find the reviews for recipe 6698 by unzipping `recipe_reviews.zip` and then searching the resulting folder.) In the folder the file `6668.txt` has many reviews. For example, consider the following: User MommyFromSeattle rated the recipe 5 stars out of 5. User mm425 gave it a 4, ALLIEA gave it a 5, foodfanatic gave it a 5, TERESAMAS gave it a 5, Ellenn gave it a 5, Stromovitch gave it a 5, iowamom gave it a 5, and so on for a total of 7565 total reviews.

Let's suppose that a new user also gives recipe 6698 (Mom's Zucchini Bread) five stars. If this is the first and only recipe preferred by the new user then a content-based or even case-based system would be a better choice. However, if this user has selected twenty recipes then we will be in a more promising position to leverage the existing reviews of the other users.

The process includes creating a matrix of all the users of the entire system. The users would label the columns and the recipes would represent the rows (although the rows and columns could be transposed and the analysis would be similar). Each cell contains the rating the user submitted for the particular recipe. If the cell is blank then the user has not yet rated the recipe.

Figure 7.1 shows a visual example of a small subset of the matrix, matching user's reviews to the corresponding recipes. There are over 1.3 million unique reviewers in the example data and over 100,000 recipes. The generation of a comprehensive matrix with over 1.3 million columns and over 100,000 rows would result in a very large, mostly empty (sparse) matrix.

	new user	MommyFromSeattle	mm425	ALLIEA	foodfanatic	TERESAMAS	Ellenn	Stromovitch	iowamom	⋮	⋮
6698	5	5	4	5	5	5	5	5	5		
6669		5				5					
6670									4		
6671	4			3			5				
6672		5									
6673			5								
6674	3					5		4			
6675				4							
6676		3					5				
6677					5						
6678	1								4		
...											
...											

Fig. 7.1 Visual example of the matrix of users as columns, recipe numbers as rows, and rated recipes as the entries in the cells. Blank cells indicate the user has not rated the recipe yet. In our example data we have over 1.3 million unique reviewers and over 100,000 recipes. (Image created by Robert Ball)

The procedure implied by the information matrix in Fig. 7.1 is to correlate the new user's preferences with all the people in the dataset to determine the people most similar to the new user. For instance, let's suppose we are trying to find the 20 people most similar to the new user. Note that settling on 20 people is completely arbitrary--we might want to determine 100 or 1000 similar users or even leverage all users in some capacity. The choice of 20 is simply a convenient number for instructional purposes.

The first step on our journey to determine similar users would be to filter the matrix to those individuals that have reviewed any of the same recipes as the new user.

In the sample matrix in Fig. 7.1 we see the new user has reviewed four recipes: 6698, 6671, 6674, and 6678. Assuming the new user has not reviewed any other recipes we can filter out all users that have not reviewed any of those four specific recipes.

For our example data, there are only 7848 users that have reviewed any of those four recipes. Now we have a matrix with 7848 columns (users, including the new user) and 4 rows (recipes). We want to further filter these users to the 20 most closely matched with the new user.

Our rating data is stored such that each user indicates their preference for the recipe on a scale from 1 to 5. This particular range of values is comparatively more extensive than those employed by many modern recommendation engines which utilize a simpler, unary approach of submitting preferences in the form of a single "like" or "thumbs up". This facilitates the determination of similar users because individuals that like the same recipe are noted as similar. Some systems also allow for a negative rating such as "thumbs down," but indicating only a "like" is the most widely accepted and simplest approach in current practice.

Given the scope of our 1 to 5 rating system, we can still find the twenty most similar people to the new user. The basic technique is to find twenty people that all marked the same responses as the new user. Ideally, we would search for twenty people that entered identical responses as the new user, specifically 5 for recipe 6698, 4 for 6671, 3 for 6674, and 1 for 6678. However, locating 20 people with those exact matches may not be possible, so we attempt to find the top twenty people that are the most similar. We could apply cosine similarity to compare each person's responses to the new user's responses in which each user's responses are mapped to a vector that is compared to the vector of the new user. Alternatively, we might use Pearson correlation to determine the set of twenty similar users.

Once we have the top 20 people of the 7848 users who are the most similar to the new user the goal is now to predict which recipes the new user will prefer. We assume the new user and these other 20 similar people are all white sheep and therefore conclude they think alike. So, if all 20 people agree with the new user and give recipe 6698 a 5 and 15 from this group give another recipe, say 8749, a 5 then most likely the new user will prefer recipe 8749 with a good chance of rating it a 5.

We can specifically calculate the predicted amount of enjoyment or likability of a recipe for the new user by averaging the amount of enjoyment that other recipes provided for the previously established users. For instance, in Fig. 7.2 on the left are the new user's rated recipes (renamed recipe 1 through 4) and all the rated recipes from the twenty people identified as most similar to the new user.

	new user	similar user 1	similar user 2	similar user 3	similar user 4	⋮	similar user 20
recipe 1	5	5	5		5		5
recipe 2	4	5	4	4			4
recipe 3	3	2	2	3	4		1
recipe 4	1	1	1		1		2
recipe 5		3	4	5	3		3
recipe 6		5		5	5		5
recipe 7		2	3	4			3
recipe 8			5		5		5
...							
recipe N		5	1		2		2

	new user	similar user 1	similar user 2	similar user 3	similar user 4	⋮	similar user 20
recipe 1	5	5	5		5		5
recipe 2	4	5	4	4			4
recipe 3	3	2	2	3	4		1
recipe 4	1	1	1		1		2
recipe 5	3.6	3	4	5	3		3
recipe 6	5.0	5		5	5		5
recipe 7	3.0	2	3	4			3
recipe 8	5.0		5		5		5
...							
recipe N	3.3	5	1		2		2

Fig. 7.2 The left matrix shows the initial values and the right matrix shows the predicted amount of enjoyment or likability that the new user would assign to the other recipes. (Image created by Robert Ball)

Note that not everyone in this group has rated each of the recipes listed, but where ratings are provided we can predict the new user's enjoyment for that specific recipe. Figure 7.2 shows that the new user has submitted ratings for the first four recipes. However, how will the new user likely rate the other recipes? By taking the average of the other users for a given recipe, we can estimate which recipes the new user will enjoy the most and then recommend those.

For instance, we can observe in the truncated tables of Fig. 7.2 that similar users 1, 2, 3, 4, and 20 all rated recipe 5. If we take the mean of the five ratings (3 + 4 + 5 + 3 + 3)/5, we compute 3.6. Doing the same calculations for all the other rated recipes produces the values displayed in red on the right side of Fig. 7.2, which are effectively predictions on how the new user would rate those recipes. We can then sort that list in descending order and recommend the following recipes in the following sequence: recipe 6, 8, 5, N, and 7.

An improved method for computing the values in red in Fig. 7.2 involves assigning weights to users. For example, suppose that when we correlate the values of new user to similar user 1 the metric calculated is 0.91. We can then use that weight in our predicted rating calculation. Assume that similar users 1, 2, 3, 4, and 20 exhibit correlation measures of 0.91, 0.89, 0.5, 0.88, and 0.6 respectively for a total of 3.78 then the calculation of (3 + 4 + 5 + 3 + 3)/5 would now become the weighted measure (3*0.91 + 4*0.89 + 5*0.5 + 3*0.88 + 3*0.6)/3.78 = 3.5.

Note that per the definition of a weighted average, the denominator in our calculation above should be the sum of the correlation measures 3.78 instead of 5 to properly emphasize the similar users. The weighted result of 3.5 instead of 3.6 highlights the values of the people that are more similar to the new user. For instance, similar user 3 submitted a 5 star rating for recipe 5, higher than any of the other users being considered. However, for the group of 5 users, the correlation metric (0.5) indicates that user 3 is the least similar to the new user, so the importance of that 5 star rating was emphasized the least.

In the case in which other users did not rate a particular recipe then they would be excluded in the average calculations. For example, in Fig. 7.2 note that similar user 3 did not rate recipes 1, 4, 8, or N and so the preference calculation would not include user 3 in predicting how much the new user would like those particular recipes.

There may be thousands of recipes that the 20 most similar users rated. However, to minimize execution time it is wise to only consider recipes that satisfy some threshold for the total number of ratings, such as a minimum of ten, to ensure that recipes only rated once or twice are not included in the final list of recommendations. For example, if only one of the similar users rated a recipe then that recipe would be ignored because it has not been vetted sufficiently by a larger portion of the group.

This common approach of only recommending established recipes or products highlights the *bubble-up problem*, which is the challenge of determining how to gather sufficient reviews for new or non-established products such that they can subsequently be considered by the recommendation engine. For example, assume a brand-new recipe that has never been reviewed. A strict collaborative-based recommendation system that does not implement any serendipity features would never recommend a recipe with no ratings. This problem underscores the fact that the new recipe could either be amazing or horrible, so how can users be incentivized to try the recipe and offer ratings if the new recipe does not appear in any recommendation list?

Overall, user-based collaborative-filtering recommendation engines are very powerful black box recommendation engines that have changed the way people shop, browse the Internet, and consume the news. These engines identify and leverage other white sheep similar to the user and typically demonstrate excellent performance in predicting what the user wants.

From an ethical standpoint, these recommendation engines also are one of the primary influences leading to tribalism, confirmation bias, and divisiveness in our communities because they show people only the narrow brands or opinions they are seeking, which may reflect neither the truth nor a balanced perspective.

Item-based collaborative-filtering recommendation engines are similar in principle to the user-based engines, but similarity measures are centered on the correlations between items rather than people belonging to a white sheep group. Both systems use the same data as a resource but perform similarity calculations on either users or items.

An example of an item-based system for recipes would be an online store where results are reported as follows: "People that liked recipe 1 generally also like recipes 5, 10, and 21." The emphasis is on the similarities among recipes rather than the similarities among users. For item-based systems, a similarity metric is calculated with respect to a specific item to produce a group of related items. Predictions about which of these items to recommend are based on survey statistics gathered across all users.

If user-based systems perform similarity comparisons on users, which were displayed as columns in Fig. 7.2, then item-based systems carry out similarity comparisons on items, displayed as rows in Fig. 7.2. In other words, the same comparisons and prediction computations are applied, but we simply switch the purpose of the rows and columns. We now consider a limited subset of related items (e.g. top 20, 50, 100) that have been similarly rated across all users to determine recommendations.

For example, if people who like *Aliens* also enjoy *Lion King* then the system can leverage a similarity metric to pair those movies together such that if a user indicates a preference for *Aliens* then the system might suggest *Lion King* as well.

In practice, user-based systems are not implemented as often as item-based systems because there are generally more users than items for a given site – recall the 1.3 million unique reviewers and over 100,000 recipes in our example data. Since items change much less frequently, the calculations to compare updated lists of items will occur less often, resulting in more responsive recommendations for users.

7.4   Specialty Types

There are many specialty types of recommendation engines. For example, some of the most powerful collaborative-based systems use matrix factorization to determine the most accurate results. *Matrix factorization* tends to produce latent variables that cannot be articulated or described but lead to very high accuracy.

The idea behind matrix factorization derives from linear algebra where the user and item (e.g., movies, recipes) matrix is decomposed into the product of two lower dimensional matrices. There are various versions of matrix factorization recommendation engines that are on the horizon of modern research.

Netflix, a movie streaming company, offered a $one million prize for whomever could create the most accurate movie recommendation engine that would outperform the company's in-house recommendation engine. Matrix factorization was one of the most effective tools utilized by the winning team and has been deemed one of the most sophisticated and accurate ways to create recommendation engines since that competition.

Another specialty type is a *demographic recommendation engine* which accounts for demographics when generating results. Demographics relate to specifics about groups of people such as residence, age, gender, religion, and other factors.

For example, a user might be identified as female, age 16, and living in Germany. For this particular user, a clothing store demographic recommendation engine would only display clothes appropriate for German teenage girls. Showing the latest popular fashion for 60-year-old Jamaican men would not be helpful.

Demographic recommendation engines are also utilized by news outlets. If you live in the outback of Australia then news about a corrupt German mayor thousands of miles distant would probably be of less interest to you than information about the corruption of the mayor in your home town.

A *context-based recommendation engine* considers the "context" of the specific request. For example, a context-based recipe recommendation system would suggest the appropriate food based on time of day. More to the point, during breakfast at a given location, the context-based recommendation system would only list breakfast recipes or foods. All other meals (e.g., brunch, lunch, dinner, tea, or supper) would not be displayed.

The successful operation of this type of system requires additional metadata for each recipe or food. For example, pancakes might be categorized as "breakfast" and "brunch" foods whereas sandwiches might be categorized as "lunch" or "tea" foods.

Hybrid recommendation engines are simply recommendation engines that combine different standard recommendation engines. For example, as mentioned above, a collaborative-based recommendation engine usually provides better overall results than a content-based system. However, collaborative-based recommendation engines require users to input significantly more preferences than the other types of recommendation engines, so a new user is disadvantaged because of the initial lack of preference data. In addition, a new user who recently signed up for an innovative trial movie streaming service might be reluctant to prime the system by spending 30 minutes entering movie preferences.

Alternatively, the system design may require that the new user submit at least one movie when signing up for the trial service. Providing the system with only one movie is a manageable request, but it is insufficient data for a collaborative-based system. A solution for this scenario is a hybrid system in which the user receives results from a content-based system at the beginning of the subscription, but later the user is switched to a collaborative-based system once there are sufficient data.

Other hybrid systems might include a mix of demographics and context as well. Any recommendation engine that does not conform to the standard textbook definition is usually considered a hybrid recommendation engine.

Exercises

1. Which group has a more positive view of recommendation engines: consumers (buyers) or businesses (sellers)? Why?
2. Why is serendipitous shopping or browsing difficult in online environments?
3. What are the societal implications of recommendation engines?
4. What is the difference between white sheep, black sheep, and grey sheep?
5. Explain the cold start problem in terms of stateful and stateless recommendation systems.
6. Look at the `mapping_ingredients_list.txt` file. What is its purpose? What are other possible solutions you can think of?
7. How important are word taxonomies (classifications and relationships) for recommendation engines? For example, the `synonym_list.txt` file lists that "brie" and "american" are both "cheese." Without some clarification of word relationships, how much can you generalize between ingredients?
8. The `synonym_list.txt` file has a very shallow taxonomy of relationships. Based on the same data, what kind of taxonomy would you suggest that might offer an improvement?
9. When would you use a constraint-based vs case-based knowledge-based recommendation engine?
10. Create a knowledge-based recommendation engine with the recipe data.
11. Create a content-based recommendation engine with the recipe data.
12. Create a collaborative-based recommendation engine with the recipe data.
13. Provide an example of a context-based recommendation engine that does not involve recipes or food.

Chapter 8
Machine Learning

Machine learning (ML) is used extensively in data science. What is it? To answer that question let us provide some context to the practice of machine learning.

In its most general form, machine learning is an algorithm or programming technique that allows a machine to learn. So, what does it mean for a machine to "learn?" Broadly speaking, to learn is to gain knowledge or become aware or informed about a certain area of expertise. This definition is intentionally expansive but so is the scope of machine learning.

If machine learning algorithms allow a machine to learn then what is artificial intelligence (AI)? Intelligence is the ability to understand information and effectively transform it into useful knowledge. Is machine learning a component of artificial intelligence?

Yes and no – it really depends on who is actively participating in your conversation about these topics. The subject of AI vs ML typically involves a fair degree of argument and contention. Essentially what will be stated here will likely invoke disagreement with many readers, but our definition is primarily intended to assist you in framing these areas of investigation within the context of data science.

Artificial intelligence occurs when any machine makes a decision based on integrating actionable information. For example, by this definition an old mercury-based thermostat may be considered a form of artificial intelligence. A given volume of mercury is both highly conductive and sensitive to temperature – it will expand when it gets warmer and shrink when it gets colder. A mercury-based thermostat works by having two electrodes connect through mercury, which acts as a switch. One electrode is fixed in place while the other can be adjusted by the human operator. Let's say that the temperature in a room is 75 °F (23.8 °C) and the thermostat is set to activate and turn on the air conditioning unit at 80 °F (26.7 °C). At this point the electrodes are not yet connecting. As the air surrounding the mercury gets hotter so too does the mercury and thus expands. When the mercury expands sufficiently then the two electrodes will connect to close the electrical circuit and activate the air conditioning for the house.

Is this thermostat example an artificial intelligence? It is most certainly a machine that "uses information" to make a decision. So, we could label the thermostat an artificially intelligent machine. Many people would argue that mercury considered in isolation cannot be deemed "intelligent." However, we could make a similar argument regarding the electrons and silicon used in microchips.

Regardless, artificial intelligence is admittedly difficult to define precisely without hundreds of people contesting the proposed definition. We rely on a convenient application of the old adage that if you gather 10 Ph.D.'s in a room and ask them to define artificial intelligence you will hear 10 different definitions.

For the purposes of this book we will define *artificial intelligence* as the ability for a computer to make decisions based on integrating actionable information. There are two main types of artificial intelligence that we will focus on: programmed intelligence and learned intelligence.

Programmed intelligence is the kind of artificial intelligence that a programmer encodes through software. For example, consider the game Tic-Tac-Toe. Can you create the artificial intelligence required for a computer to play this game with a person? In Tic-Tac-Toe, there are only 9 positions to choose from with any three X's or O's in a row resulting in a win. Given

R. Ball, B. Rague, *The Beginner's Guide to Data Science*, https://doi.org/10.1007/978-3-031-07865-1_8

the resources, most professional computer scientists could create the artificial intelligence required to always win or draw, but never lose at Tic-Tac-Toe. In 1997, an IBM supercomputer named Deep Blue became the first system to defeat a reigning world champion in chess, considered an historically significant moment for artificial intelligence. Regardless of the complexity of the game, non-adaptive computer opponents trained solely on game rules and strategies are programmed intelligence.

Alternatively, *learned intelligence* or *machine learning* is the idea that a computer learns from or adapts to data and does not rely entirely on stepping through a series of programming instructions to demonstrate intelligence. For example, a computer might be exposed to millions of images of cats and millions of images of dogs and then be able to distinguish between **new** images of these two animals in future applications.

Figure 8.1 illustrates the general relationship in which programmed intelligence and machine learning are both part of artificial intelligence. To complicate communications about this topic even further, when someone refers to artificial intelligence they might be addressing programmed intelligence, machine learning, or both. In most cases it can be difficult to discern what is truly being discussed without asking follow-up questions.

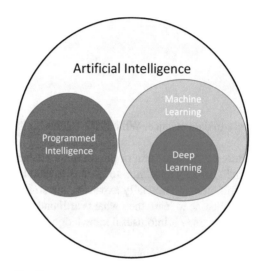

Fig. 8.1 A visual comparison of programmed intelligence vs machine learning. (Image created by Robert Ball and Brian Rague)

Figure 8.1 also includes "deep learning" as part of machine learning. Deep learning is a subset of machine learning that leverages extensive use of neural networks (see Sect. 8.6). Deep learning often involves large amounts of data and high-end hardware. We will not be investigating the details of deep learning in this book beyond describing where it falls in the taxonomy of artificial intelligence.

Both programmed intelligence and machine learning have their place and purpose. The good and bad aspect about programmed intelligence is that once code is written we know the intelligence is effectively "hard-wired" and may be regarded as predictable.

Conversely, for either the supervised or unsupervised cases machine learning requires a large amount of training data before useful and accurate predictions can be realized. The general process with machine learning is to choose an algorithm, provide large amounts of training data then determine how well the algorithm learned by examining the outcome from a test set of known predictions. As an example, suppose we have one million images of cats and one million images of dogs for a total of two million images. Typically, we would train the machine learning algorithm with 80% of the data (1.6 million images) then test how well the algorithm performs with the other 20% of the data (0.4 million images).

In data science the subset of programmed intelligence techniques is not utilized very often. We more commonly use machine learning strategies because the data we need to analyze changes regularly. If the data were constant, like the game theory for winning a chess match, then programmed intelligence might be preferred. Of course, if the data did not change frequently there would be no real need for the discipline of data science.

8.1 Machine Learning Overview and Terminology

There are many basic concepts that generally apply to most machine learning algorithms.

As mentioned above, training a machine learning algorithm using data is something that applies to all machine learning algorithms. If an algorithm is not trained, and subsequently does not learn, can it truly be considered a machine **learning** algorithm? In other words, without training data, a machine learning algorithm by definition does not learn.

As stated earlier, the standard practice for most machine learning algorithms is to train the algorithm with 80% of the data and test it with the other 20% of the data. However, it is common to see the train/test ratio range anywhere from 70/30 to 85/15. An essential aspect of machine learning is that the training data and the testing data should be different. If these sets overlap even by one sample then we are essentially asking the algorithm to learn something and then simply recall or "remember" the original training data.

For an algorithm to remember or recognize what it has already processed may be useful, but this functionality is not the crux of machine learning. Consider the example of an algorithm that scans images of cats. If it were shown one of the original training images then the algorithm would only have to recall the image from storage and recite the classification previously provided for that sample. This is **not** machine learning but instead is a form of file system or database retrieval and may be considered a version of AI *expert system*.

A well-known example of an expert system is IBM Watson – a computer system that emerged victorious on the live television quiz show *Jeopardy!* Expert systems are often written in the logic programming language Prolog to enable the system to create new meaningful logical associations based on known fundamental axioms and statements. Expert systems are an important subfield of AI, but not heavily utilized in data science.

Instead, given a test observation the point of machine learning is to be able to classify or predict outcomes correctly without having previously processed that test instance. For example, assume a machine learning algorithm were trained on separate cat and dog images and then offered a new scanned image of a cat that it had not encountered before. The performance of this machine learning algorithm would be considered a success if it correctly classified the new image as a cat.

Training the machine learning algorithm should involve selecting training samples randomly from the original dataset. This is important because one set of training data might be biased toward different values of features (the inputs) when compared to another set of training data. A feature is an individually measurable property or attribute in your data. For example, height is a feature because it can be measured independently from other features, such as age, weight, and eye color.

The following function from the *sklearn* package will automatically randomize and split your data into train and test datasets (the default for `test_size` is 0.25 [25%]):

```
X_train, X_test, y_train, y_test = train_test_split(X, y, test_size=0.2)
```

It is standard to use an uppercase X for the data and a lowercase y for the predictor. This convention derives from mathematics where X stands for a multidimensional data matrix and y represents a one dimensional predictor array.

The `train_test_split` function takes all the original data in X and the corresponding predictors in y, then randomly selects (while maintaining the rows that correspond to the predictors) and splits the data into appropriate training and test sets. See Fig. 8.2 for a visual example.

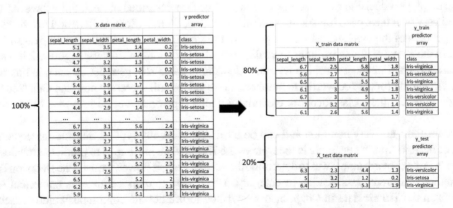

Fig. 8.2 Example of the train_test_split function results in an 80/20 percent split. The X and y data are randomized then split into training and testing data. (Image created by Robert Ball)

The following is a canonical example of a typical machine learning training and testing set of instructions applied to the Iris dataset, regardless of the type of machine learning algorithm used:

```
File: machine_learing_introduction.py
from sklearn.datasets import load_iris
from sklearn.model_selection import train_test_split
from sklearn.metrics import confusion_matrix, accuracy_score
from sklearn.tree import DecisionTreeClassifier

iris = load_iris()
X = iris.data
y = iris.target

classifier = DecisionTreeClassifier()

X_train, X_test, y_train, y_test = train_test_split(X, y, test_size=0.2)

classifier.fit(X_train, y_train)
y_pred = classifier.predict(X_test)

print(confusion_matrix(y_test, y_pred))
print(f'Accuracy = {str(accuracy_score(y_test, y_pred))}')
```

Which **sometimes** produces the following output:

```
[[12  0  0]
 [ 0 10  1]
 [ 0  1  6]]
Accuracy = 0.9333333333333333
```

The reason the algorithm sometimes produces the above output is because the `train_test_split` function randomizes the data used for training and testing on each invocation. So, for one calculation the accuracy is 93.34% as shown above and for another execution it might be 100%.

The above code analyzes the well-known Iris dataset described in detail in Chap. 4. The output displays accuracy and a confusion matrix. Accuracy, confusion matrix, and classification reports are common metrics for determining how well a machine learning algorithm performed (see Sect. 6.3 for more information).

The above code employs a decision tree machine learning algorithm described in Sect. 8.2. However, the machine learning technique could very well have been logistic regression (Sect. 8.4), SVM's (Sect. 8.5), neural networks (Sect. 8.6), or one of many more. Regardless of the machine learning algorithm, the same training and testing split pattern is utilized.

The Iris dataset in the above example includes three different classifications of flowers. Figure 8.2 indicates the predictors for this example are strings. For most machine learning algorithms these strings must be encoded as numbers (see Sect. 3.3.11). Even when the predictors are converted to numbers, these values still represent nominal data (see Sect. 5.2) and their values are analyzed to determine a classification, essentially which type of flower corresponds to the particular petal and sepal observations.

Classification algorithms are very common and can be used to predict any type of discrete label such as identifying an image as dog, cat, or person. Another example is to leverage the *Titanic* data in Chap. 5 to predict a traveler's passenger class. Because the objective in the above example was to determine the category of Iris flower, we used the DecisionTreeClassifier class.

If classifiers predict a discrete classification or category, a regressor predicts continuous numerical values. Proceeding with the example from the *Titanic* data in Chap. 5, if we wished to predict the continuous value passenger fare (price of ticket) then we would use a regressor. In this case with decision trees, we would instantiate the DecisionTreeRegressor class.

When should you use a classifier as opposed to a regressor? This determination is not based on the **inputs** (X input), but the **predictor** (y predictor array). For example, the Iris dataset has all continuous values as input features, but classification types (i.e., versicolor, setosa, and virginica) represent the predictor. In summary, if what you are trying to predict is a classification then it is a classification problem, and if what you are trying to predict is a continuous number then it is a regressor problem.

As for terminology, the expression 'fit' (like `classifier.fit(X_train, y_train)`) means 'learn' or 'fit the model.' Because machine learning is based heavily on statistics, we use the statistics term 'fit the model.' To fit the model means to adapt the inner workings of the statistical (or machine learning) model to the data. The model is simply the functional math and statistics that make the algorithm work. Because the internal operations of many machine learning algorithms work differently, to fit the model simply means to apply the appropriate learning and refinement procedures so that the algorithm will be able to predict future values.

Once we have fitted the model then we can use the classifier for prediction purposes. The code `y_pred = classifier.predict(X_test)` utilizes the classifier that has previously been fitted (or taught) and asks the model (the classifier) for predictions based on the `X_test` matrix. We then can store the predictions in the output variable `y_pred` based on the input of `X_test`.

One of the general assumptions about the training of machine learning algorithms is they are provided with balanced datasets. A *balanced dataset* is one that contains the same number of representations per category. For example, with the Iris dataset each category of Iris flower has exactly 50 example entries.

With unbalanced datasets you run the risk of having the machine learning algorithm always choose the majority class for each prediction. The *majority class* is the one that has the most representative samples in the training set. For example, suppose that for the Iris dataset we have 1000 examples of virginica and only 50 examples each of setosa and versicolor. Given such an unbalanced dataset and the extremely high occurrence rate of virginica compared to setosa and versicolor, many machine learning algorithms might simply predict virginica and be correct most of the time if the frequencies of the test set classifications mirror those of the training set.

Combatting unbalanced datasets can be tricky. The most common approaches to remedy unbalanced datasets are undersampling, oversampling, and creating synthetic data. *Undersampling* obtains a subset of the majority class. For example, if we truly had a 1000/50/50 representation for the individual Iris flowers, then we could undersample the majority class (the 1000 examples) down to 50 so the dataset is balanced.

Oversampling is the opposite of undersampling. Instead of having fewer of the majority class, we include repeated samples from the minority classes. Using oversampling in our Iris example, we would retain the 1000 samples from the virginica class and repeat or clone the other 50 samples from the other classes until all three classifications had the equivalent 1000/1000/1000 representation.

Creating synthetic data can be achieved in various ways. For example, SMOTE (Synthetic Minority Oversampling Technique) is a form of oversampling that maintains the majority class samples, but instead of simply replicating the minority classes, it generates additional samples based on the current feature space. For example, for the setosa class, SMOTE would look at the existing feature space of setosa petals and sepals and create new synthetic samples by interpolating between random selections of existing setosa samples and their nearest neighbors.

Although machine learning is very powerful, paradoxically too many input features can lead to less accuracy. *The Curse of Dimensionality* occurs when there are so many features (or dimensions) that people and machine learning algorithms have a hard time deciphering the core information or "signal" of a dataset.

What is the concept behind dimensions? What does that mean? A "dimension" is simply a column of data or an individual feature. As your input includes more and more features the predictive power of your machine learning algorithms become less and less accurate. For a more detailed discussion of this issue, including ways to combat the curse of dimensionality, see Sect. 8.9.

There are many machine learning packages and frameworks available. Regardless of the one you choose, the fundamental principles are the same. For this book we will use *sklearn* (Scikit-learn) primarily for its ease of use and teachability. There are many other Python packages and frameworks including PyTorch, TensorFlow, Theano, Keras, and MXNet. Due to the various application purposes and constantly evolving nature of machine learning software, our previous list does not imply any particular order of preference or endorsement.

To help with any potential confusion, *sklearn* and Scikit-learn refer to essentially the same library even though they have different names.

Generally, one installs the Python package with the following command:

```
pip install scikit-learn
```

In a Python program, the framework is imported with the following command:

```
import sklearn
```

Using `import scikit-learn` will generate an error. However, you can also install the package with the following command:

```
pip install sklearn
```

Behind the scenes, the above command will install a dummy package whose only purpose is to install scikit-learn.

As discussed in Sect. 1.9, big data is often confused with the term 'data science.' However, *big data* often implies a large volume of data and this concept is important with regard to machine learning. Many machine learning algorithms do not do well with small amounts of data. So, when working with big data machine learning will often be an effective tool that will help you elucidate the patterns and nuances in your data which cannot be easily consumed and understood by an individual person.

However, its commonly understood that processing a large amount of data can be time consuming. This leads to the topic of high-end hardware components, specifically GPUs. A *GPU* is a *Graphics Processing Unit*. GPU's are designed with high-end visualizations in mind. They are engineered for the specific purpose of rendering high-end 3D graphics very quickly. People that play high-end video games often rely on computers with expensive GPU's.

GPU's work by managing matrices of data and manipulating that data rapidly in a non-linear (parallel) way. Matrix data structures are used because monitors essentially display a matrix of pixels to the user. In other words, when you see an image on a monitor or projector what you really see are millions of little color squares that your brain processes into larger objects and shapes. When those pixels are generated along the pipeline from the operating system to the monitor the GPUs are leveraged to speed up rendering.

If you happen to have a matrix, such as the input X data, and can transform it into a form that a GPU expects then you can take advantage of the increased power of a GPU over a general CPU (Central Processing Unit) that typically runs computational devices. Since all input data to a machine learning algorithm are structured in matrix form then machine learning algorithms can be accelerated considerably if they take advantage of high-end GPU's.

Another important general aspect of machine learning algorithms to discuss at this point are the terms supervised and unsupervised. A *supervised* machine learning algorithm is one that includes a specific predictor or target class associated with the training set. In our Iris example, given certain petal and sepal measurements, a particular target class is the setosa Iris.

An *unsupervised* machine learning algorithm does not have such an explicit classification associated with each of the training set observations. One common type of unsupervised algorithm is *clustering*. The fundamental concept behind clustering is to group similar things together. For example, based on features describing different dog attributes we can determine breeds that are the most similar? The results of this analysis may create separate groups of dogs based on size.

It's important to note that clustering algorithms can be both supervised or unsupervised. For example, *k-means* is an unsupervised clustering algorithm where the objective of the algorithm is to create k clusters. If k-means were given example data on dogs and k = 2 then two clusters of dogs would be created based on scaled distance metrics between the individual features. The value of k establishes the number of groups to be created by the k-means algorithm. It is unsupervised because the dogs are not labeled or categorized prior to the learning phase. *Hierarchical Clustering* is another common unsupervised clustering algorithm that starts by assigning each training sample to its own cluster of size one and proceeds to build larger clusters in a bottom-up fashion.

Two additional important concepts about machine learning implementations are overfitting and underfitting. *Overfitting* can be roughly described as memorizing a specific training set. This situation will likely compromise the accuracy of the algorithm because the generated model may not be able to generalize appropriately to non-training data. Consider a machine learning model trained on a specific breed of dog (beagle) and a specific breed of cat (calico). Although the model may do very well in distinguishing the difference between these two animals, subsequent testing with images of a wide variety of dog and cat breeds would reveal severe shortcomings in the predictive power of the model.

Underfitting denotes the opposite scenario where the model does not learn very well from the training data. This situation often occurs when there is insufficient training data.

Overfitting and underfitting can be more formally characterized by the errors known as bias and variance. *Bias* is the difference between the predictions generated by the machine learning model and the correct values. Bias is sometimes referred to as *bias error* or *error due to bias* as it provides a measure of the discrepancy between the machine learning model's predicted values and the true values.

Variance is the variability of the model predictions based on the training set. As mentioned above, it is important to randomize our training data. The degree to which our accuracy changes each time we rerun the algorithm is a measure of variance. In the previous code example in which we applied the DecisionTreeClassifier to the Iris dataset the accuracy is often between 93-100% – this range is a demonstration of variance.

Underfitting is often described as generating a high bias, low variance model. This means that the model is oversimplified, does not capture underlying patterns in the data well, and has a higher error rate.

Conversely, overfitting may be described as generating a low bias, high variance model which tracks the vagaries and fluctuations of the training set, essentially capturing the noise along with the underlying pattern in the dataset. As mentioned previously, this model does not generalize effectively.

You cannot have both low bias and low variance because of the inverse, competing relationship between these two errors. Finding the balance between bias and variance is difficult. However, one common approach to reduce both bias and variance is to increase the training dataset. Generally, the more data available for training the better the model.

Now that we have more vocabulary related to the machine learning paradigm, we can revisit the term deep learning from Fig. 8.1. Deep learning is a big data machine learning technique generally accomplished with unsupervised neural networks using high-end GPUs.

Proponents of deep learning often envision the future powered by deep learning algorithms and powerful supercomputers. Skeptics say that deep learning is simply a hyped-up characterization much like "Big Data" and is merely another type of machine learning.

For data scientists, machine learning algorithms are simply tools to assist in your analysis. Like a car, you can focus too much on surface features like how shiny it is and how fast it goes, but realistically if it does not transport you from one place to another then it does not matter how nice it looks or how well it runs. Similarly, you can focus too much on optimizing machine learning algorithms, such as balancing bias and variance, but if the model does not help you with your task of capturing actionable insights then it is essentially useless.

Clearly, the more you know about machine learning algorithms, the better, but ultimately it comes down to how you apply them in practice. Similarly, an automotive engineer, someone who designs and builds cars, may be a world's leading expert on designing and creating automobiles. However, that person may not in reality be a good driver. The goal of machine learning algorithms with data science is knowing how and when to responsibly apply these algorithms to solve your data problems.

It is both impractical and impossible to learn **all** the machine learning algorithms. It is advisable to be familiar with the most common ones, but since researchers are constantly creating new ones assimilating all the machine learning algorithms becomes a moving target and an impossibility.

We strongly advise studying one machine learning algorithm at a time. It is far better to understand a few machine learning algorithms well than to be familiar with dozens but without any practical knowledge of how and when to use them for a given data science project.

The following is a list of the most common machine learning algorithms used in data science:

- Linear Regression
- Neural Networks
- Decision Trees
- Logistical Regression
- Random Forest
- SVM (Support Vector Machines)
- Naïve Bayes
- KNN (K-Nearest Neighbor)
- K-Means
- Gradient Boosting & AdaBoost

8.2   Decision Trees

One of the simplest machine learning algorithms is the decision tree. A trained (fully realized) decision tree is essentially a series of *if* and *else* statements corresponding to the branches of the tree.

Let's build a simple representative decision tree together based on a narrative. The objective is to build a decision tree that predicts a specific outcome based on the different input features. The point of our example decision tree is to predict if a man will go shopping with his wife, which requires a binary response of either yes ("Go shopping") or no ("Stay home").

To inform the construction of the decision tree, we rely on the following narrative: "The man will always go shopping with his wife if it is Christmas Eve unless his mother-in-law accompanies them at which point he will not go. If it is not Christmas Eve, he will go shopping if the sun is shining outside or it is not raining. However, if his favorite candy is 50% or more on sale then he will go shopping no matter what other conditions exist (including if his mother-in-law is going)."

Before we construct the decision tree, let's summarize the different input features of interest based on the above description. Note the state of each feature is characterized as either *present* or *absent*:

- Christmas Eve
- Mother-in-law accompanies
- Sun is shining
- Rain
- Favorite candy is 50% or more on sale

Figure 8.3 illustrates the design of a possible decision tree based on the narrative and list of key features. The decision tree figure depicts a series of *if* and *else* statements.

Fig. 8.3 Visual decision tree depiction of the shopping/staying home narrative given in the text. (Image created by Robert Ball)

One of the more powerful properties of decision trees is they can be easily visualized and comprehended by an observer. *Explainable AI (XAI)* is the idea of an observer achieving a clear understanding why a machine made its decision. Decision trees are often used when XAI is a requirement. Decision trees are in the category of "white box" machine learning algorithms in which the observer can clearly view the mechanisms by which predictions are generated. These are contrasted with "black box" machine learning algorithms (described below) which obscure the complexity of the computational process employed to produce the output results.

The alert reader will realize there are several ways to construct decision trees that modify the feature set and achieve similar results. One easy optimization applied to the above tree is to prune or remove the "sun is shining node" under the assumption that it is mutually exclusive with the "raining node." As with most programming strategies and design, there are often many ways to produce the same results, as shown in the comparison between Figs. 8.3 and 8.4.

Fig. 8.4 Same decision tree depicted in Fig. 8.3 except the sunshine node has been pruned (removed). (Image created by Robert Ball)

However, the decision trees that we created in Figs. 8.3 and 8.4 are derived with programmed intelligence, essentially our narrative directs the definition of *if* and *else* branches programmed into the computer. This approach provides a straightforward demonstration of the usefulness of decision trees, but it does not show how they are created with machine learning.

Consider the following dataset (sledding.csv) that records conditions influencing a child's decision of whether or not to go sledding:

	snowing	temperature	wind	time	predictor(sledding)
0	yes	below_freezing	high	morning	yes
1	yes	freezing	high	morning	yes
2	yes	above_freezing	low	afternoon	yes
3	yes	below_freezing	low	afternoon	yes
4	yes	freezing	medium	morning	no
5	yes	below_freezing	low	afternoon	yes
6	no	freezing	medium	afternoon	yes
7	no	above_freezing	medium	morning	no
8	no	above_freezing	medium	morning	no
9	no	below_freezing	low	afternoon	yes
10	no	freezing	low	afternoon	no
11	no	above_freezing	high	afternoon	no

Based on the above data, what would the resulting decision tree look like? The following code loads the data, encodes the string data to integers using the LabelEncoder class (see Sect. 3.3.11 for more details), and then trains the decision tree:

```
File: decision_tree.py:
from sklearn import preprocessing, tree
from sklearn.tree import export_text
import matplotlib.pyplot as plt
import pandas as pd

df = pd.read_csv('sledding.csv')

df = df.apply(preprocessing.LabelEncoder().fit_transform)  # convert all strings to integers
features = ['snowing', 'temperature', 'wind', 'time']
X = df[features]  # the data
y = df[['predictor(sledding)']]  # the predictor

# print(X)
# print(y)

decision_tree = tree.DecisionTreeClassifier(criterion='entropy')  # 'gini' is the default
decision_tree = decision_tree.fit(X, y)

# print(decision_tree.feature_importances_)

print(export_text(decision_tree, feature_names=features))

fig = plt.figure(figsize=(5, 5), dpi=300)
tree.plot_tree(decision_tree, filled=True, feature_names=features)
plt.savefig('decision_tree.png')

print(decision_tree.feature_importances_)
```

The above code produces the following text representation of the generated decision tree on the left and the equivalent visualization on the right (Fig. 8.5).

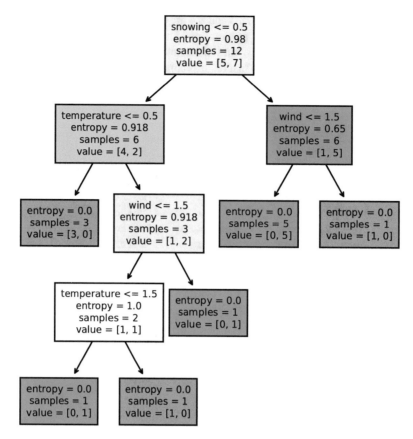

Fig. 8.5 Visual depiction of the decision tree from the plot_tree function. Compare to the text on the left. (Image created from decision_tree.py)

There are various competing algorithms for creating decision trees. The simplest algorithm usually presented first is the ID3 (Iterative Dichotomiser 3) algorithm. In practice, ID3 is **not** typically the most optimal (in terms of memory, accuracy, and execution speed), but it is easily conceptualized for those learning the fundamental behavior of decision trees even if most modern machine learning algorithms do not utilize it.

The C4.5 and C5.0 algorithms both possess improvements over the ID3 algorithm. The above decision tree from the *sklearn* package uses an optimized version of the Classification and regression trees (CART) algorithm, which is considered an improvement over the C4.5 algorithm.

Understanding the decision nodes in the above text representation and visualization requires knowledge of the label encodings itemized here:

- Snowing: no (0), yes (1)
- Temperature: above_freezing (0), below_freezing (1), freezing (2)
- Wind: high (0), low (1), medium (2)
- Time: afternoon (0), morning (1)
- Predictor(sledding) (this is referred to as 'class' in the textual output and delineated with specific "leaf" node colors in the visualization): no (0), yes (1)

Also note in the visualization displayed in Fig. 8.5 a "true" response at an individual node in the decision tree will lead to the left child node and a "false" response branches to the right child node. The branch navigated depends on the resolution of the conditional expression given in the node. The textual representation mirrors an indented *if-else* ladder construct with conditionals clearly stated at each level.

Let's go through two specific pathways to better understand the output. Starting at the top (root) node, we ask "Is it snowing?" If the response is Yes, then based on our encoding, the "snowing" value is equal to 1. Since 1 is greater than 0.5 and the conditional in the root node is not satisfied, we follow the right branch. Now we ask "What is the wind?" If the wind is high or low (0 or 1, which are both less than or equal to 1.5) then the conditional in the node is satisfied and we follow the left branch to a terminal "leaf" node indicating the child sleds (class = 1, dark blue node). However, if the wind is medium

(2 is greater than 1.5) we follow the right branch to a terminal "leaf" node indicating the child does not sled (class = 0, brown node). This decision sequence corresponds precisely with the data; the only time when it is snowing that the child does not sled is when the wind is medium (see the record with index 4 in the original dataset).

What about the cases when it is not snowing? If it is not snowing (0 is less than or equal to 0.5) we follow the left branch from the root node and then check the temperature. If the temperature is above freezing (0 is less than or equal to 0.5) we follow the left branch to a terminal "leaf" node indicating the child does not sled (class = 0, brown node).

Now that we have seen results generated from the *sklearn* decision tree using the CART algorithm, we will create a competing decision tree using the ID3 algorithm. As mentioned above, the ID3 algorithm is primarily used for its instructional value and explaining the processing steps associated with ID3 is often included as a job interview question. ID3 demonstrates the core auto-generation mechanisms for decision trees.

A decision tree built by machine learning relies on the concept of information gain to automate the construction of a decision tree. Information gain quantifies the degree of usefulness associated with a given feature relative to the other features used in the decision making process. For example, in the above example, how important is the snowing feature when compared to the time feature? Before we begin with the mathematics associated with information gain, we can determine directly what the *sklearn* decision tree indicates are the most important features. When creating decision trees, *sklearn* utilizes a function called the *criterion* to measure the quality of a split at a particular node. This criterion function can be specified as either 'gini' or 'entropy'. Using the following code, *sklearn* determines the importance of each feature by returning the normalized total reduction of this criterion per feature, typically known as Gini importance:

```
Partial code: full program in decision_tree.py:

print(decision_tree.feature_importances_)
```

Which produces the following output:

```
[0.19973045 0.40438132 0.39588822 0.        ]
```

These values may vary for a given execution of the above code based on the structure of the decision tree solution. The order of the numbers correspond to the order of the input features. Consequently, snowing has a relative importance of about 0.2, with the remaining measures for each feature as follows: temperature 0.4, wind 0.4, and time 0.0. In other words, using the CART algorithm, the *sklearn* decision tree ranks the features from the most important to least important as temperature (by a slim margin), wind, snowing, and time.

To answer the question of which features provide the most informational value, decision trees typically use either Gini impurity or entropy. Gini impurity measures the variety of a particular feature. For example, if all the recorded values of a feature such as sledding time were identical (e.g. only afternoon), then the corresponding Gini index for that feature would show its minimum value of 0.0 indicating no impurity. Conversely, if half of the listed sledding times were morning and the other half were afternoon then the Gini index would be its maximum value of 0.5.

Alternatively, entropy is popularly understood as a measure of disorder or chaos. More random numerical values reflect a higher level of chaos and disorder as indicated by entropy's maximum value of 1.0. For a feature with uniform values throughout, there would be no disorder or chaos resulting in the minimum entropy value of 0.0. Entropy is defined mathematically for each feature by the following formula where S is the set of feature data and p_i is the frequency (or probability) of the element in the set:

$$Entropy = H(S) = -\sum p_i \log_2 p_i$$

To determine the quality of a split at a decision tree node, entropy is traditionally used with the ID3 and C4.5 algorithms while Gini impurity is normally used with the CART algorithm.

The above entropy equation may be too abstract for some readers, so here is an illustrative example. To determine the impact on entropy associated with the snowing feature, we consider the root node of our decision tree which lists 12 total inputs, 5 negative and 7 positive. Note that branching at the root node depends solely on the snowing value (no-0, yes-1). The entropy value at the root node is calculated based on the total predictor(sledding) outcomes of the original dataset: 5 "no" and 7 "yes" often depicted as [5-, 7+] or [5,7] as shown in the visualization of the decision tree in Fig. 8.5. The "samples = 12" listed in the root node in Fig. 8.5 are the total inputs being considered at that level.

The following specific formulas might provide greater insight to the above equation:

$$Entropy(snowing) = Entropy([5-,7+])$$

Since snowing is the deciding feature at the top node, we refer to the initial entropy of the decision tree system using the snowing parameter. In the following formula the $\ominus$ symbol represents the negative probability and the $\oplus$ symbol represents the positive probability:

$$Entropy([5-,7+]) = -(p_\ominus \log_2 p_\ominus + p_\oplus \log_2 p_\oplus)$$

Substitution produces the following:

$$Entropy([5-,7+]) = -\left(\frac{5}{12}\log_2\frac{5}{12} + \frac{7}{12}\log_2\frac{7}{12}\right) = 0.98$$

The same calculation can be applied to determine the resulting entropy for each decision branch. For example, Entropy(snowing$_{yes}$) = Entropy([1-,5+]) = 0.65 and Entropy(snowing$_{no}$) = Entropy([4-, 2+]) = 0.92. We now look at entropy calculations for the remaining features and their associated values. For each case, we assume the feature is used to make the initial branching choice from the root node. For example, instead of using snowing as our initial decision feature, we use temperature.

Table 8.1 displays the results for the entropy calculations for all four features and their respective values (attributes). These calculations can be found in the `entropy_and_information_gain.py` file.

Table 8.1 Entropy results for the four features and their respective attributes

	Snowing	Yes	No
Count	[5-, 7+]	[1-, 5+]	[4-, 2+]
Entropy	0.98	0.65	0.92

	Temperature	Above_freezing	Below_freezing	Freezing
Count	[5-, 7+]	[3-, 1+]	[0-, 4+]	[2-, 2+]
Entropy	0.98	0.81	0.00	1.00

	Wind	Low	Medium	High
Count	[5-, 7+]	[1-, 4+]	[3-, 1+]	[1-, 2+]
Entropy	0.98	0.72	0.81	0.92

	Time	Morning	Afternoon
Count	[5-, 7+]	[3-, 2+]	[2-, 5+]
Entropy	0.98	0.97	0.86

These calculations can be found in the entropy_and_information_gain.py file

The summary information gain formula for each feature is based on separating out the different attributes of a feature and measuring the entropy of each attribute. In the following formula, S is one of the four features: snowing, temperature, wind, or time. A is the set of attributes for each feature. S_v is the number of outcomes attached to attribute v. For example, for the wind feature, the three attributes are high, medium, and low.

$$Gain(S,A) = Entropy(S) - \sum_{v \in A} \frac{S_v}{S} Entropy(S_v)$$

From this formula we can now derive the information gain for the snowing feature. The snowing feature has two attributes: yes and no. From above, we note that there are 5 negatives and 7 positives at the root decision node: [5-, 7+]. What about snowing$_{yes}$? In other words, how many times does following the yes attribute for snowing correspond to a negative outcome (child will not sled) and how many times does it correspond to a positive result (child will sled) for the predictor? As demonstrated above, snowing$_{yes}$ = [1-, 5+] and snowing$_{no}$ = [4-, 2+]. Based on these data, we produce the following results by substituting into the information gain formula:

$$Gain\left(snowing,\{yes,no\}\right) =$$
$$Entropy\left(snowing\right) - \left(\frac{6}{12} Entropy\left(snowing_{yes}\right) + \frac{6}{12} Entropy\left(snowing_{no}\right)\right)$$

Using the values from Table 8.1, we obtain the following information gain for the *snowing* attribute:

$$Gain\left(snowing,\{yes,no\}\right) = 0.98 - \left(\frac{6}{12} 0.65 + \frac{6}{12} 0.92\right) = 0.196$$

Following similar steps we compute the information gain for the remaining three features to derive the results shown in Table 8.2. The calculations can be found in the `entropy_and_information_gain.py` file.

Table 8.2 Information gain results for the four features. The calculations can be found in the entropy_and_information_gain.py file

	Information gain
Snowing	0.196
Temperature	0.376
Wind	0.179
Time	0.072

The relative importance of each feature is directly related to the resulting values of information gain. Following the ID3 algorithm the information gain for each feature is shown in Table 8.2 where we discover that for splitting the root node the list of important features in order of importance are temperature, snowing, wind, and finally time.

Recall the CART algorithm found the following order of overall importance: temperature, wind, snowing, and time, which is close but not an exact match between the two algorithms. An important observation is that for combinatorial and efficiency reasons the CART algorithm employed by *sklearn* is constrained to generate only binary decision trees which ultimately influences the information gain calculations for the individual features. Consequently, although the information gain for temperature emerges as the highest value in our Table 8.2 calculations, that result was based on allowing a three-way split for that feature, which would not be permitted in the design of a binary decision tree.

With this caveat in mind, given the information gain calculations in Table 8.2 and following an algorithm that permits multi-way (more than two) splits for the nodes in our decision tree we will set temperature as the root node as shown in Fig. 8.6.

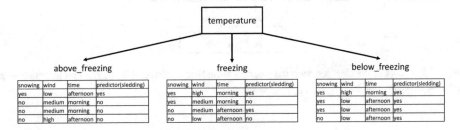

Fig. 8.6 Initial decision tree split with temperature as the root node. All entropy and information gain calculations are included in entropy_and_information_gain.py

The three attributes for temperature allow the dataset to be divided into three smaller parts. We repeat the process again for each of these three subtrees by examining the information gain for the remaining three features.

Specifically, for the above_freezing data – the leftmost table – we need to determine the information gain for snowing, wind, and time. We find that snowing and wind both have the same information gain and both are greater than the information gain value for time. With an ensuing tie, from these top two features we can arbitrarily select the one to be used to split the node, so we randomly choose snowing.

Examining the data in the leftmost table in Fig. 8.6 we find the values for snowing are identical to the predictor (sledding) values. In other words, if snowing is 'yes' then the child sleds, but if snowing is 'no' then the child does not sled. At this point the other features do not matter and may be ignored. The updated decision tree is shown in Fig. 8.7.

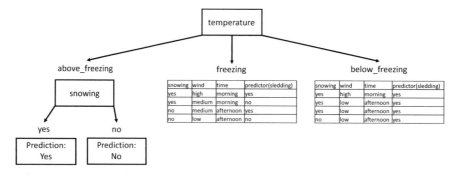

Fig. 8.7 The next step of decision tree construction with the above_freezing attribute considered. All entropy and information gain calculations are included in entropy_and_information_gain.py

For the freezing data – the center table – we find the wind feature has the highest information gain. For the wind feature the single 'high' value corresponds to a prediction of 'yes', and the single 'low' attribute corresponds to a prediction of 'no'.

Since the 'medium' attribute is associated with both a positive and negative outcome, we need to investigate the information gain for snowing and time to further discriminate our decision tree. Once again these two features have the same information gain and so we randomly choose the snowing feature. Figure 8.8 provides the current status of our decision tree.

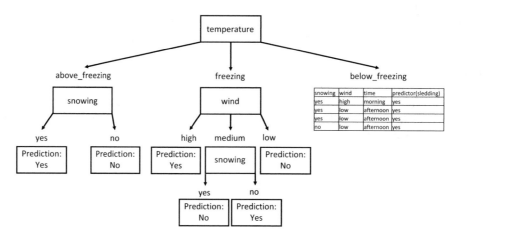

Fig. 8.8 The next step of decision tree construction with the freezing attribute considered. All entropy and information gain calculations are included in entropy_and_information_gain.py

The last attribute to consider is below_freezing – the rightmost table. No matter the conditions for the snowing, wind, and time features, the predictor(sledding) column in the rightmost table shows that if the temperature attribute is below_freezing then the child always sleds. Figure 8.9 shows the final decision tree that we created using entropy and the multi-way ID3 algorithm.

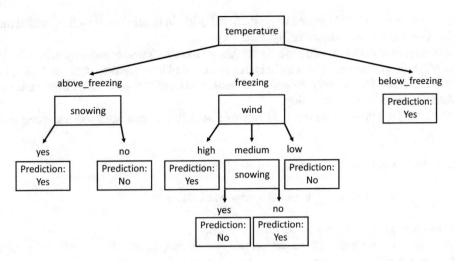

Fig. 8.9 The final step of decision tree construction with the below_freezing attribute considered. All entropy and information gain calculations are included in entropy_and_information_gain.py

Comparing Figs. 8.5 and 8.9 reveals similar complexities, but different pathways. In addition, we used entropy as the criterion when we could have opted for Gini impurity instead. Debating or enumerating the pros and cons of the different decision tree construction algorithms (e.g., ID3, C4.5, CART) and criteria (entropy vs Gini impurity) is beyond the scope of this book.

However, knowing that decision trees are transparent and explainable (i.e. you can trace the path of each decision no matter how complex) and are considered an overall effective machine learning tool is very important for data science. In the *sklearn* package there is both a decision tree classifier (DecisionTreeClassifier) and decision tree regressor (DecisionTreeRegressor).

All entropy and information gain calculations performed to create the decision trees in this section can be found in the `entropy_and_information_gain.py` file.

8.3 Linear Regression

Linear regression is one of the most fundamental yet powerful tools available to a data scientist. The main objective for linear regression analysis is to extract a pattern or meaning from chaos. Figure 8.10 shows random data on the left and a visually recognizable pattern on the right.

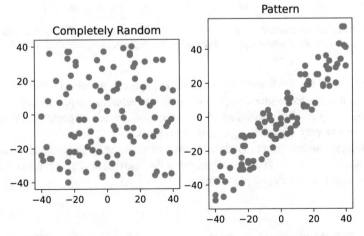

Fig. 8.10 Comparison of random data (on the left) compared to an obvious pattern (on the right). (Image created from linear_regression_random_vs_pattern.py)

The purpose of linear regression is to both find the pattern that exists in linear data as well as predict the pattern if it were to extend beyond the X-axis boundaries shown in Fig. 8.10.

One of the key assumptions about linear regression is the data possess the linearity property. *Linearity* denotes that a given set of data can be sufficiently characterized by a meaningful visual trend line. Random data, as depicted on the left side of Fig. 8.10, does not demonstrate the linearity property, but most real multivariate data, as illustrated on the right side of Fig. 8.10, reveal some degree of linear relationship.

The following code analyzes the data on the right side of Fig. 8.10 and generates the linear regression line shown in Fig. 8.11:

```
Partial code: full program in linear_regression.py:

classifier = LinearRegression()  # create the classifier

# LinearRegression is expecting X to be a matrix.
# Since it is not, we reshape it to match expected shape.
classifier.fit(np.array(X).reshape(-1, 1), y)  # Train the model

y_pred = classifier.predict(np.array(X).reshape(-1, 1))

mse = mean_squared_error(y, y_pred)
r2_score = r2_score(y, y_pred)
```

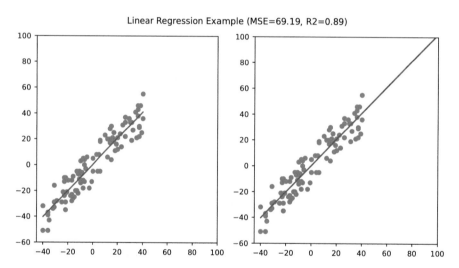

Fig. 8.11 The left figure shows the linear regression over the data on the right side of Fig. 8.10. The right side shows how the same function that creates the red line segment can be used to determine (or predict) any other future point. (Image created from linear_regression.py)

The left side of Fig. 8.11 shows the red linear regression superimposed on the original data. The right side of Fig. 8.11 depicts an extrapolated version of the same linear regression up to the value of x equal to 100.

The MSE (Mean Squared Error) and R^2 (coefficient of determination) are also reported in the title for Fig. 8.11 – for more information on measuring the accuracy of linear regression see Sect. 6.2.

A linear regression algorithm is an algorithm that determines coefficient values of a linear equation that solves for the best fit. The resulting line generated by a linear regression algorithm is the line that best fits all the existing data points. A bivariate linear regression solution is written in the following form:

$$Y = a + bX$$

In the above formulation, Y is the dependent result, X is the independent variable plotted on the X-axis, a is the intercept (the value of Y when X = 0), and b is the slope of the fitted line.

The most common approach for solving the equation is the method of *least squares*. The first step is to find $\bar{y}$ and $\bar{x}$ (the sample averages of y and x respectively) using the following:

$$\bar{y} = \frac{\sum y}{N}, \quad \bar{x} = \frac{\sum x}{N}$$

The next step is to find the slope of the line:

$$b = \frac{\sum(x_i - \bar{x})(y_i - \bar{y})}{\sum(x_i - \bar{x})^2}$$

The last step is to determine the intercept:

$$a = \bar{y} - b\bar{x}$$

The following Python code demonstrates an example implementation of determining the line produced by a linear regression analysis using the least squares method and compares these results to the implementation provided by *sklearn*:

```
Partial code: Full program in file: linear_regression_calculations.py:

from sklearn.linear_model import LinearRegression

classifier = LinearRegression()  # create the classifier

# LinearRegression is expecting X to be a matrix.
# Since it is not, we reshape it to match expected shape.
classifier.fit(np.array(X).reshape(-1, 1), y)   # Train the model

print(f'sklearn results: Y = {classifier.intercept_:.2f} + {classifier.coef_[0]:.2f}X ')

y_hat = np.mean(y)
x_hat = np.mean(X)

b_numerator = 0
b_denominator = 0
for i in range(len(X)):
    b_numerator += (X[i] - x_hat) * (y[i] - y_hat)
    b_denominator += (X[i] - x_hat) ** 2

b = b_numerator / b_denominator
a = y_hat - b * x_hat
print(f'Example results: Y = {a:.2f} + {b:.2f}X ')
```

Which produces the following output:

```
sklearn results: Y = 0.39 + 1.02X
Example results: Y = 0.39 + 1.02X
```

Although our above example only shows the result of using a single feature array (X) as input, a linear regression can be applied using any number of dimensions (features), a technique called multiple regression. A multiple regression plane is an extension of the linear expression above and is written in the following form:

$$Y = a + b_1 X_1 + b_2 X_2 + \cdots + b_n X_n$$

Sometimes a single linear regression model does not work for a given dataset. For example, when modeling the wealth of individuals, the wealth model for multi-billionaires is distinctly different from those in abject poverty. For these cases, *piecewise* linear regression is recommended.

Piecewise linear regression is simply linear regression but with the data broken into more logical units that exhibit local consistency. For instance, the wealth of individuals in a particular country might be separated into different models for the abject poor, poor, lower class, and middle class. However, employing too many models risks overfitting the data, whereas applying too few models risks underfitting the data.

Frequently for many machine learning algorithms, you need to transform the data for the algorithm to work more effectively. Linear regression models are no exception and often work best when applied to data that are normally distributed. The residuals in linear regression models are also assumed to be independent and normally distributed. More information about distributions can be found in Chap. 4.

The top-left plot in Fig. 8.12 shows how a linear regression is a noticeably poor fit for an exponential distribution of the data. The other plots demonstrate how various transformations applied to the data can substantially improve the linear regression fit. MSE (Mean Squared Error) values are provided in the plot titles – the smaller the MSE the better the fit.

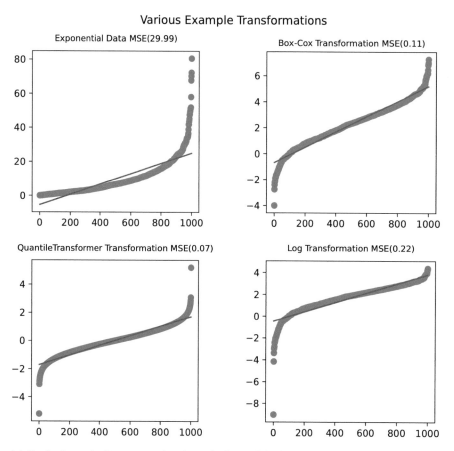

Fig. 8.12 An exponential distribution and a linear regression shown in the top-left plot. Various other transformations applied. The MSE (Mean Squared Error) is shown – the smaller the MSE the better the linear regression fit. (Image created from exponential_to_normal_distribution.py)

8.4 Logistic Regression

Logistic regression or binary logistic regression predicts what statisticians call dichotomous outcomes. *Dichotomous* means two distinct values, such as yes/no or on/ off. In other words, logistic regression excels at predicting binary results or answers limited to one of two alternatives. Also, although it is called a regression, this technique actually performs a classification by leveraging what's known as a logit regression in its underlying algorithm.

However, you are not limited to only dichotomous outcomes if you choose multinomial logistic regression instead. Fortunately, modern machine learning packages like *sklearn* will automatically select binary logistic regression vs multinomial logistic regression based on the number of predictor targets in the problem. For example, if you used the Iris dataset that has three classifications then multinomial logistic regression would automatically be utilized as the following code demonstrates:

```
Partial code: Full program in: logistic_regression.py

scaler = StandardScaler()
X = scaler.fit_transform(X)

classifier = LogisticRegression()

X_train, X_test, y_train, y_test = train_test_split(X, y, test_size=0.2)

classifier.fit(X_train, y_train)
y_pred = classifier.predict(X_test)

print(confusion_matrix(y_test, y_pred))
print(f'Accuracy = {str(accuracy_score(y_test, y_pred))}')
```

Which produces the following output:

```
[[ 8  0  0]
 [ 0 14  0]
 [ 0  0  8]]
Accuracy = 1.0
```

One of the key assumptions about logistic regression is that, similar to linear regression, the data conforms to some type of linear behavior. Before using logistic regression you will want to visualize the data and transform it if necessary, as is demonstrated in Fig. 8.12.

Also, note in the above example code all the input data is scaled. This transformation is important to ensure that each feature is granted equal weighting. The StandardScaler used in the above program normalizes all the dimensions using the principles of the z-score (see Sect. 4.9 for more information).

Figure 8.13 provides a visual representation of how logistic regression divides the feature space of the Iris dataset into three distinct areas based on the input feature set. The sample points are distinguished using different colored shapes based on the three different flower types. The diagram on the left plots the training data, and the diagram on the right shows the test data. As revealed in the test data visualization, the logistic regression analysis was able to completely separate one of the Iris flowers (purple) from the other two, however, a few flowers represented by blue and yellow were misclassified. Overall, this logistic regression model achieved a 92.11% accuracy (for one run of the program – accuracy varies based on the randomness of the `train_test_split` function).

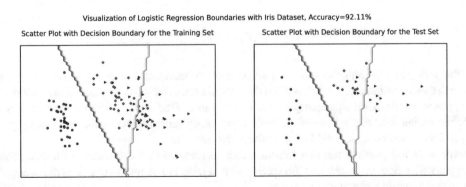

Fig. 8.13 Visualization of the dividing boundary of the Iris dataset. (Image created from logistic_regression_visualization.py)

Logistic regression is founded on two main concepts: probabilities and the sigmoid function.

The sigmoid function is a function integral to many applications from microchips to neural networks. For any input value, the sigmoid function limits the output to a value between 0 and 1. By definition this function is asymptotic, which means it never produces an output exactly equal to 0.0 or 1.0. The following is the traditional formula for the sigmoid function and is visualized in Fig. 8.14:

$$S(x) = \frac{1}{1 + e^{-x}}$$

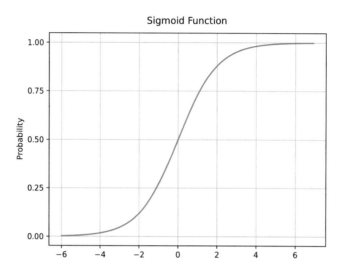

Fig. 8.14 Visual depiction of the sigmoid function. (Image created from sigmoid_visualization.py)

With the sigmoid function the narrow transition range ensures your results are effectively biased or "pushed" either toward 1.0 or 0.0. Logistic regression models leverage the sigmoid function where the exponent x is a linear expression. For example, the equation for a binary logistic regression (binary dependent variable) with two predictors (x_1 and x_2) is the following:

$$Binary\ Logistic\ (x_1, x_2) = \frac{1}{1 + e^{-(a + b_1 x_1 + b_2 x_2)}}$$

The above equation is very similar to linear regression but applied to the sigmoid function instead. A more generalized equation for multinomial logistic regression (dependent variable has more than two levels) with n predictors and accompanying weights is very similar:

$$Multinomial\ Logistic\ (x_1, x_2, \cdots, x_n) = \frac{1}{1 + e^{-(a + b_1 x_1 + b_2 x_2 \cdots + b_n x_n)}}$$

The core idea underlying logistic regression is the prediction of probabilities. For example, imagine the task of assigning a dish to either French cuisine or Indian cuisine based on the Scoville scale, which determines pungency of chili peppers, and the pyruvate scale, which determines the pungency of onions and garlic. Traditionally, French food has low pungency scores (not very spicy) while Indian food has a some of the highest pungency scores in the world (extremely spicy). The Scoville scale would be x_1 and the pyruvate scale would be x_2 in the above equation.

The *Scoville scale* is an independent test that returns a measure related to the spiciness of a given food. For example, compare a bell pepper with a Scoville score of 0 (no spiciness) to a jalapeño pepper with a Scoville score of about 3000 and a cayenne pepper with a Scoville score of about 30,000.

If we focus only on the Scoville score as our input going forward and train our logistic regression with example French and Indian dishes and their associated Scoville scores then we may determine the probability that a new dish is either French or Indian based on the Scoville score. For instance, for a dish with a Scoville score of 5000, we might find a very low probability that this dish belongs in the French category. Conversely, the probability that we can assign the dish to the Indian cuisine would be very high.

For our example, logistic regression would utilize the sigmoid function in conjunction with probability to determine if a dish should be classified as French or Indian. The probability threshold for selecting a particular target result is typically set at 0.5. For instance, if the logistic regression returns a 0.6 probability that the dish is Indian then the logistic regression algorithm returns Indian as its answer, otherwise we predict French.

How do we solve for the intercept (a) and the coefficients (b's) in the above equations? The technique used to resolve logistic regression employs *parameter estimation* or guessing.

This approach first determines the "loss function." The loss function is a measure of the quality of our current guess for the intercept and coefficients. For instance, if you were solving the intercept and coefficients by hand and decided to increase the value of the intercept, how would you know if the resulting sigmoid function used for prediction is now better or worse?

Similar to error functions described in Sect. 6.2, a loss function is an indicator of how well the current intercept and coefficients perform with regard to predicting the targets provided by the training set. In *sklearn* you can designate one of several types of loss functions using the penalty attribute. There are two main loss functions used: L1 and L2 (the default for *sklearn*).

The *L1 loss function*, or *LAD (Least Absolute Deviations)* obtains a sum of the absolute differences between every true point compared to every predicted point. The equation for the L1 loss function is shown here:

$$L1 = \sum \left| y_{true} - y_{predicted} \right|$$

Alternatively, the *L2 loss function*, or *LS (Least Squared Errors)* calculates the sum of the squared differences between the true and predicted values.

$$L2 = \sum \left(y_{true} - y_{predicted} \right)^2$$

The L2 function is default because it is generally preferred and typically results in better outcomes. However, the L1 function performs better with outliers that should be ignored, as L1 does not magnify outlier differences by squaring them. In effect, the L2 function squares any differences of outliers and makes them more important whereas the influence of outliers on the L1 function is less significant.

Once a loss function has been identified then an iterative process is applied until the best results are discovered. However, this rough description naïvely suggests that all combinations of intercepts and coefficients are attempted, which is an intractable, time intensive strategy. In other words, a significant amount of execution time is wasted using a brute force approach that attempts infinitely available combinations of potential intercept and coefficient values.

In the machine learning domain, each iteration cycle is often labeled an *epoch*.

Parameter estimation is frequently utilized with optimization techniques like gradient descent. *Gradient descent* works by randomly setting initial values to the variables (the intercept and coefficients in our case) then systemically and incrementally modifying these values by calculating the slope and moving the estimated values in a direction that improves the optimization. In other words, if we modify a parameter in a certain direction, say increasing the intercept by 10, does the loss function generate an answer closer to zero? If so, keep going, otherwise, change direction.

There are three main types of gradient descent: batch gradient descent, mini batch gradient descent, and stochastic gradient descent (SGD). *Batch gradient descent* considers the entire dataset for every modification. Consequently, batch gradient descent is the slowest, but most precise way of minimizing loss. *Mini batch gradient descent* is similar to batch gradient descent, but focuses on a specific subset of the dataset, an approach that is faster, but not as precise. *Stochastic gradient descent (SGD)* is the fastest algorithm and uses random samples each iteration to asymptotically approach the optimized loss.

Regardless of the type of gradient descent used, there is always a risk of reaching a *local minimum* rather than a *global minimum* which will significantly skew the results. Consider an expansive range of mountains and valleys. If you were trying to find the lowest of all the valleys, the global minimum, how would you do it? Since logistic regression starts with random values, the analogy is that different locations in the mountain range would be initially selected. If you start the optimization problem in a deep valley, how do you know that your current valley location is the lowest of all the valleys, the global minimum, or just the lowest in the area where you started, the local minimum?

To summarize the process of logistic regression, first select a loss function (e.g., L2), then select an optimizing function (e.g., SGD), at which point the computer iterates until either a local minimum or (hopefully) a global minimum is determined. Once the parameters of interest, the intercept and coefficients, have been estimated then the model is considered trained.

Although there are many steps involved, logistic regression is highly efficient in terms of memory usage and execution time. Also, logistic regression is not prone to overfitting, like decision tree algorithms, although overfitting can easily become a problem with logistic regression for high-dimensional data (the curse of dimensionality!).

Recall that the input feature data should be standardized before the logistic regression model is created. Also, logistic regression is not easily conceptualized in terms of explainable AI (XAI) because the intercept and coefficient values have no interpretable real world meaning for investigators in the context of the broader classification problem.

8.5 SVM (Support Vector Machine)

SVM (Support Vector Machine) is a supervised learning model primarily used for binary classification, but it can also be used for regression as well. SVM is similar to logistic regression in the sense that it is also trying to find the best boundary between datasets. SVM specifically works by finding the maximum distance, called the *margin*, between datasets.

The main idea of SVM is that this margin between datasets in feature space is characterized by a *hyperplane*. A hyperplane is simply a line in 2D space, a plane in 3D space, and more generally a N-1 dimensional surface in N dimensional feature space. The SVM algorithm leverages all the features in the dataset: if there are only two features then SVM works in 2D space.

A Support Vector Machine is based on *support vectors*. Support vectors are the data points that are the closest to the hyperplane and most heavily influence the orientation and position of the hyperplane. In other words, support vectors are the samples that define the hyperplane.

Figure 8.15 depicts linearly separable data with the support vectors identified by green arrows. The margin is defined by a double-pointed arrow and the optimal hyperplane is the single solid line.

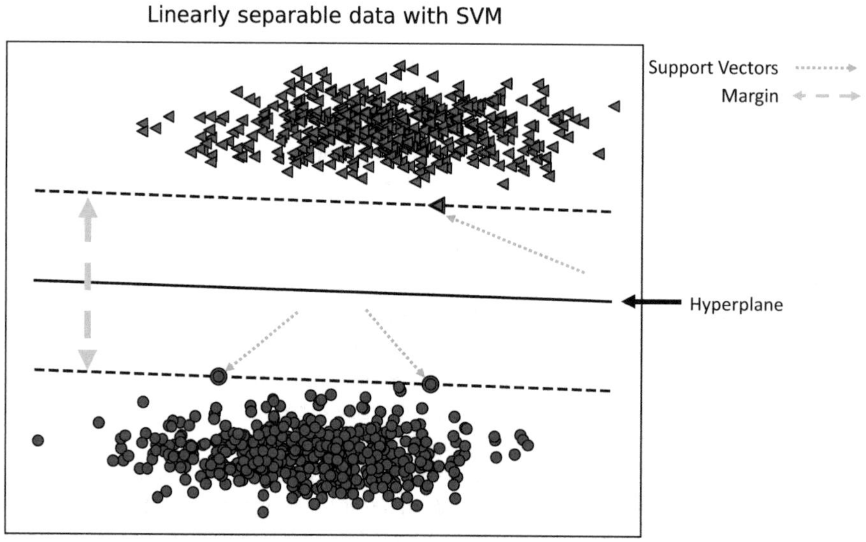

Fig. 8.15 Visualization of a hyperplane from a SVM. The main visualization was created in svm_basic_figure.py and then subsequently annotated

SVM's are especially noteworthy for their ability to perform the *kernel trick*. The gist of the kernel trick is to map lower-dimensional data into a higher dimensional space where points become more easily separable. The trick part of the kernel trick is to select a special kernel function or mapping that captures a key relationship between the datapoints as if they were in higher dimensions without actually transforming these points into a higher dimension.

In brief, the kernel trick reduces the amount of computation required to actually transform the data to a higher dimension. For data that is not linearly separable, as shown in Fig. 8.16, this higher dimensional mapping is necessary to establish a reliable decision boundary and subsequently achieve higher degrees of prediction accuracy.

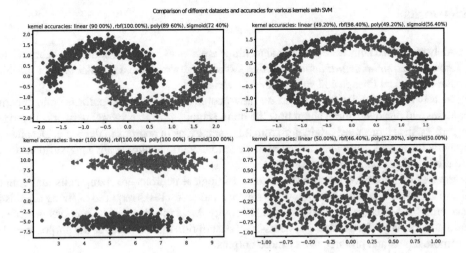

Fig. 8.16 Comparison of various datasets and accuracies with different kernels with SVM. The code for this figure is in svm_visualization.py

Figure 8.16 shows four different datasets. The title for each graph compares the prediction accuracy based on the kernel used with SVM. The left-top dataset is a traditional half-moon example; the right-top is a traditional concentric circles dataset; the left-bottom displays distinct linear separability; and the bottom-right is a collection of random data. For each of the four graphs, the accuracy of the respective linear, rbf, poly, and sigmoid kernels are reported.

The RBF kernel is the default kernel for the *sklearn* SVM implementation and represents the Radial Basis Function. The Radial Basis Function is an exponential function whose value depends on the distance between two feature vector in the input space and ranges between one (equal feature vectors) and zero (infinite distance between feature vectors). The following generic code shows how to employ SVM with *sklearn*:

```
Partial code: full program in svm_visualization.py:

from sklearn import svm
from sklearn.preprocessing import StandardScaler

#read in data... put the data into X and y respectively...
ss = StandardScaler()
X = ss.fit_transform(X)

classifier = svm.SVC()  # rbf is the default
X_train, X_test, y_train, y_test = train_test_split(X, y)
classifier.fit(X_train, y_train)
y_preds = classifier.predict(X_test)
```

As the name suggests, SVM fundamentally leverages vectors and is highly influenced by linear algebra. SVM employs dot products and matrices to compute results. Each choice of kernel (e.g., linear, rbf, poly, and sigmoid) specifies different approaches to calculating the margin and hyperplane.

Aside from selecting the kernel, SVM implementations can also be tuned by choosing different values of *C* and *Gamma*. The C parameter, or regulation, is the penalty parameter. A smaller C creates a smaller margin and a larger C creates a larger margin. In practical terms, the value of C defines the influence of outliers in the calculations for the SVM. The larger the value of C, the more the SVM will ignore outliers.

Gamma is used only with the RBF kernel. Gamma dictates the degree of curvature we allow in a decision boundary. Figure 8.15 provides a simple example with a straight line hyperplane. However, when dealing with higher-dimensional data the boundary of the hyperplane will include curvature. The higher the gamma, the higher the curvature.

For more discussion on setting the optimized values of C and Gamma, see Sect. 8.8.

The SVM technique performs well when classifying high-dimensional data into one of two categories and often achieves high accuracy. However, this approach may not be appropriate with large datasets due to arbitrarily long training time, depending on parameter tuning and kernel selection.

8.6 Neural Networks

Among the most widely recognized machine learning algorithms are *neural networks*, sometimes abbreviated as *neural nets*, or more formally known as *artificial neural networks (ANN)*. Neural networks are a popular type of machine learning algorithm with an extensive historical lineage and vast number of variations.

Originally, neural networks were designed based on how people initially conceptualized brain function. In reality, as neuroscience progresses and discovers more about how the brain actually operates, neural networks begin to appear less like human brains. Regardless, the term *neural* originated from this analogy, as in neurons in the brain. Despite any dissimilarities between neural network machine learning and actual brain function, neural networks are a powerful set of machine learning algorithms and are the launching point for deep learning.

The brain is composed of a massive number of neurons. Biological neurons are living cells that connect to each other through electrical and chemical interfaces to manage the sensory and motor tasks required of living organisms. The phenomenon of intelligence emerges fundamentally from neuronal activity. Most machine learning books provide a brief biological lesson on how neurons work. We offer a concise systems-level description that suffices for the purposes of this book: neurons may receive one or more inputs and produce one or more outputs.

The behavior of an artificial neuron is similar to that of a biological neuron, except the activity of the artificial neuron is defined and regulated by a function called an *activation function*. There are many different activation functions that dictate how artificial neurons respond. The mechanisms by which biological neurons work to generate output in response to inputs involve an interesting and complex series of biochemical events.

There are many common and effective artificial neuron activation functions to choose from. For example, the sigmoid function, which we explained in detail in Sect. 8.4, is a popular activation function. Activation functions provide the core "intelligence" for the individual artificial neurons. Other common artificial intelligence functions include, but are not limited to, *ReLU (Rectified Linear Unit) function* and *hyperbolic tangent function*.

For the *sklearn* MLPClassifier class (the neural network classifier class), the default activation function is ReLU. When instantiating the classifier, the activation function can be specified using the `activation` keyword argument. For instance, to explicitly declare the ReLU activation function define the keyword `activation='relu'`. The sigmoid activation function is specified by the keyword value `logistic`, and the hyperbolic tangent activation function is selected using the keyword value `tanh`. The following is an example of applying the neural networks classifier MLPClassifier to the Iris dataset:

```
Partial code: full program in nueral_networks.py:

from sklearn.neural_network import MLPClassifier
from sklearn.preprocessing import StandardScaler
...

scaler = StandardScaler()
X = scaler.fit_transform(X)

classifier = MLPClassifier()

X_train, X_test, y_train, y_test = train_test_split(X, y, test_size=0.2)

classifier.fit(X_train, y_train)
y_pred = classifier.predict(X_test)

print(confusion_matrix(y_test, y_pred))
print(f'Accuracy = {str(accuracy_score(y_test, y_pred))}')
```

Which produces the following output:

```
[[ 9  0  0]
 [ 0  7  0]
 [ 0  0 14]]
Accuracy = 1.0
```

ReLU, the default activation function in *sklearn* and considered the most common in general practice, has a very straight-forward design in that any positive input passes through to the output or, for the case in which the input is negative, zero is returned. ReLU behavior is formalized by the following mathematical formula:

$$f(x) = \max\{0, x\}$$

The hyperbolic tangent activation function is often used for natural language processing (see Chap. 9) and speech recognition and is very similar to the sigmoid function. Whereas the range of the sigmoid function is (0, 1), the range of the hyperbolic tangent function is modestly extended to (−1, 1).

$$f(x) = 2 * sigmoid(2x) - 1$$

There are also many other activation functions in the machine learning research literature. For example, the Leaky ReLU is the same as the original ReLU, but instead of returning 0 for all negative numbers it returns a small negative number like −0.001 that decreases very slowly as the negative input decreases. This specialized behavior helps combat the "dying ReLU" problem in which neurons with 0 output are unlikely to recover and play no significant role in discriminating and processing the input.

At this juncture in our discussion, an important observation is that an isolated artificial neuron is not very useful. The act of combining individual neurons together as a network with *weights* assigned to the interconnections is instrumental in achieving the prediction power expected from these algorithms.

The fundamental operational structure for neural networks is called a *perceptron*. A single-layer perceptron takes one or more inputs, modulates these inputs with assigned weights, sums the transformed inputs together, adds a bias, transfers the resulting sum as input to an activation function, and then returns the output of the activation function. This process is shown schematically in Fig. 8.17.

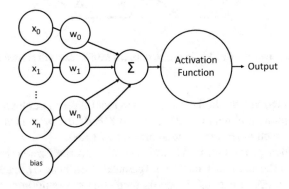

Fig. 8.17 Visual representation of a single-layer perceptron with weighted inputs, and a bias. (Image created by Robert Ball)

The *bias, or bias neuron* as it is sometimes called, is important for various reasons. The bias represents a constant that effectively shifts the activation function and allows for more flexibility in the tuning of the neuron specifically and the performance of the network more broadly.

For a single input perceptron without a bias, the behavior of the neuron is constrained by the shape of the activation function, most evident when there is zero input. A negative bias can help delay the triggering of the activation function until there is a sufficient level of excitation as represented by the single input value and assigned weight. As the model becomes multi-dimensional with more than one input then the bias allows the model to achieve more flexibility in the decision space.

Overall, the single-layer perceptron is a very powerful computational unit but is only useful for linearly separable problems with a binary target. The classic problem that a single-layer perceptron **cannot** solve is the XOR problem. Figure 8.18 provides a visualization of the XOR problem with a schematized solution by a single-layer perceptron depicted on the left and a successful multilayer perceptron solution on the right.

Fig. 8.18 Classic XOR problem. A single-layer perceptron cannot solve the XOR problem because the problem is not linearly separable (on the left), but a multilayer perceptron can solve it (on the right). (Image created by Robert Ball)

In contrast to a single-layer perceptron, a *multilayer perceptron (MLP)* contains one or more hidden layers apart from the input and output layer. The first hidden layer nodes receive the dot product of the input layer values and the weights associated with each node plus any bias. Activation functions are employed at each node in the layer generating results which are subsequently combined using the same dot product plus bias operation as they are being passed forward to the next layer.

Figure 8.19 is an abstract representation of a MLP. Note the number of neurons can vary at each of the different layers and, as stated above, the total number of hidden layers can be greater than one. Although not shown in Fig. 8.19, the presence of a bias neuron is implied for every layer except the output layer.

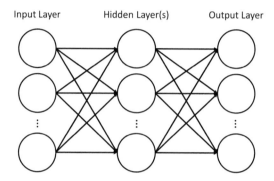

Fig. 8.19 Abstract model of a multilayer perceptron (MLP). As depicted here, a MLP consists of at least three layers: an input layer, at least one hidden layer, and an output layer

Since all directed edges are oriented in a forward direction from input layer to output layer and there are no cycles in the above physical network, this configuration of neural nodes is called a *feed forward neural network*. There are other types of neural networks that exhibit cycles, such as recurrent neural networks (RNN).

Backpropagation is a computational process utilized to train neural networks. Similar to the logistic regression procedure described in Sect. 8.4 that guesses the optimal parameters (parameter estimation), backpropagation is a technique that dynamically adjusts the weights in a neural network based on prediction errors through a series of iterations or epochs to ultimately increase the accuracy of the results.

Backpropagation simply means backward propagation of errors. The reason it uses the term "back" is because backpropagation initiates at the output layer and changes the weights of that layer before working backward to the next adjoining inner layer. This process continues until all weights for each node have been tuned based on the magnitude of the prediction errors. The same concepts related to gradient and loss functions from Sect. 8.4 apply here. Technically, a single weight associated with a node in a given layer is modified by using the chain rule from calculus to optimize that weight.

Backpropagation is considered fast and simple, generating solutions that optimize for the training data.

Single-layer perceptrons and multilayer perceptrons are often considered the core networks from which all other neural networks are constructed. There are a large variety of neural networks one can explore. Aside from single-layer perceptrons, MLPs, and RNNs, there are also Radial Basis Function (RBF) neural networks, Convolutional neural networks (CNN), Modular neural networks (MNN), Long/Short Term Memory (LSTM), Markov Chain (MC), Deconvoluational Networks (DN), and many more.

We recommend hat before learning about all the different types of neural networks you first take some time to understand how MLPs work, when to use them, and how to use them. There is little to be gained by digging deeper into the weeds of neural networks until you have a firm grasp of MLPs.

From a practical perspective, neural networks represent a black box machine learning algorithm, so explainable AI (XAI) for this particular approach is not possible. Also, neural networks require scaled data and the encoding of strings to numbers.

8.7 Ensemble Algorithms

Ensemble algorithms are algorithms that involve the collaborative effort of more than one machine learning model. For example, one of the most popular ensemble algorithms is random forest.

Random forest acquires its name by combining the predictions generated from two or more decision trees. The different trees each provide a prediction at which point the random forest ensemble algorithm either returns the majority vote (typically for classification) or returns the average vote (typically for regression).

The 'random' quality of random forest is embodied by the strategy that not all the decision trees in the random forest are trained the same, which is accomplished at the tree learning algorithm level by allowing only a subset of features to be considered when determining how to split a node in a given tree.

Random forest leverages an algorithm called *Bagging (**Bootstrap Aggregating**)* as a pre-processing step. *Bagging* is the process of providing machine learning models, decision trees in this case, different bootstrap samples of the training data. Bootstrapping is a selection method that uses random sampling with replacement. The replacement property allows multiple bootstrap samples to have the same observations. The aggregation step is combining the outcomes of each tree into a single prediction.

The bagging operation effectively provides the decision trees in the random forest different training subsets. The primary objective of this approach is to allow for a better balance of bias and variance for the resulting predictions as opposed to a single decision tree trained with the full, original training set and consequently prone to overfitting.

There are dozens of different variations of ensemble algorithms each with their own unique approach. For example, extra trees (Extremely Randomized Trees) ensemble is similar to random forest except that it uses the entire training set for each tree instead of a random subsample with replacement, and it configures the decision trees to select the split points for each feature at random.

AdaBoost (Adaptive Boosting) is also a common ensemble algorithm that adapts a series of weak classifiers sequentially into a strong one, successively improving model predictions. Technically, AdaBoost can be implemented with any classifier, but is often associated with decision trees.

Boosting methods attempt to achieve low bias predictions by iteratively combining weak learner classifiers to produce a strong learner that predicts outcomes with increasing accuracy. A weak learner by definition classifies observations slightly better than random guessing. For decision tree models weak learners are typically characterized by a single split, or decision stump.

The core concept in boosting involves training an initial decision tree on unweighted observations. This first decision tree, a weak learner, will provide predictions that will necessarily contain misclassifications. These errors will be weighted more heavily when used as training input to a second subsequent classifier designed to correct the previous errors. Misclassifications from the second decision tree will ensue, which will become emphasized in the training data that supplies the third decision tree. This process continues iteratively until the specified maximum number of weak learners is reached (n_estimator keyword parameter for *sklearn* AdaBoostClassifier.) A final prediction is generated from the weighted average outcomes from each weak learner.

The following code shows an example of a random forest and AdaBoost:

```
File: ensemble_examples.py
from sklearn.datasets import load_iris
from sklearn.model_selection import train_test_split
from sklearn.ensemble import RandomForestClassifier, AdaBoostClassifier
from sklearn.metrics import accuracy_score

iris = load_iris()
X = iris.data
y = iris.target

X_train, X_test, y_train, y_test = train_test_split(X, y, random_state=13)

random_forest = RandomForestClassifier(random_state=13)
random_forest.fit(X_train, y_train)
y_preds = random_forest.predict(X_test)
print(f'Random forest accuracy = {accuracy_score(y_test, y_preds)*100:.2f}%.')

AdaBoost = AdaBoostClassifier(random_state=13)
AdaBoost.fit(X_train, y_train)
y_preds = AdaBoost.predict(X_test)
print(f'AdaBoost accuracy = {accuracy_score(y_test, y_preds)*100:.2f}%.')
```

Which produces the following output:

```
Random forest accuracy = 94.74%.
AdaBoost accuracy = 92.11%.
```

8.8   Cross Validation, Hyperparameter Tuning, and Pipelining

Cross Validation is the practice of assessing the quality of your model by rotating through all the data for training and testing purposes and averaging the output from each of the train/test cycles. The objective of this analysis goes back to the balance of bias and variance – we want to determine the overall generalized performance of our trained model, ensuring that we are neither overfitting nor underfitting the data.

By splitting the data into train and test sets we are randomly assigning some of the data for training and some of it for testing. However, a potential problem exists in randomly determining these datasets, as different sets of randomly split data might elicit model performance that is either better or worse.

The key idea of cross validation is to test all the data, not just a fractional part. For example, we might measure exceptional performance with our particular choice of train and test data. However, using a different part of the data for testing might reveal much poorer performance. How do we know if the test data we use is a representative sample or not? Cross validation addresses that concern.

Strictly speaking, doing any testing at all is a form of cross validation. Employing the traditional approach of applying the `train_test_split` function where we randomly split the data into training and testing sets is called the *hold-out approach*. This conventional technique is convenient, fast, and easy to implement, but it has the problem described above that it arbitrarily selects a test set that might not be the best candidate when assessing the model. It is likely that this particular test data may not generalize to the broader observation set yet to be collected.

The *Leave One Out Cross Validation (LOOCV)* approach tests every single observation separately. With *n* observations, such as the 150 observations included with the Iris dataset, the model is fitted with training data of 149 observations and the test data comprises 1 single observation.

For the Iris dataset example, this train/test analysis would be repeated for a total of 150 times so that each observation has a chance to be the one observation in the test data. The accuracy of each of the 150 models is averaged together to determine an overall accuracy.

Compared to the traditional hold-out approach, the accuracy of the LOOCV technique has less bias and is a better test to determine if the model is production ready. However, it can take a long time to perform the LOOCV test. With 150 observations, the LOOCV approach is not computationally intensive, but with a modest dataset containing hundreds of thousands of observations it is impractical.

The *k-Fold* approach combines the strategies of both the hold-out approach and the LOOCV approach. Instead of simply testing one time with the hold-out approach, the k-fold approach divides the data into many folds or sections. The standard number of folds is k = 5, but any integer value larger than 1 can be specified.

Using the standard k = 5, the concept underlying k-fold is to divide the data into 5 equal sizes and then iterate so that each fold is assigned to be the test dataset, as in LOOCV. Instead of *n* folds of fit and predict iterations in LOOCV, k-fold has much fewer iterations, in this case, only five. The average accuracy results may not be as representative as LOOCV, but it is much more practical when applied to larger datasets. Figure 8.20 illustrates a visual example of the train/test division for k-fold with k = 5.

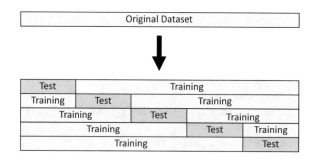

Fig. 8.20 Visual representation of splitting the original data for the k-fold approach. For the standard k = 5, the dataset is split into five equal parts and then each part has a separate turn as the test data. (Image created by Robert Ball)

Aside from the hold-out method, the k-folds approach is the most common technique for cross validation. However, for imbalanced datasets k-fold does not work as well as the *stratified k-fold* approach. Stratified k-fold is the same as k-fold, but it maintains the same balance for each fold. For example, if the overall dataset observations are 60% class A and 40% class B, then the stratified k-fold will maintain that same ratio (60/40) for each fold.

There are a number of other lesser known cross validation methods. For example, *repeated random train-test splits* is just as its name implies: repeatedly split the original dataset into training and test datasets an arbitrary number of times. The difference between the repeated random train-test splits approach and k-folds is when the number of random splits equals k, then k-folds is more systematic and tests every observation whereas repeated random train-test splits is likely to test several of the same observations and miss some altogether.

To implement the k-folds approach with cross validation separate the data into different folds, then iteratively train and test models in a loop *k* times. Alternatively, you can use a prebuilt function that performs these steps, such as `cross_val_score`. The following code provides an example:

```
Partial code: Full program in cross_validation.py:

tree = DecisionTreeClassifier()

X_train, X_test, y_train, y_test = train_test_split(X, y, random_state=rs)

tree.fit(X_train, y_train)
y_pred = tree.predict(X_test)

print(f'Traditional Hold Out Accuracy = {(accuracy_score(y_test, y_pred)*100):.2f}%.')

result = cross_val_score(tree, X, y)   # defaults to 5 folds

print(f'5-fold Cross Validation Average Accuracy: {(result.mean())*100:.2f}%.')
```

Which produces the following output:

```
Traditional Hold Out Accuracy = 97.37%.
5-fold Cross Validation Average Accuracy: 96.67%.
```

The above output indicates that the model performance is fairly good. With the traditional hold out approach the model achieves about 97% accuracy. After cross validating the model with the number of folds equal to 5, the resulting accuracy is approximately 96-97%.

If cross validation seeks to assess the quality of a model given different test sets then *hyperparameter tuning* seeks to determine the best parameters for a given model.

Consider that some machine learning algorithms come equipped with many different loss functions, optimizers, and other parameters that can have a significant influence on performance.

Recall in Fig. 8.16 the accuracy of four different kernel transformations for SVM was calculated and presented. The idea of hyperparameter tuning is to establish the best parameter choices through repeated training/testing and cross validation so that uncertainty can be minimized regarding which parameters might be the most effective for a machine learning algorithm.

There are various ways to perform hyperparameter tuning. We show an example below using the GridSearchCV class from *sklearn*. Briefly, the GridSearchCV class searches for the best set of hyperparameters from a grid (thus the name) of possible hyperparameter values. The CV in GridSearchCV means cross validation.

To use the GridSearchCV class and similar classes a list of hyperparameters to assess must be provided. As expected, the longer the list of hyperparameters, the more execution time required to find the best ones. The following code is an example that uses the half-moon data in the top left quadrant of Fig. 8.16 to find the best *kernel*, *gamma*, and *C* parameters:

```
Partial file: full program in hyperparameter_tuning_with_svm.py:

# arange provides 15 different values of C from 0.0 to 1.5 with 0.1 steps
possible_C_values = np.arange(0.1, 1.6, 0.1)
```

```
# hyperparameters:
parameters = {"kernel": ['linear', 'rbf', 'sigmoid', 'poly'],
              "gamma": ['scale', 'auto'],
              "C": possible_C_values}

clf = GridSearchCV(svm.SVC(), parameters, n_jobs=-1)   # -1 means to use all available CPUs

clf.fit(X=X, y=labels)
svm_model = clf.best_estimator_   # the best trained SVM model

# print out the best score and best parameters:
print(f'Average (mean) r2 scores from cross validation for the best parameters:
{clf.best_score_}.')
print(f'Best parameters: {clf.best_params_}')
```

Which produces the following output:

```
Average (mean) r2 scores from cross validation for the best parameters: 0.999.
Best parameters: {'C': 0.8, 'gamma': 'scale', 'kernel': 'rbf'}
```

The above example demonstrates that a very impressive average R^2 score of 0.999 can be achieved with a C of 0.8, the gamma set to `scale` and the kernel set to `rbf`. The resulting best trained model in now available through the `clf.best_estimator` attribute.

Overall, cross validation and hyperparameter tuning can validate your machine learning models and make them more accurate. However, since cross validation and hyperparameter tuning may require significant time, these techniques can wait until after the data has been sufficiently explored and understood.

8.9   Dimensionality Reduction and Feature Selection

Dimensionality reduction and feature selection are both techniques that address a prevailing problem in data science referred to as The Curse of Dimensionality. *The Curse of Dimensionality* occurs when there are so many features (or dimensions) that people and machine learning algorithms have a hard time understanding the main points or informational value of a dataset. What is the idea behind dimensions? What does that mean? A "dimension" is simply a column of data or a feature.

For example, consider metadata about people. If we refer to people only by their height then it is easy to compare them based on adjectives such as tall, average, or short. Alternatively, if we evaluate people by their weight we can utilize adjectives like fat, average, or skinny. More convoluted, but still manageable descriptors emerge when evaluating people using both height and weight with phrases like "short and fat", "short and skinny", and "tall and skinny". However, how do you describe people succinctly (in as few words as possible) when you consider height, weight, age, and intelligence? Finding a single word for older, intelligent, fat, short people is difficult.

If we have a hard time with only four features, what happens when we add many more such as income level, debt, education, race, and gender? The prospect of adequately describing and understanding a dataset of people with 10 or more features is so demanding for both people and machine learning that dimensionality reduction and feature selection techniques were created.

Feature selection is the easier of the two techniques to understand. Feature selection is literally selecting which features to retain and which features to remove. Although this process can be done by hand, in practice we rely on the automation provided by algorithms. Although there are many approaches, the general idea is to identify the feature that has the least amount of variance or information content and remove it from the list. The algorithm then repeats as necessary.

The "least amount of variance" pertains to features that exhibit the least amount of change. For example, if all the people in the dataset are the same height then the height feature does not help to distinguish between them. More specifically, given ten people who are all 183 cm (6 feet) but with various weights then describing any of the ten people as 183 cm does not help draw any significant distinction among them. So, if a dataset had ten people with height and weight and the heights of all the people were the same then the height feature could easily be dropped with little or no penalty to our subsequent analytical and prediction methods.

Dimensionality reduction, in contrast, is more powerful than feature selection, but is often very confusing to students that have never been exposed to the idea. The basic strategy in dimensionality reduction is not to drop any specific feature, but to reduce and combine features together into something called components.

For example, if we were to merge weight and height together into one component then instead of thinking of a person by their separate measures of weight and height we now have a combined weight-height feature to describe a person. Conceptually, this may not appear like a helpful process when attempting to describe an observation in the dataset, but computationally dimensional reduction has several advantages.

Effectively combining or merging features relies heavily on the variability in the original features. For example, if there is no variability in height because everyone is 183 cm then the two features of weight and height "merge" by simply dropping height. For this particular case feature selection and dimensionality reduction are equivalent and would both result in essentially the same result. However, let's say that for our set of observations height explains 25% of the variance and weight explains 75% of the variance, which means there is some variance in the height, but three times more in the weight. Whereas feature selection would simply drop the height property because of its lower relative variance, dimensionality reduction merges the two features together, but with 3 times the emphasis on the weight feature than the height feature (75% = 3 × 25%). In other words, dimensionality reduction takes the most important constituent parts as measured by their variance contribution and creates something new.

For both dimensionality reduction and feature selection functions, the target number of features (or components in dimensionality reduction) can be specified. Alternatively, the desired percentage of explained variance can be stipulated and these routines would return the number of features or components that account for that amount of variance.

Let's continue with our simple example using the original features of height and weight. One dimension is height and the other dimension is weight. Figure 8.21 displays the different heights (cm) and weights (kg) of 8 people in a scatterplot, a two-dimensional visualization.

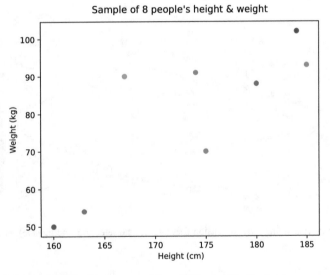

Fig. 8.21 Scatterplot of 8 people's height and weight. (Image created from PCA_visualization1.py)

An examination of Fig. 8.21 shows that in general shorter people weigh less than taller people. There is at least one main outlier (a person at 167 cm [5 feet 5 inches] weighing 90 kg [198 pounds]), but overall, the trend is neither difficult to understand nor surprising. In this case we are working with two dimensions or features, height and weight.

When we add a third dimension, say age, then the visualization requires a bit more effort to comprehend. Figure 8.22 offers a 3D (3-dimensional) scatterplot of height, weight, and age. What is the insight gained from this visualization? Can you understand now that the outlier, the person at 167 cm weighing 90 kg, is older?

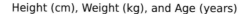

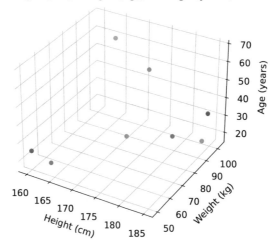

Fig. 8.22 Scatterplot of 8 people's height, weight, and age. (Image crated from PCA_visualization2.py)

The tricky thing about 3D visualizations is they are hard to read without rotating the chart and viewing the visualization from many different angles. Generally, a person needs to interact dynamically with 3D visualizations to sufficiently understand them. Unfortunately, the visualization in Fig. 8.22 resides in a book and is not interactive.

What happens if we add a fourth dimension? Visualizing the data becomes even more difficult, though we are not implying that 4D visualizations cannot be achieved. The data will need to be presented and contextualized using advanced visualization techniques. Figure 8.23 shows a parallel coordinates plot for the four dimensions of height, weight, age, and intelligence.

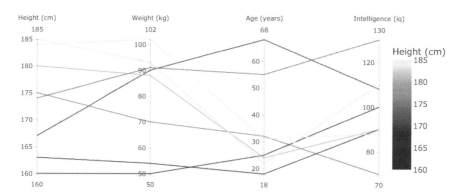

Fig. 8.23 A parallel coordinates plot visualizing four dimension (height, weight, age, and intelligence). (Image created from PCA_visualization3.py)

The main problem with this rendering is that a parallel coordinates plot takes some training and time to understand how to use and, as mentioned above for a 3D visualization, bestows the most value when the user can directly interact with the data. A parallel coordinates plot is a multivariate plot that allows a person to view how the different dimensions interact with each other. The best implementations allow the user to dynamically rearrange the various feature axes (e.g., height, weight) in customized order from left to right.

Adding dimensions increases the difficulty in capturing the main insights or fundamental meaning of the data. As explained above, this is an illustration of The Curse of Dimensionality because the more dimensions (the more features) we include in a dataset, the more difficult it is to grasp its complexity.

The primary objective of dimensionality reduction and feature selection is to employ fewer features (or components) in the model, which ultimately uses less space, can improve execution speed, and enhance the accuracy of machine learning algorithms. The exciting news is that many machine learning algorithms improve both their speed AND accuracy when they operate on datasets with fewer features or components. What's not to like about this approach?

The next subsections explain feature selection and dimensionality reduction, providing an example of the two with a side-by-side comparison. The feature selection subsection will be illustrated with an example of recursive feature elimination (RFE) and the dimensionality reduction section will demonstrate an example of principal component analysis (PCA).

8.9.1   *Feature Selection with RFE (Recursive Feature Elimination)*

Feature selection is also referred to as *feature reduction* or *feature elimination* because this accurately describes the goal of the procedure, to eliminate or remove the least essential features. This operation can also be considered a form of feature selection because at the same time the best features for a given problem are being identified and culled.

Feature selection is often considered a subdiscipline of *feature engineering*. For a given problem, feature engineering includes both testing for features to retain or eliminate as well as specifying features to create. The goal of feature engineering is to preserve and utilize the correct features, no more and no less, that are required for an optimal model given a specific data science problem.

Recursive feature elimination (RFE) is one of the most common feature selection techniques. With RFE a person defines an external model, usually a machine learning algorithm like decision trees or support vector machines to iteratively construct a model to remove features with low weights or influence. In the end, only the most predictive features remain. Some people consider RFE a wrapper function because it requires the external model to do all the hard work while the RFE function just provides the model with smaller and smaller subsets of the dataset.

In other words, given an external model, like a decision tree, RFE first trains the decision tree on the dataset with all the features included. Then, the decision tree ranks the features in terms of importance. The least important feature is removed by RFE then this process repeats. See Sect. 8.2 regarding how decision trees evaluate feature importance.

Consider an example in which we are predicting if an individual visiting a sidewalk lemonade stand will purchase a drink based on features such as the customer's height, weight, age, and iq plotted in Fig. 8.23. To determine the accuracy of the model, an important thing to point out is that RFE requires a target feature used for prediction. As a result, we added another feature to our working example: "bought_lemonade," a Boolean variable that indicates if the person bought from the local lemonade stand. Essentially, from the individual input explanatory variables height, weight, age, and iq, we would like to predict if that person will buy from the local lemonade stand.

The following code creates the data, establishes the classifier (in this case a decision tree), and then defines the X matrix (the features to be considered) and y list (the target feature – person purchases lemonade). The code then runs the RFE function specifying the number of features (n_features_to_select=3) to select (return as results):

```
Partial code: full program in rfe.py:


classifier = tree.DecisionTreeClassifier()
rfe = RFE(estimator=classifier, n_features_to_select=3)


rfe_fitted = rfe.fit(X, y)


print(f"Returns the data of the results without the names removing the eliminated feature:\
n{rfe.fit_transform(X, y)}")
print(f"\nReturns the ranking of the features to keep with a '1' and a '2' for the features to
eliminate:\n{rfe_fitted.ranking_}")
print(f"\nReturns the same idea as 'ranking_' but with True and False:\n{rfe_fitted.support_}")
# print(features)
print(f"\nUses the above Boolean array from 'support_' to only have the features that were not
eliminated:\n{np_features[rfe_fitted.support_]} "
      f"\n*Note that you have to use Numpy arrays to use Boolean arrays to filter.")
```

Which produces the following output:

```
Returns the data of the results without the names removing the eliminated feature:
[[ 54  18   90]
 [ 93  22  110]
 [ 90  68  108]
 [102  31   88]
 [ 88  24   90]
 [ 50  25  100]
 [ 70  32   70]
 [ 91  55  130]]


Returns the ranking of the features to keep with a '1' and a '2' for the features to eliminate:
[2 1 1 1]
Returns the same idea as 'ranking_' but with True and False:
[False  True   True   True]


Uses the above Boolean array from 'support_' to only have the features that were not eliminated:
['weight' 'age' 'iq']
*Note that you have to use Numpy arrays to use Boolean arrays to filter.
```

With `n_features_to_select` set to 3, you can see that the `fit_transform` function determines which of the features is the least useful, eliminates that feature, and returns the top 3 columns of data. However, with access to the `rfe_fitted` variable, which only fits (or processes) the data using RFE, we can view the `ranking_` and `support_` attributes. The `ranking_` attribute orders the original features by importance. All '1' results are retained. All other results are ranked in order of less usefulness. The `support_` attribute is similar and stores a Boolean index to signify which features to keep and which to eliminate.

By applying the Boolean `support_` index to a Numpy array that holds the names of the features, we can filter out the one eliminated feature and see that weight, age, and iq, are the prevailing predictors for determining if a person is going to buy lemonade from the local lemonade stand.

If you change the parameter `n_features_to_select` to 1 and run the code again then you will discover that age is the most useful feature in predicting if someone is going to buy lemonade from the local lemonade stand. If you take the time to study the data then you will notice that the people that are older (55 and higher) bought the lemonade, whereas the younger people did not. All the other features have less predictive power for this particular outcome.

There are various other feature selection techniques beyond RFE. For example, *Recursive Feature Elimination with Cross-Validation (RFECV)* ranks features with recursive feature elimination and leverages cross-validated selection to determine the best number of features; *SelectKBest* selects the K best features using statistical models (e.g., Chi-Squared, ANOVA); and *SelectPercentile* selects features based on percentile results from statistical models.

8.9.2 Dimensionality Reduction with PCA (Principal Component Analysis)

Using the original data from above (without the "bought_lemonade" feature) we will now investigate how dimensionality reduction with *PCA (Principal Component Analysis)* operates to reduce or change the number of feature dimensions. How can we reduce the number of dimensions and preserve approximately the same amount of useful information when we have fewer dimensions? For our lemonade stand example, how can we reduce four dimensions down to two?

Before we delve into the details of how PCA works, let's run the code that transforms the four dimensions of our 8 people (height, weight, age, and intelligence) visualized in Fig. 8.23 down to two dimensions. The following code is used to create Fig. 8.24 below:

```
Partial code: full program in PCA_example1.py:

# the following will scale all the values
scaled_numpy_array = StandardScaler().fit_transform(df)
```

```
# create the PCA object and set it so that it will reduce the features
# (columns) down to two components:
pca = PCA(n_components=2)
# the result is 8 pairs of components
principal_components = pca.fit_transform(scaled_numpy_array)
```

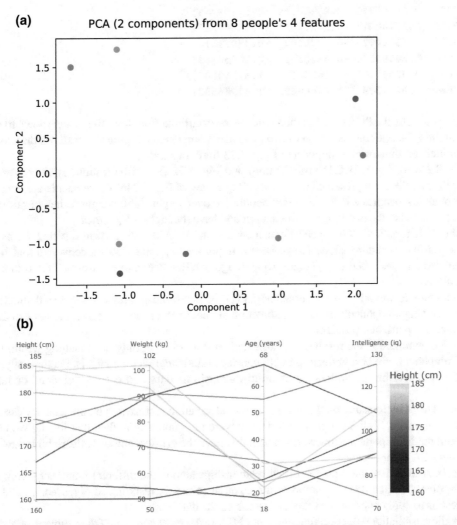

Fig. 8.24 (**a**) Visualization of the PCA results from four dimensions down to two dimensions for the 8 people. (Image created from PCA_example1.py). (**b**) Duplicate of Fig. 8.23 for comparison

The code in PCA_example1.py first creates the dataframe then scales each of the features. Scaling or normalizing the data is an important step for many algorithms like PCA and neural networks. Normalizing converts each of the four dimensions into the same range. With the StandardScaler class, all dimensions are normalized using the principles of the z-score (see Sect. 4.9 for more information). The following output prints the dataframe before being normalized (print(df)):

```
     height    weight    age    iq
0       163        54     18    90
1       185        93     22   110
2       167        90     68   108
3       184       102     31    88
4       180        88     24    90
5       160        50     25   100
6       175        70     32    70
7       174        91     55   130
```

Note the range for each individual data column is not uniform across all features. For example, the range for height is 25 (160 – 185) and the range for weight is 52 (50 – 102). However, the following normalized data (`print(scaled_numpy_array)`) does exhibit consistent ranges across all features:

```
[[-1.19080075 -1.42465916 -0.98972978 -0.48831317]
 [ 1.30421035  0.73307704 -0.74796373  0.69547633]
 [-0.73716237  0.56709734  2.03234589  0.57709738]
 [ 1.19080075  1.23101617 -0.20399011 -0.60669212]
 [ 0.73716237  0.4564442  -0.6270807  -0.48831317]
 [-1.53102954 -1.64596544 -0.56663919  0.10358158]
 [ 0.17011439 -0.53943405 -0.14354859 -1.67210266]
 [ 0.0567048   0.6224239   1.24660622  1.87926582]]
```

After normalizing the data, the PCA function then reduces the dataframe from four dimensions down to two. Figure 8.24.a shows the results of the PCA after the reduction to two generated components. Figure 8.24b shows the original four features rendered in the parallel coordinates plot (duplicate of Fig. 8.23 for comparison).

Comparing Fig. 8.24a to Fig. 8.21 (2D visualization) and Fig. 8.22 (3D visualization) is an interesting exercise. For example, consider the two people represented in the top-left quadrant of Fig. 8.24a. These people are clustered together and separate from the other six people. Why? These two people are overweight (high weight to height ratio) compared to the other six people. They are also the two oldest in the dataset and have the highest IQ scores.

So, what did the PCA function accomplish? Fundamentally, the PCA transformation distilled the essence of our four dimensional dataset down to two dimensions. However, it is important to point out that it converted four **features** that could be concisely articulated and described (e.g., height, weight, age, intelligence) and transformed them to a pair of nebulous, ill-defined **components**.

A *feature* is a measurable variable. For example, height can be measured and understood as a distinct, independent feature. However, the resulting components of the PCA function are not independently measurable nor can they be understood outside the context of this particular problem.

In other words, a component is the result of a mathematical function that only has meaning within the context of the original dataset. Components are best understood with respect to statistical factor analysis. Factor analysis is an advanced topic whose goal is to describe the variability of multi-dimensional data and extract "hidden" or latent features and patterns.

In summary, PCA works by focusing on the essence of the information represented by the data. By essence, we mean the variability inherent in the data. If all 8 people in our dataset were the same height, then that feature would not possess any variability and would not be important. However, features that have the greatest amount of variability have the most importance in the resulting PCA components.

Mathematically PCA identifies the most important relationships among the different features by computing the covariance matrix, performing an eigen decomposition of the covariance matrix resulting in eigenvectors and eigenvalues, and leveraging these results to apply appropriate transformations on the data.

There are many dimensionality reduction functions, but PCA is the most common. Other dimensionality reduction functions include *Kernel PCA (KPCA)*, *Linear Discriminant Analysis (LDA)*, *Singular Value Decomposition (SVD)*, *Locally Linear Embedding (LLE)*, and *TSNE (t-distributed Stochastic Neighbor Embedding)*. However, the primary purpose for all these techniques is to understand the most important relationships between features and express those relationships as components.

Although each of the aforementioned functions perform dimensionality reduction differently, let's explore a visual overview of the concept. For example, let's consider converting Fig. 8.21 from two dimensions (weight and height) to one component. Figure 8.25 plots the same data as Fig. 8.21 but with a residual drawn from every person's point to a best-fit line.

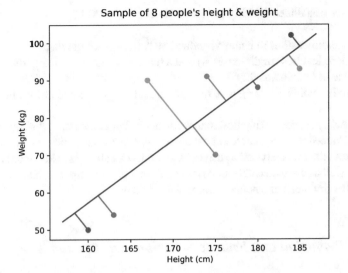

Fig. 8.25 Visualization of each person's height & weight with the shortest distance line from each point to the best-fit line. Note: if this were not primarily a visual example then the axes would need to be scaled before being processed. (Image created from PCA_explanation1.py)

Figure 8.25 displays a best-fit line that depicts the general trend of the relationship between weight and height for the eight people in the dataset. If we then project every point along its residual onto this trend line we obtain Fig. 8.26. Note: recall for actual PCA the axes in our example would need to be scaled before they are processed, but we are using this visual example for instructional purposes only.

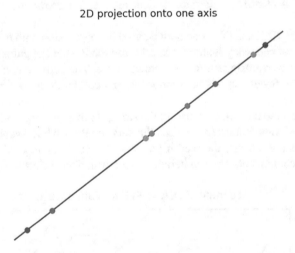

Fig. 8.26 Projection of the 8 points from Fig. 8.25 onto the best-fit line. The transition from Fig. 8.25 to this figure illustrates the concept of reducing two-dimensional data to one-dimensional data. (Image created by Robert Ball)

The resulting image is then rotated to present a single horizontal axis, as depicted in Fig. 8.27.

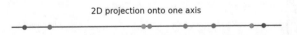

Fig. 8.27 The finished results of mapping two-dimensional data in Fig. 8.25 to a single dimension. (Image created by Robert Ball)

Figures 8.25, 8.26, and 8.27 demonstrate the general progression involved in dimensionality reduction with functions like PCA. Note there is no label associated with the axis in Fig. 8.27 because this component is derived from a combination of the two features weight and height. You might name the axis "weight-height component" as appropriate because it incorporates both features, but it is better practice to simply call it "component 1."

What are the main use cases of a dimensionality reduction function like PCA?

- Visualization: To reduce the dimensions to better visualize the "essence" of the data.
- Machine Learning: The more features (or dimensions) that a dataset contains, typically the worse machine learning algorithms perform – The Curse of Dimensionality.
- Clustering: To cluster together points that might be difficult or near impossible to cluster in a higher dimensional space.

Choosing between dimensionality reduction functions can be difficult. For example, in the case of data uniformly distributed across the target variable (balanced dataset), LDA almost always performs better than PCA. However, if the data is highly skewed (irregularly distributed) then it is advised to use PCA since LDA can be biased towards the majority class. PCA can be applied to both labeled as well as unlabeled data since it doesn't rely on the output labels. In contrast, LDA requires output classes to determine linear discriminants and hence requires labeled data.

8.9.3   Dimensionality Reduction and Feature Selection with Examples

Dimensionality reduction and feature selection are related. There are times when feature selection is preferable, such as when you need to know which specific features are the best predictors. In this case, you can use feature selection to determine the best features both for greater insight and to save money.

With the Iris dataset, most of the variance can be explained by the single feature petal length. Consequently, if you were a botanist measuring sepals and petals then you could drop sepal measurements entirely, focus only on measuring petal length, and still achieve a 97% prediction accuracy. If the goal of the botanist is to identify which types of Iris flowers were in a selected meadow then only measuring petal lengths when compared to measuring sepal length, sepal width, petal length, and petal width, is effectively four times faster and almost as accurate.

This money saving technique of identifying the most salient features for predicting outcomes can be applied to many disciplines.

In our next example, we see how RFE and PCA can both be used in conjunction with machine learning to realize the goal of discovering the precise number of necessary features. Here we use the Palmer Penguins dataset named `penguins_no_nulls.csv` file with null entries previously removed for convenience. The code contained in the file `pca_vs_rfe.py` progressively reduces the dataset to fewer features or components and computes the accuracy based on the total number of remaining features or components.

More specifically, the code reads the dataset, converts all the strings to ints (see Sect. 3.3.11 for more details), assigns the explanatory features to X and the target feature to y, scales the data for PCA while keeping non-scaled versions for RFE, splits the data for training and testing (change the seed if you wish to see different random splits), executes a for-loop to reduce the number of features or components on each iteration, then visualizes the accuracy as shown in Fig. 8.28.

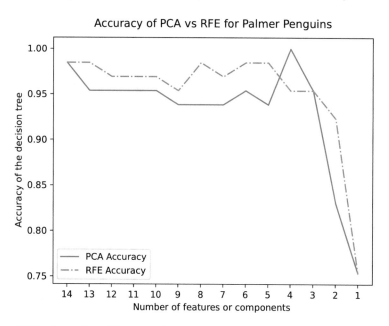

Fig. 8.28 Accuracy of PCA vs RFE as the number of features and components decreases. (Image created from pca_vs_rfe.py)

A careful examination of Fig. 8.28 reveals that utilizing between 3 to 14 features or components results in an accuracy between 94% and 100% for both methods. Dropping incrementally from 3 to 1 feature or component induces a significant decrease in accuracy. In other words, we may conclude there are three or four main features that represent most of the variability and the rest of the features are consequently not very useful predictors.

With regards to an altogether different data science application, dimensionality reduction is preferred for analysis techniques such as NLP (Natural Language Processing – see Chap. 9) where you might encounter hundreds or thousands of unique words in a given corpus (the input text). Normally each unique word would be considered its own feature. Using feature selection on thousands of words is out of the question, so we turn to dimensionality reduction.

Figure 8.29 plots the results of a two component PCA applied to the TFIDF vectorization of the words from 344 books each of which was written by one of six authors. For more details on TFIDF see Sect. 9.2. Note in the figure that different authors are to some degree visually separate from each other. However, PCA is not necessarily the best technique for so many dimensions. Also consider that PCA is converting the TFIDF of tens of thousands of words into only two components.

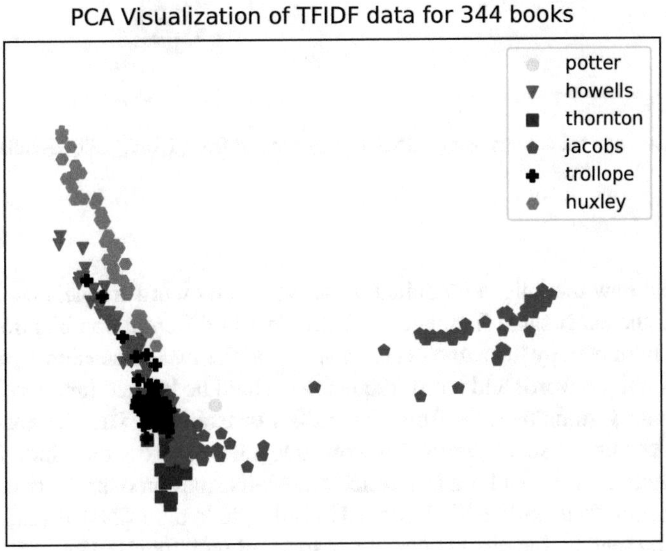

Fig. 8.29 Visualization of 344 books from six authors using PCA. (Image created from PCA_tfidf_visualization_of_books.py. The books can be found in books.zip_

Figure 8.30 illustrates the same conceptual approach but with another technique called *TSNE (t-distributed Stochastic Neighbor Embedding)*. TSNE or t-SNE is a dimensionality reduction technique that excels with high-dimensional data such as unique words in a collection of books. Comparing Figs. 8.29 and 8.30 clearly impart the reason to become familiar with more than one technique for a given machine learning task. Both Figs. 8.29 and 8.30 demonstrate the utility of converting the words contained in 344 books down to only 2 components, but for this particular dataset, TSNE does a better job of distinguishing authorship.

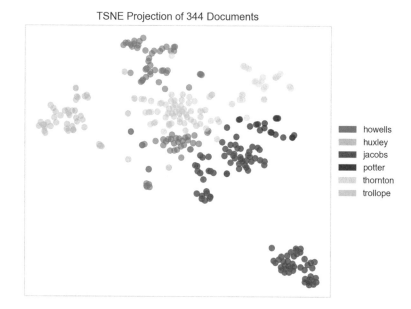

TSNE Projection of 344 Documents

howells
huxley
jacobs
potter
thornton
trollope

Fig. 8.30 Visualization of 344 books from six authors using TSNE. (Image created from TNSE_tfidf_visualization_of_books.py. The books can be found in books.zip)

Figures 8.29 and 8.30 reveal how the unique combination of words and word frequencies in each book compares to the other books. If two books had the exact same frequency of words but in different combinations (compare "I am Sam" and "Sam I am") then they would both occupy the same decision space on the two dimensional plot. In contrast, books that are very different both in terms of unique words and word frequencies would be located far away from each other.

We can recognize and extract various patterns from Fig. 8.30. For example, Thomas Huxley (green dots) and Beatrix Potter (purple dots) employ a consistent set of words and word frequencies across their individual works and can be easily distinguished from the other authors who exhibit a less concentrated distribution over the two dimensional space. However, other authors like William Jacobs (red dots) and William Howells (dark blue dots) reveal primarily two main ways of expressing themselves as can be seen by the distinct clusters associated with their books.

Exercises
1. What is the difference between machine learning and programmed intelligence?
2. How would you define artificial intelligence?
3. What is the purpose in training a machine learning algorithm?
4. Why are expert systems not heavily used in data science?
5. What is XAI? When would it be important in data science projects?
6. For all machine learning algorithms listed in this chapter, indicate which of them is a black box or a white box.
7. In your own words, how does the ID3 algorithm construct decision trees?
8. What are the main concepts underlying the SVM algorithm?
9. When would you use neural networks?
10. What is the difference between decision trees and random forest?
11. What is feature selection?
12. What is dimension reduction?
13. How are feature selection and dimension reduction similar and how are they different?

Chapter 9
Natural Language Processing (NLP)

Natural Language Processing (NLP) is a combination of computer science, artificial intelligence / machine learning, and linguistics (sometimes referred to as computational linguistics). The primary objective of NLP is to gather knowledge or insight from textual information expressed in a human-readable language. Whether the text resource is from social media, religious works, or the latest novel, the application of NLP is the same.

Given the scope of this topic, this chapter – as with many others in this text – could easily be developed into a stand-alone book. In fact, there are many dedicated references that focus exclusively on NLP. However, we will emphasize the essential aspects of this discipline related specifically to data science so you may begin immediately with your text processing and analysis tasks. As with all the other subjects in this book, if you want to know more about NLP there are a wide selection of resources to choose from to increase your knowledge on this subject.

NLP at its core is importing a *corpus* (collection of text), determining and assessing the meaning of the text, then performing some action based on the information provided by the text. Recalling the motivation for this book described in Chap. 1, we will now provide two examples based on how to effectively apply NLP to either generate or save money.

Our first example involves the stock market. Although there are many reasons why a person might invest in stocks, most often the primary purpose is to make money. To achieve this goal, the classic approach is to purchase a stock at a lower price then sell that same stock at a higher price ("buy low/sell high"). However, how do we know if the price of the stock is likely to increase before we buy it?

One approach is to create a mathematical time series model, as described in Chap. 10. Another approach is to perform web scraping (see Sect. 2.7) of social media websites that discuss stocks. By obtaining and analyzing the blogs and posts that people compose about stock purchasing and trends, and then applying NLP *sentiment analysis* on that text, we can deduce a general positive or negative mood (known as sentiment) about the prospects of a particular stock. If our investigation reveals that people are expressing an overall positive sentiment about a stock then we would buy the stock, but if the general sentiment leans negative or neutral then we would likely forego any investment in the stock, at least in the short term.

Another example of the application of NLP within a data science context that is becoming more popular among call centers is the use of automated chat bots that answer questions from clients or customers. There are now many companies implementing the theories and techniques from NLP to create automated chat bots.

For chat bot operation, the fundamental functionality is a user writes or speaks a question to the chat messaging system at which point another software system converts the speech to text (if it was spoken), which is then processed to determine semantics and sometimes tone. The system is designed to analyze the question from the user and to respond appropriately. During the system learning phase, the response may sometimes be simultaneously hilarious and embarrassing in its inaccuracy, but typically responses are relevant and adequately answer the client's question. Using automated chat bots, call centers have been able to decrease their work force by as much as 50% since the year 2000, thus saving companies a significant amount of money.

R. Ball, B. Rague, *The Beginner's Guide to Data Science*, https://doi.org/10.1007/978-3-031-07865-1_9

We now detail the most common data science related uses of NLP then gradually expand to more sophisticated and less known techniques. We highly recommend you read and understand at least the first few sections pertaining to Bag of Words, TFIDF, and Naïve Bayes. The other sections are also helpful, but we have found generally that most employers are seeking employees with at least some practical working knowledge of NLP basics.

There are various open source NLP libraries available such at NLTK, spaCy, Gensim, and Hugging Face. We will primarily use examples from the NLTK library.

9.1 Bag of Words

One of the most basic applications of NLP within the field of data science is to predict the author of a book. Consider how you would approach the task of determining the unknown author of a book.

To achieve even mild success at this task, you would need some expertise in comparative literary analysis. At a minimum, you should have read a wide variety of books. Even though you cannot possibly read all books ever written, the more you read the more expertise you acquire. While reading through a collection of works, you would likely notice the writing techniques of some authors are similar whereas other authors display a very distinct and easily recognizable writing style.

If you were to read roughly 10,000 books, something that many literature professors have done over a lifetime of study, and then picked up an entirely new book you would likely be able to compare and to contrast that book with the ones you previously read. Without looking at the cover or jacket copy (the text written on the book flaps), you might also be able to effectively predict the type of book, the author's gender, the author's nationality, and other attributes related to the work.

However, what if you do not have sufficient discretionary time to devote to the study of literature? NLP can assist us in this situation. Let's employ a more modest example of trying to predict authorship based on the works of only three classic writers: Jules Verne (considered the father of science fiction), Jane Austen (an important author in developing the modern novel), and L. Frank Baum (author of the *Wizard of Oz* series).

There are myriad ways to compare the books of Verne, Austen, and Baum. However, one of the simplest and most reliable techniques is to observe the frequency of words included in the books they wrote.

We have provided four example books from each of the three authors in the `books` folder with the files that correspond with this chapter.[1] If you take the time to read the 12 books you will find that overall word usage among each of the authors is distinct. Jules Verne was French and his works were translated to English. Jane Austen was English and typically used formal language. L. Frank Baum was American and wrote children's books.

If we had access to experts on all three authors, we could create a small artificial intelligence program to quickly predict the author. Given an unknown book from one of the authors this panel of literary authorities could assign the author with overwhelming confidence. If the word "radioactive" shows up in the unknown book the panel would likely predict the author as Jules Verne. If the word "Oz" appears in the book the predicted author is L. Frank Baum. If the word "Lady" shows up at least five times Austen is the predicted writer. However, this approach does not use NLP since it leverages domain experts with previous knowledge of author styles and word usage.

Our above approach capitalizes on foreknowledge of what distinguishes the authors and leverages that information to predict accordingly. However, what would we do if faced with a prediction task about authors in which a panel of specialists were not available? We use a computer to determine the common words.

The most widely used approach to determine how often a word occurs is called Bag of Words. The idea behind *Bag of Words* is simple: for every word encountered create a dictionary entry that tracks the number of occurrences for that word. Let's look at two examples:

1. "Add three eggs to the bowl. Mix the eggs with a whisk. With the bowl, add the sugar."
2. "Go to the hotel, then take get a room if there is one available. Otherwise, let's keep going."

For Bag of Words we count how many times each word appears. From the first sentence, we derive the following Bag of Words results:

"Add": 1, "three": 1, "eggs": 2, "to": 1, "the": 4, "bowl.": 1, "Mix": 1, "with": 1, "a": 1, "whisk.": 1, "With": 1, "bowl,": 1, "add": 1, "sugar": 1.

---

[1] The 12 books are from Project Gutenberg (https://www.gutenberg.org/). All 12 books can also be found with the book files from this chapter.

Notice there are two main problems with the above bag of words: capitalization and punctuation. The word "Add" appears once and the word "add" appears once. Also, "bowl." appears once and "bowl," appears once. Convention dictates that we convert all words to lowercase and all punctuation be removed. Using NLTK's version of Bag of Words, called CountVectorizer, we note this implementation changes all words to lowercase and removes all punctuation as demonstrated in the following code:

```
File: bag_of_words1.py:
from sklearn.feature_extraction.text import CountVectorizer

corpus = ['Add three eggs to the bowl. Mix the eggs with a whisk. With the bowl, add the sugar.',
'Go to the hotel, then take get a room if there is one available. Otherwise, let\'s keep
going.']

vectorizer = CountVectorizer()
X = vectorizer.fit_transform(corpus)

print(vectorizer.get_feature_names())
print(X.toarray())
```

Which produces the following output:

```
['add', 'available', 'bowl', 'eggs', 'get', 'go', 'going', 'hotel', 'if', 'is', 'keep', 'let',
'mix', 'one', 'otherwise', 'room', 'sugar', 'take', 'the', 'then', 'there', 'three', 'to',
'whisk', 'with']
[[2 0 2 2 0 0 0 0 0 0 0 0 1 0 0 0 1 0 4 0 0 1 1 1 2]
 [0 1 0 0 1 1 1 1 1 1 1 1 0 1 1 1 0 1 1 1 1 0 1 0 0]]
```

The first array produced by the `get_feature_names()` method is a set of all the words from the two sentences. The second array is two-dimensional (2×25) where each row represents the number of times each word appears in the respective sentence. For example, the word "add" corresponding to the first entry in each row indicates this word appears twice in the first sentence and zero times in the second sentence.

Looking at the output there are several words that overlap between the two sentences. However, overlap typically occurs with the most common words in the English language, such as "to," and "the." The most common words in any language are called *stop words*. Fortunately, it is very easy to remove stop words. We can either flag the CountVectorizer class to remove stop words automatically or customize our list of stop words and perform the task separately. Changing the constructor to `vectorizer = CountVectorizer(stop_words='english')` and running the new program (`bag_of_words2.py`) produces the following output:

```
['add', 'available', 'bowl', 'eggs', 'going', 'hotel', 'let', 'mix', 'room', 'sugar', 'whisk']
[[2 0 2 2 0 0 0 1 0 1 1]
 [0 1 0 0 1 1 1 0 1 0 0]]
```

The above output shows that with stop words removed our two sentences now have no words in common. If you want to remove the stop words separately then download the most common stop words from NLTK using the following code:

```
File: stop_words_example.py:
from nltk.corpus import stopwords

# the following two lines need to only be run once:
import nltk
nltk.download('stopwords')  # needs to be run only once!

stop_words = stopwords.words('english')
print(stop_words)
```

The output from the above code is a list of the most common English words. If your particular text has other common words you can expand this list as needed.

Note: You need to run the code `nltk.download('stopwords')` only **once** to download the list of stopwords to your computer. You can comment that line out after you have executed that statement.

9.2   TFIDF (Term Frequency-Inverse Document Frequency)

The previous techniques utilizing Bag of Words, capitalization, punctuation, and stop words represent a great start to managing and analyzing textual data. When trying to predict book authorship, these ideas are more than sufficient and work effectively a majority of the time.

However, consider what happens when comparing works of different lengths, specifically a long text to a short text? Assume we want to predict which of two sample texts is most related to cooking. If the first text is a long novel where the characters in the book cook in the kitchen in only one chapter and the second text contains brief instructions from a cookbook then the longer novel will usually be selected as the one most associated with cooking by prediction algorithms.

The reason is the longer text will contain a large number of occurrences for each word and the shorter text will have a relatively smaller number of occurrences for each word. More to the point, longer texts are selected over shorter texts simply because of sheer length, leading to the logical conclusion that each word of interest will occur more frequently.

To remedy this situation, for a given text we focus not on the raw number of occurrences of a word, but on the **normalized number of occurrences** of that word relative to the number of occurrences of all other words for a given text. *Term frequency*, the TF in TFIDF, defines how often a term or word appears in a document.

The next part, IDF in TFIDF, stands for *Inverse Document Frequency*. The IDF calculation considers whether a word is used across all or some of the texts under consideration. The IDF factor emphasizes words that are particular to individual works by computing the logarithm of the quantity Q, where Q equals the total number of documents divided by the number of documents that include that particular word. A word used in only one document out of 100 will be weighted more heavily by IDF compared to a word used in all 100 documents.

TFIDF (Term Frequency-Inverse Document Frequency) values are calculated by multiplying the results of the TF function and the IDF function and then normalizing the results for each document. The following code provides an example employing TfidfVectorizer using the same sentences from the previous subsection:

```
File: tfidf_example.py:
from sklearn.feature_extraction.text import TfidfVectorizer

corpus = ['Add three eggs to the bowl. Mix the eggs with a whisk. With the bowl, add the sugar.',
'Go to the hotel, then take get a room if there is one available. Otherwise, let\'s keep
going.']

vectorizer = TfidfVectorizer(stop_words='english')
X = vectorizer.fit_transform(corpus)

print(vectorizer.get_feature_names())
print(X.toarray())
```

Which produces the following output:

```
['add', 'available', 'bowl', 'eggs', 'going', 'hotel', 'let', 'mix', 'room', 'sugar', 'whisk']
[[0.51639778 0.         0.51639778 0.51639778 0.         0.
  0.         0.25819889 0.         0.25819889 0.25819889]
 [0.        0.4472136  0.         0.         0.4472136  0.4472136
  0.4472136 0.         0.4472136  0.         0.         ]]
```

The above output demonstrates that TFIDF is generally preferred to Bag of Words because it determines the normalized term frequency of each word relative to each individual document and also across all documents through the scaling imposed by IDF. The next section will show how to use these concepts in the context of predictive analytics.

9.3 Naïve Bayes

One of the most common computational approaches to prediction with NLP is Naïve Bayes. *Naïve Bayes* is a statistical technique for predicting different outcomes based on the prior probability of those outcomes. The naïve qualifier in Naïve Bayes arises from the assumption that given a particular outcome *y*, the probabilities of all measurable explanatory variables used in the calculation are independent of each other. This assumption is naïve and rarely satisfied because the predictors or features of interest are typically correlated in some way. However, this assumption of feature probability independence for a particular outcome makes the math much easier and computationally efficient!

When specifically applied to NLP, Naïve Bayes is about predicting a classification given a collection of new words without considering or completely understanding the interconnections and correlations of the words used as predictors. Figuring out the various correlations and context among words in a single book is both computationally intensive and unnecessary from the standpoint of acceptable prediction accuracy.

For example, the words "to" and "be" are often paired sequentially to form the phrase "to be," indicating a correlation or dependance. However, the Naïve Bayes algorithm assumes the words "to" and "be" are not correlated and does not utilize any information on word relationships when calculating prediction results, thus substantiating its classification methodology as naïve.

The independence assumption of Naïve Bayes is essential because this classifier primarily processes high-dimensional data. If each word represents a dimension and a typical medium-sized book contains about 5000–8000 unique words then by definition we have a very high-dimensional dataset!

Given a specific outcome *y*, the formula for Naïve Bayes derives from Bayes' original rule:

$$P\left(y\mid x_1, x_2, \cdots, x_n\right) = \frac{P\left(x_1, x_2, \cdots, x_n \mid y\right) P\left(y\right)}{P\left(x_1, x_2, \cdots, x_n\right)}$$

The Naïve Bayes formula performs classification by selecting the outcome *y* that generates the greatest probability given the values of feature set x_1, x_2, ...x_n. For example, suppose we have 1000 apples with three possible classifications (y): Red Delicious, Pink Lady, and Honey Crisp. Each apple has three binary features: Red (x_1), Sweet (x_2), and Juicy (x_3). Counts from our sample of 1000 apples for each type and feature are shown in Table 9.1:

Naïve Bayes determines the classification (*y*) given the three features (x_1, x_2, x_3). Let's start with an example A_1 of an apple that is red, sweet, and juicy. Given those three features, the algorithm finds the most likely apple type for A_1. More formally what is the most likely classification (*y*) given features of Red, Sweet, and Juicy *P(y | Red, Sweet, Juicy)*?

To calculate *P(y| x_1, x_2, x_3)* for each possible classification, the right hand side of Bayes' rule above tells us we need to initially determine the probability for each *P(y)*, *P(x_n)*, and *P(x_n | y)*. For the three classifications Red Delicious, Pink Lady, and Honeycrisp, we calculate the prior probabilities *P(y)* using the far right column in Table 9.1:

- P(Red Delicious) = (500/1000) = 0.5
- P(Pink Lady) = (300/1000) = 0.3
- P(Honeycrisp) = (200/1000) = 0.2

Table 9.1 Example data for Naïve Bayes computation

Feature/Type	Red	Not red	Sweet	Not sweet	Juicy	Not juicy	Total
Red Delicious	400	100	350	150	450	50	500
Pink Lady	0	300	150	150	300	0	300
Honey Crisp	100	100	150	50	50	150	200
Total	500	500	650	350	800	200	1000

For the features Red, Sweet, and Juicy, we calculate the summary probabilities of each feature *P(x_n)* using the bottom row in Table 9.1:

- P(Red) = (500/1000) = 0.5
- P(Sweet) = (650/1000) = 0.65
- P(Juicy) = (800/1000) = 0.8

Now we compute $P(x_n \mid y)$, the probability of each feature given a specific apple type:

- P(Red | Red Delicious) = (400/500) = 0.80
- P(Sweet | Red Delicious) = (350/500) = 0.70
- P(Juicy | Red Delicious) = (450/500) = 0.90

Because of the Naïve Bayes assumption of feature independence we can state the following two equalities: (1) $P(x_1, x_2, x_3) = P(x_1)P(x_2)P(x_3)$ and (2) $P(x_1, x_2, x_3 \mid y) = P(x_1|y)P(x_2|y)P(x_3|y)$. Therefore:

- P(Red, Sweet, Juicy | Red Delicious) = (0.8 * 0.7 * 0.9) = 0.504

Calculations for the remaining classifications follow:

- P(Red | Pink Lady) = (0/300) = 0
- P(Sweet | Pink Lady) = (150/300) = 0.50
- P(Juicy | Pink Lady) = (300/300) = 1
- P(Red, Sweet, Juicy | Pink Lady) = (0 * 0.5 * 1) = 0
- P(Red | Honeycrisp) = (100/200) = 0.50
- P(Sweet | Honeycrisp) = (150/200) = 0.75
- P(Juicy | Honeycrisp) = (50/200) = 0.25
- P(Red, Sweet, Juicy | Honeycrisp) = (0.5 * 0.75 * 0.25) = 0.09375

We are now able to perform the final calculation $P(y \mid x_1, x_2, x_3)$ for each possible classification by leveraging the naïve assumption of feature independence. For example, for the type Red Delicious, the explicit formula becomes:

P(Red Delicious| Red, Sweet, Juicy) = P(Red | Red Delicious)*P(Sweet | Red Delicious)*P(Juicy | Red Delicious)*P(Red Delicious) / P(Red)*P(Sweet)*P(Juicy). A similar formula is employed for the other apple types.

The results for each of the apple classifications are shown here:

- P(Red Delicious | Red, Sweet, Juicy) = (0.504 * 0.5) / (0.5*0.65*0.85) = 0.91
- P(Pink Lady | Red, Sweet, Juicy) = (0 * 0.3) / (0.5*0.65*0.85) = 0
- P(Honeycrisp | Red, Sweet, Juicy) = (0.09375 * 0.2) / (0.5*0.65*0.85) = 0.068

Based on these three Naïve Bayes computations derived from the input values listed in Table 9.1, the classification with the highest probability (0.91) given a sample apple that is red, sweet, and juicy is Red Delicious.

There are many nuances associated with the Naïve Bayes method. For example, most Naïve Bayes implementations use the Laplace Correction which sets the floor for any value in Table 9.1 to 1 instead of 0. For instance, *P(Pink Lady | Red, Sweet, Juicy) = 0* because Pink Lady is always pink and never red. This result emerges because of the 0 entry in the Red column for the Pink Lady apple type in Table 9.1, which produces a value of 0 for *P(Red | Pink Lady)*. However, with the Laplace Correction applied, the 0 entry would be replaced by a minimum value of 1, which would revise the calculation such that *P(Red | Pink Lady) = (1/300) = 0.003*.

Also, Gaussian Naïve Bayes assumes a continuous Gaussian, normal distribution for each of the features and is most suitable for general classification tasks. The Bernoulli Naïve Bayes is useful for independent binary inputs like the example given previously in which the features are either present or absent.

Multinomial Naïve Bayes is the most commonly employed approach for data science and NLP investigations. Multinomial refers specifically to the multinomial distribution, a generalization of the binomial distribution. The binomial distribution refers to a distribution used in discrete probability with the "bi" in binomial emphasizing two key measures, the probability of success and the probability of failure.

Naïve Bayes calculations rely on carefully selected probability distributions. The binomial distribution works great when flipping a coin but is not sufficient for datasets with a large number of unique words. For these cases, the multinomial distribution is a better fit since it involves probabilities for more than two possibilities.

To illustrate the application of Naïve Bayes to NLP, let's continue with a code example. Using the `books` and `test_books` directories in the book files for this chapter, we first read in 12 books from the `books` folder, four books each from the authors Jules Verne, Jane Austen, and L. Frank Baum. We use TFIDF to determine the normalized frequency of each word then proceed with a multinomial Naïve Bayes analysis to predict three anonymized books in the `test_books` directory as demonstrated in the following code:

```
File naive_bayes_prediction.py:

from sklearn.feature_extraction.text import TfidfVectorizer
from sklearn.naive_bayes import MultinomialNB
import os

directory = 'books'
# get only the directories in the books directory:
dirs = [d for d in os.listdir(directory) if os.path.isdir(os.path.join(directory, d))]

corpus = []
labels = []
for directory in dirs:
    books = os.listdir(f'books/{directory}')
    for book in books:
        labels.append(directory)
        all_words = ''
        with open(f'books/{directory}/{book}', 'r') as f:
            for line in f:
                all_words += line  # TfidfVectorizer expects a list of strings
        corpus.append(all_words)

vectorizer = TfidfVectorizer(stop_words='english')
tfid_vect_results = vectorizer.fit_transform(corpus)

# MultinomialNB stands for multinomial Naïve Bayes
clf = MultinomialNB().fit(tfid_vect_results, labels)

books = os.listdir('test_books')
for book in books:
    print(f"For {book}, the predicted author is: ", end='')
    all_words = ''
    with open(f'test_books/{book}', 'r') as f:
        for line in f:
            all_words += line  # TfidfVectorizer expects a list of strings
    tfid_vect_results = vectorizer.transform([all_words])  # vectorizer is expecting a list
    prediction = clf.predict(tfid_vect_results)
    print(prediction[0])
```

Which produces the following output:

```
For anonymous_verne_book.txt, the predicted author is: Verne
For anonymous_austen_book.txt, the predicted author is: Austen
For anonymous_baum_book.txt, the predicted author is: Baum
```

Without investing any time and effort to read the 12 training books and the 3 sample books yourself, you can see the Naïve Bayes algorithm correctly predicted the author for each of the three anonymized books 100% of the time.

Naïve Bayes is not the only available tool for text prediction. Many other machine learning algorithms work as well. In the file `text_identification_multiple_algorithms.py` we have a comparison of five different algorithms that are used for text prediction tasks. The file also shows how to use the PipeLine class from *sklearn* to improve code size and readability.

Also, consider that you might not always have the same predictive power when analyzing different authors. Notice the clustering plotted in the two dimensional PCA visualization in Fig. 9.1 reveals that Jules Verne appears to have two distinct styles of writing, whereas L. Frank Baum and Jane Austen only have one. Because the clusters are so easily distinguishable, most machine learning algorithms should not have much trouble deciding between authors. To understand more about the visualizations associated with dimensionality reduction (like PCA) see Sect. 8.9.

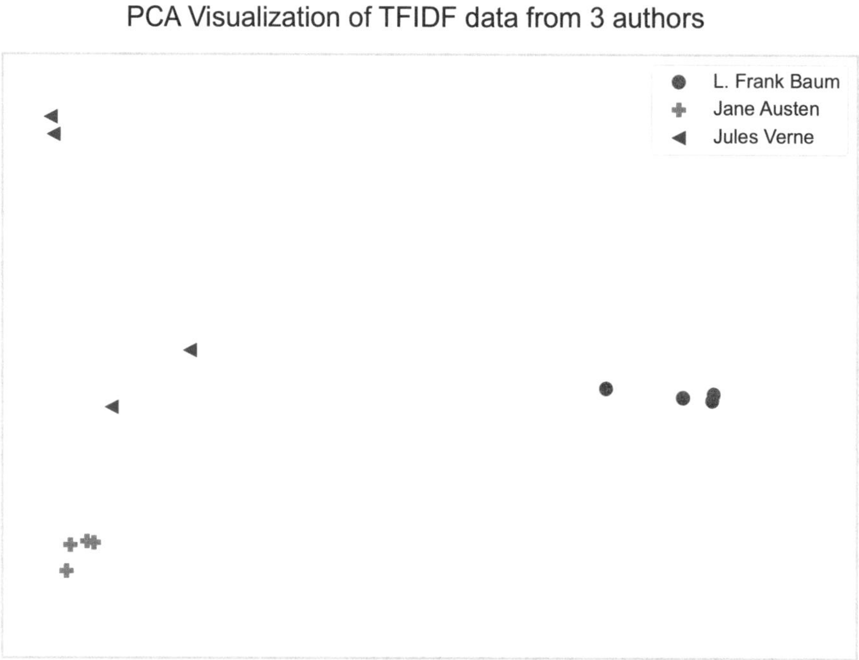

PCA Visualization of TFIDF data from 3 authors

Fig. 9.1 PCA visualization of the TFIDF data from the 12 Baum, Austen, and Verne books. (Image created from tfidf_visualization.py)

9.4 Stemming, Lemmatization, and Parts of Speech

The first three sections of this chapter covered the fundamentals of NLP that all data scientists should know, namely, Bag of Words, punctuation removal, capitalization, stop words, TFIDF, and basic text prediction.

We are now in a prime position to get a little more sophisticated with our textual analysis. As we saw with Naïve Bayes, the likelihood of assigning the creation of an anonymous document to a given class (or author) is based on the frequency of words used by that class (or author) as reflected in selected book samples. For example, Jane Austen invokes the word "Lady" quite often in her writings. However, TFIDF and Naïve Bayes are not sophisticated enough to determine that "Ladies" and "Lady" are closely related and are conceptually the same word.

The key concept behind *stemming* is the removal of different endings from related words. This text preprocessing strategy is especially important when working with verbs. Consider the verb "love" which can be expressed in variant forms such as love, loved, loves, and loving. There may be contexts in which these different forms of the word "love" are important. However, if our analysis deems the core essence of the word is of primary importance rather than its alternative forms then the application of stemming will be helpful. The following code illustrates the idea of stemming with various forms of the work "love."

```
File: stem_example.py:
from nltk.stem import PorterStemmer

words = ['love', 'loved', 'loves', 'loving', 'lady', 'ladies', 'strawberry', 'strawberries']
ps = PorterStemmer()

for word in words:
    print(f'{word}: {ps.stem(word)}')
```

Which produces the following output:

```
love: love
loved: love
loves: love
loving: love
lady: ladi
ladies: ladi
strawberry: strawberri
strawberries: strawberri
```

Careful observation of the output listed above indicates that stemming works well some of the time. All variants of "love" became *love*, but all forms of "lady" became *ladi*, and all forms of "strawberry" became *strawberri*. If your input text represents the data for a predictive machine learning algorithm like Naïve Bayes then we need to ensure that all words have undergone the stemming operation. However, if humans are part of the processing pipeline then the mapping from lady to *ladi* will be problematic since the resulting word stem can lead to some confusion.

So, stemming possesses the drawback of producing words that may not exist. However, some benefits of stemming include fast execution, low memory consumption, and compatibility with predictive algorithms. Generally speaking, stemming may be considered another naïve deterministic algorithm that truncates words with no context or knowledge informing the transformation from the input string to the word stem result.

However, if human readability and word precision are required, then *lemmatization* is recommended. The operational concept behind lemmatization is similar to that used in stemming, but instead the mapping is performed using a lookup table. This implementation strategy means that lemmatization demands more memory and more processing cycles than stemming, but with the associated advantage that the result is a legitimate, recognizable word in the language of interest.

To fully understand lemmatization and obtain correct results, we need to understand parts of speech (POS). *Parts of speech (POS)* is defined as the traditional classification of words into categories such as nouns, pronouns, verbs, and conjugations. The following code demonstrates an example of POS:

```
Partial file: full program in pos_example.py:

text = "I like to eat yummy strawberries on Sundays!"
tokens = nltk.word_tokenize(text)
print(tokens)
tag = nltk.pos_tag(tokens)
print(tag)
```

Which produces the following output:

```
['I', 'like', 'to', 'eat', 'yummy', 'strawberries', 'on', 'Sundays', '!']
[('I', 'PRP'), ('like', 'VBP'), ('to', 'TO'), ('eat', 'VB'), ('yummy', 'JJ'), ('strawberries',
'NNS'), ('on', 'IN'), ('Sundays', 'NNS'), ('!', '.')]
```

The different capitalized labels PRP, VBP, and VB are defined in the current index from NLTK, which can be acquired using the following code:

```
File: get_tagset.py:
import nltk
nltk.download('tagsets') # this only has to be done once!
nltk.help.upenn_tagset()
```

The result generated by the above code is an alphabetical listing of the POS description for each label. Some examples include PRP for personal pronoun, VBP for present tense verb, and VB for a verb in its base form.

POS is important because different words will have different meanings based on their category. For example, again consider the word "love." Love has different meanings based on whether it is used as a noun ("The couple held a special kind of love") or a verb ("They loved their child.").

Lemmatization works by using a mapping (in this case WordNet – see Sect. 9.5) and POS to determine the exact base word for any input word. Consider the following code:

```
Partial file: Full program in lemmatization_example.py:

lemmatizer = WordNetLemmatizer()

words = ['love', 'loved', 'loves', 'loving', 'lady', 'ladies', 'strawberry', 'strawberries']

for word in words:
    print(f'{word}:\tnoun:{lemmatizer.lemmatize(word)}\tadjective:{lemmatizer.lemmatize(word,
    pos="a")}\t verb: {lemmatizer.lemmatize(word, pos="v")}')
```

Which produces the following result:

```
love:      noun: love     adjective: love      verb: love
loved:     noun: loved    adjective: loved     verb: love
loves:     noun: love     adjective: loves     verb: love
loving:           noun: loving   adjective: loving   verb: love
lady:      noun: lady     adjective: lady      verb: lady
ladies:           noun: lady     adjective: ladies    verb: ladies
strawberry:       noun: strawberry     adjective: strawberry      verb: strawberry
strawberries:     noun: strawberry     adjective: strawberries    verb: strawberries
```

The program output demonstrates the POS classification directly determines the lemmatization return value. Given the first four words in the example (love, loved, loves, and loving) were all verbs, the verb form ("love") returned by the lemmatization of each of these words was accurate. Similarly, the next four words (lady, ladies, strawberry, and strawberries) were all nouns, and the correct noun form was returned for each call to lemmatize.

Overall, lemmatization returns a more satisfactory result than stemming but with the additional overhead of greater memory requirements, slower speed, and specifying the correct POS.

9.5 WordNet

WordNet is a lexical database developed by Princeton University (wordnet.princeton.edu). Lexical means relating to words and vocabulary. However, it would be more appropriate to describe WordNet as a lexical graph or network in the traditional mathematical and computer science sense. Note that WordNet only contains nouns, verbs, adjectives, and adverbs, so it will not contain articles, prepositions, pronouns, conjunctions such as "the," "a," and "because.".

To appreciate WordNet's full functionality let's start with the word "strawberry." At first glance, WordNet appears to be identical to a dictionary because it lists "strawberry" as a noun with several possible definitions. However, what distinguishes WordNet is the word strawberry is graphically **connected** to other related words in the database.

For each definition of the word "strawberry" WordNet provides a series of synsets (synonym sets) which comprise sister terms, hypernyms, hyponyms, holonyms (for nouns), meronyms (for nouns), troponyms (for verbs), entailment (for verbs), and derivationally related forms. Not all word definitions have entries for every type of synset.

Sister terms are similar to synonyms. For example, strawberry has many sister terms that each come with their own definitions and connections. For example, the first definition of strawberry – "sweet fleshy red fruit" – relates to sister terms that include bilberry, huckleberry, blueberry, cranberry, boysenberry, and blackberry. Note that all sister terms are similar fleshy fruit that can be consumed.

The *hypernyms* associated with the "strawberry" definition "sweet fleshy red fruit" are a series of terms that traverse a range from the most concrete to the most abstract. There may exist one to many branches of hypernyms. For our particular case, WordNet shows two branches of hypernyms. The following is one of the example hypernyms chains for our word: strawberry ➔ berry ➔ edible fruit ➔ produce, green goods, green groceries, garden truck ➔ food, solid food ➔ solid ➔ matter ➔ physical entity ➔ entity.

The hypernym example demonstrates how WordNet depicts relationships with words, in this case from specialized to generalized descriptors. Every word in WordNet is connected through a type of relationship with another word.

Hyponyms, by contrast, represent more narrowly defined subtypes of words, essentially navigating the opposite direction of hypernyms toward more specificity. From these descriptions, we can see that based on the word's definition hypernyms divulge the more general or abstract types whereas hyponyms emphasize the more specific descriptors.

The *holonyms* of "strawberry" explore a different dimension from hypernyms by essentially classifying the word as a member of a group, akin to the taxonomies used in the life sciences. To illustrate, the listed holonyms for "strawberry" with definition "sweet fleshy red fruit" are the following: garden strawberry, cultivated strawberry, Fragaria ananassa (widely cultivated strawberry). From this list of holonyms, we can state that strawberry is an instance of garden strawberry, cultivated strawberry, or Fragaria ananassa.

A *meronym* denotes a part of the whole. A *part meronym* in text analysis refers to the different parts or characteristics of the word. For example, for the word "leaf" with definition "leaf, leafage, foliage (the main organ of photosynthesis and transpiration in higher plants)" there are three part meronyms listed: venation (the veins in the leaf), lobe (how leaves are divided), and leaf shape or leaf form.

There is also a *substance meronym* that denotes composition. For example, the substance meronym for leaf lists the following: "parenchyma (the primary tissue of higher plants composed of thin-walled cells that remain capable of cell division even when mature; constitutes the greater part of leaves, roots, the pulp of fruits, and the pith of stems)." In other words, a "leaf" is composed of parenchyma.

Troponyms are a particular manner or way of doing something. For example, whispering and shouting are both a manner or method of talking, and thus are appropriate troponyms.

Entailment is similar to troponyms but establishes logical extensions to other actions. More specifically, entailment means if you are doing one thing then by logical deduction we can assume you are also simultaneously doing something else. Relevant examples are that to be snoring you must also be sleeping, or walking implies you are using your legs.

The entries for derivationally related forms provide the appropriate variants of the word based on parts of speech usage. For example, the word "leaf" produces the following derivationally related forms: (adj.) leafy, (n.) leaflet, (v.) leaf, and (v.) foliage.

Overall, WordNet is a very powerful tool that has attained a significant level of maturity since its inception in 1985. There are a wide range of interesting linguistic analyses possible given WordNet's scope and functionality. From being able to retrieve definitions to comprehensively capturing broad lexical relationships, WordNet is a useful NLP resource. Like most things in life, how you choose to leverage WordNet is left to your imagination.

The following is a brief example demonstrating the use of WordNet through the NLTK library:

```
File: WordNet_example.py:
from nltk.corpus import wordnet as wn

# According to the documentation: There are four possible parts of speech (POS):
# verb, noun, adjective, and adverb. Synsets are identified in a 3-part form: word.pos.nn
data = wn.synset('data.n.01')
print(f'Definition: {data.definition()}')
print(f'Examples: {data.examples()}')
print(f'Hyponyms: {data.hyponyms()}')
print(f'Member holonyms: {data.member_holonyms()}')

tab = ''
hypernym = data.hypernyms()
print(f'\nHypernyms of {data}. From {data} to {data.root_hypernyms()[0]}:')
while len(hypernym) > 0:
    print(tab, hypernym[0])
    tab += '\t'
    hypernym = hypernym[0].hypernyms()

print(f'The similarity score between {data} and {wn.synset("science.n.01")} is {data.path_
similarity(wn.synset("science.n.01"))}.')
print(f'The similarity score between {data} and {wn.synset("metadata.n.01")} is {data.path_
similarity(wn.synset("metadata.n.01"))}.')
```

Which produces the following output:

```
Definition: a collection of facts from which conclusions may be drawn
Examples: ['statistical data']
Hyponyms: [Synset('accounting_data.n.01'), Synset('metadata.n.01'), Synset('raw_data.n.01')]
Member holonyms: []


Hypernyms of Synset('data.n.01'). From Synset('data.n.01') to Synset('entity.n.01'):
  Synset('collection.n.01')
          Synset('group.n.01')
                  Synset('abstraction.n.06')
                          Synset('entity.n.01')
The similarity score between Synset('data.n.01') and Synset('science.n.01') is 0.1.
The similarity score between Synset('data.n.01') and Synset('metadata.n.01') is 0.5.
```

Aside from a general demonstration of using WordNet programmatically, the last two statements in the above code provide examples of the `path_similarity` function. The `path_similarity` metric is based on the shortest path in relation to the is-a (hypernym/hyponym) hierarchy. Other similarity functions are available based on different properties of WordNet's lexical graph.

Overall, WordNet represents a major milestone for computational linguistics and has rightfully deserved many awards and commendations. The ability to programmatically analyze inferences and complex relationships among words in documents rather than applying painstaking manual techniques is a huge achievement for the NLP/data science community.

9.6   Natural Language Understanding, and Natural Language Generation

Both *Natural Language Understanding (NLU)* and *Natural Language Generation (NLG)* are considered subdisciplines of NLP. NLP is a broad term while NLU and NLG are more specific.

NLU refers to strategies and implementations that allow a computer to understand human language naturally (thus the 'N'). The term "naturally" is primarily descriptive of the manner of communication occurring between human and computer. More specifically, from the perspective of the human user conversations with a computer would be experientially indistinguishable from the verbal exchanges that normally take place between two persons. A key implication of NLU is that during these conversations a computer would comprehend intention and meaning comparable to or better than another human.

The holy grail for NLU developers is the successful *Turing test* scenario in which an interrogator chatting (either verbally or through text) with either a computer or another person over the Internet or some other channel of communication would not be able to reliably distinguish the identity of their conversation partner. In other words, if you are chatting with someone over the Internet, are you chatting with another person or a computer?

NLG represents the other side of the coin from NLU. NLG refers to the generation (thus the 'G') or the creation of language designed to be understood by a human recipient. In our example of chatting over the Internet, the human-generated input to the computer would be interpreted through NLU and the computer's response or output would be the product of NLG. An apt analogy is NLU corresponds to the computer's ears and NLG represents the computer's voice.

The hard problem of NLP is refining and perfecting NLU and NLG to the point where a person could not determine if the entity with whom they are communicating is a computer or a person (thus passing the Turing test). Some computer scientists claim this challenge is one of the most difficult tasks currently confronting artificial intelligence scientists.

The ultimate goal envisioned by the AI community is commonly depicted in works of science fiction as a casual, natural conversation between a person and computer. In this perfect scenario neither the person nor the computer would require any specialized training prior to the conversation. Instead, they would just chat naturally back and forth. This vision may be realized 1 day, but technological and linguistic challenges remain.

The simplest versions of NLU are delivered in the form of easy, concise commands. These straightforward verbal commands have clear, distinguishable boundaries and are becoming more prevalent with voice activation systems such as Google speech, Apple's Siri, and Amazon's Alexa. These personal digital assistant systems are constantly improving and have worked their way into mainstream, day-to-day usage.

Admittedly, the actual comprehension capabilities of these systems are far from ideal. Consider the subtle mechanisms and finely tuned algorithms that would be required for a computer employing perfect (or at least human equivalent) NLU to understand the subtleties of sarcasm, or to summarize text as a person does when providing a short synopsis of a book recently read. Consider the computational complexities inherent in performing sentiment and emotional analysis (see Sect. 9.9), and providing sentence and grammar correction.

If the ideal of NLU is to understand humans as they speak naturally without any formalized syntax or training, then the analogous ideal of NLG is expressed in Roald Dahl's short story *The Great Automatic Grammatizator* about a mechanical genius that creates a machine that generates novels of any genre. The machine is so adept and efficient that highly-profitable novels can be produced in just minutes. The cautionary message put forth by this short story is yet another example of automation replacing human workers, in this case authors.

Similar to the current state of NLU, NLG is far from creating a massive volume of highly profitable books, but NLG technologies have been utilized to effectively respond to people's inquiries. Chatbots responding to customers are getting more sophisticated and saving companies time and money. Highly specific automated responses from voice and text systems are both more prevalent in modern life and improving every day.

Although NLU and NLG may never reach the pinnacle imagined by science fiction authors, they are becoming more entrenched in current mainstream activities.

For the developer interested in these technologies, there are various NLU and NLG libraries and programs including rasa, snip-nlu, Luis.ai, and IBM Watson.

9.7 Collocations/*N*-Grams

What are collocations? In a nutshell, a *collocation* is a clearly defined thought unit or semantic concept expressed by two or more consecutive words. A collocation example using two words is "strong tea" whereas an 8 word collocation instance is something like "The President of the United States of America."

Another commonly used term for collocations is *N-grams*. There are many varieties of *N*-grams defined by the value of N: unigram (single word), bigram (two words), trigram (three words), quadgram (four words) and so on. However, because referencing an octogram (eight words) in conversation might be confusing and awkward, people will refer to these word groups as *N*-grams where a single word is a 1-gram, two words is a 2-gram, and so on. **Note:** unigrams are not actually collocations because they only involve one word.

Given the origin of the terminology associated with N-grams, why are these particular collections important?

Since one of the primary goals of NLP is to gain some level of insight into text structure and semantics, as data scientists we need to be cognizant of a fundamental property of natural language. Although a single word in isolation may have a certain meaning, a combination of two or more words will often represent a concept vastly different from the individual meanings of its constituent words.

For instance, consider the word "baby." "Baby" has a clear definition that can be found in any dictionary. However, for our purposes, we consider the popular understanding of the word as "a human infant." A "toy" generally means "a physical thing that people play with" such as a ball, Rubik's Cube, or doll.

Focusing solely on single words can easily lead to missing a great deal of context and meaning. For example, does "baby toy" possess two separate but simultaneous definitions: a human infant **and** a physical thing that people play with? No, combining these two words invokes deeper, assorted meanings.

The phrase "baby toy" could mean a physical thing designed to engage and amuse human infants. However, it also might have a more pejorative connotation, such as, "Company X's latest car is nothing more than a baby toy."

Regardless of the explicit or implied meaning of "baby toy," the sequence "baby toy" is clearly not interpreted or understood the same as "baby" and "toy" considered separately and independently. Statistically speaking, this means that the words comprising the bigram "baby toy" are dependent on each other for meaning.

An *N*-gram can be any length, but has only one perceived meaning. For example, the 8-gram "The President of the United States of America" is referring to a single person. In other words, although there are eight words that make up this collocation, it only has one clearly defined meaning, not eight. Interestingly, because of the sheer length of this title there is also an acronym for this person who holds the highest office in the United States: POTUS. Whether someone refers to "POTUS" or "The President of The United States of America" the idea is the same and understood by the hearer as equivalent expressions.

Thus, collocations are a key tool for gaining broader and deeper insights into textual meaning. Collocations embody a combination of words and produce a single, self-contained, more contextual thought rather than simply representing individual, disjoint words when deciphering meaning.

Figure 9.2 illustrates the bigrams for a given sentence: [The watermelon], [watermelon is], [is considered], [considered ripe], [ripe when], [when the], [the vine], [vine that], [that it], [it connects], [connects to], [starts to], [to wither.].

Fig. 9.2 Depiction of all the bigrams (or 2-grams) for a single sentence

Figure 9.3 shows the trigrams for the same sentence: [The watermelon is], [watermelon is considered], [is considered ripe], [considered ripe when], [ripe when the], [when the vine], [the vine that], [vine that it], [that it connects], [it connects to], [connects to starts], [starts to wither.].

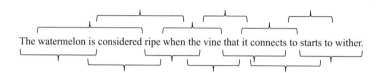

Fig. 9.3 Depiction of all the trigrams (or 3-grams) for a single sentence

The following Python code retrieves N-grams of all lengths for a given input text:

```
Partial file: full file in n_gram_production.py:

def add_collocation(collocation, collocation_length, results):
    if collocation_length not in results:
        results[collocation_length] = dict()
    if collocation not in results[collocation_length]:
        results[collocation_length][collocation] = 1
    else:
        results[collocation_length][collocation] += 1

def get_collocations(text):
    results = dict()
    words = text.split()
    if len(words) == 0:
        return
    for i in range(1, len(words)):
        beg = 0
        end = beg + i
        while end <= len(words):
            collocation = " ".join(words[beg:end])
            add_collocation(collocation, i, results)
            beg += 1
            end += 1
    return results
```

The above code will compute all *N*-grams that are possible. With our example sentence "The watermelon is considered ripe when the vine that it connects to starts to wither," the code will produce all the 1-grams to 14-grams (15-grams and above are not possible with this short sentence). However, in practice it is not generally necessary to compute collocations beyond the quadgram (4-gram). The previous example highlighting the special importance of the 8-gram "The President of the United State of America" is more the exception than the rule. Significant collocations of five, six, or even seven words are rare.

Also, the NLTK package in python will obtain all *N*-grams of size *N*. The following Python code example retrieves all collocations of size 2 through 5:

```
File: nltk_n_gram_production.py:
import nltk
import string
text = "This is example text. Getting collocations from a small amount of text is not very
valuable. Collocations are usually performed on thousands to millions of words."
# this next step removes all punctuation from the text and
# makes all the text lowercase:
text = text.translate(str.maketrans('','',string.punctuation)).lower()
# this next step breaks the string into individual words:
text2 = nltk.word_tokenize(text)
for i in range(2, 6): # for loop that generates all 2-grams to 5-grams
        print(f"{i}-grams: \n{list(nltk.ngrams(text2, i))}")
```

Which produces the following results:

```
2-grams:
[('this', 'is'), ('is', 'example'), ('example', 'text'), ('text', 'getting'), ('getting',
'collocations'), ('collocations', 'from'), ('from', 'a'), ('a', 'small'), ('small',
'amount'), ('amount', 'of'), ('of', 'text'), ('text', 'is'), ('is', 'not'), ('not', 'very'),
('very', 'valuable'), ('valuable', 'collocations'), ('collocations', 'are'), ('are', 'usu-
ally'), ('usually', 'performed'), ('performed', 'on'), ('on', 'thousands'), ('thousands',
'to'), ('to', 'millions'), ('millions', 'of'), ('of', 'words')]
3-grams:
[('this', 'is', 'example'), ('is', 'example', 'text'), ('example', 'text', 'getting'),
('text', 'getting', 'collocations'), ('getting', 'collocations', 'from'), ('collocations',
'from', 'a'), ('from', 'a', 'small'), ('a', 'small', 'amount'), ('small', 'amount', 'of'),
('amount', 'of', 'text'), ('of', 'text', 'is'), ('text', 'is', 'not'), ('is', 'not',
'very'), ('not', 'very', 'valuable'), ('very', 'valuable', 'collocations'), ('valuable',
'collocations', 'are'), ('collocations', 'are', 'usually'), ('are', 'usually', 'performed'),
('usually', 'performed', 'on'), ('performed', 'on', 'thousands'), ('on', 'thousands', 'to'),
('thousands', 'to', 'millions'), ('to', 'millions', 'of'), ('millions', 'of', 'words')]
4-grams:
[('this', 'is', 'example', 'text'), ('is', 'example', 'text', 'getting'), ('example',
'text', 'getting', 'collocations'), ('text', 'getting', 'collocations', 'from'), ('getting',
'collocations', 'from', 'a'), ('collocations', 'from', 'a', 'small'), ('from', 'a', 'small',
'amount'), ('a', 'small', 'amount', 'of'), ('small', 'amount', 'of', 'text'), ('amount',
'of', 'text', 'is'), ('of', 'text', 'is', 'not'), ('text', 'is', 'not', 'very'), ('is',
'not', 'very', 'valuable'), ('not', 'very', 'valuable', 'collocations'), ('very', 'valu-
able', 'collocations', 'are'), ('valuable', 'collocations', 'are', 'usually'), ('colloca-
tions', 'are', 'usually', 'performed'), ('are', 'usually', 'performed', 'on'), ('usually',
'performed', 'on', 'thousands'), ('performed', 'on', 'thousands', 'to'), ('on', 'thousands',
'to', 'millions'), ('thousands', 'to', 'millions', 'of'), ('to', 'millions', 'of', 'words')]
5-grams:
[('this', 'is', 'example', 'text', 'getting'), ('is', 'example', 'text', 'getting', 'collo-
cations'), ('example', 'text', 'getting', 'collocations', 'from'), ('text', 'getting',
'collocations', 'from', 'a'), ('getting', 'collocations', 'from', 'a', 'small'), ('colloca-
tions', 'from', 'a', 'small', 'amount'), ('from', 'a', 'small', 'amount', 'of'), ('a',
'small', 'amount', 'of', 'text'), ('small', 'amount', 'of', 'text', 'is'), ('amount', 'of',
'text', 'is', 'not'), ('of', 'text', 'is', 'not', 'very'), ('text', 'is', 'not', 'very',
'valuable'), ('is', 'not', 'very', 'valuable', 'collocations'), ('not', 'very', 'valuable',
'collocations', 'are'), ('very', 'valuable', 'collocations', 'are', 'usually'), ('valuable',
'collocations', 'are', 'usually', 'performed'), ('collocations', 'are', 'usually', 'per-
formed', 'on'), ('are', 'usually', 'performed', 'on', 'thousands'), ('usually', 'performed',
'on', 'thousands', 'to'), ('performed', 'on', 'thousands', 'to', 'millions'), ('on', 'thou-
sands', 'to', 'millions', 'of'), ('thousands', 'to', 'millions', 'of', 'words')]
```

Note the large number of different collocations produced from a small amount of text! Which of these N-grams should we emphasize and subsequently analyze? Fortunately, the next section describes helpful techniques to determine which collocations are important and which ones are not.

9.8   Scoring Collocations

The previous section presented the concept of collocations, which are essentially N-grams constituting N number of words that express a single human phrase, concept, or thought. However, the core challenge arising from the use of N-grams in textual analysis is automatically determining which collocations are the most important from a large corpus.

There are many different methods available to determine which collocations are important. The common approaches include Pointwise Mutual Information (PMI), t-test, and Chi-Squared test. Since the t-test and Chi-Squared test are both covered in Chap. 5, in this section we will examine in depth only PMI.

Pointwise Mutual Information (PMI) derives from statistics and measures the association of independent events. The main idea behind the use of PMI in NLP is to compare the likelihood of words found together (dependently) against the likelihood that these same words are found separately and independently in the corpus under analysis. The following represents the generalized formula for PMI where x and y are two events being measured:

$$PMI(x,y) = \log_2 \frac{P(x,y)}{P(x)P(y)}$$

The above formula is generic and not very applicable for NLP purposes. Instead, if we substitute w_1 for x and w_2 for y, where w_1 is the first word and w_2 is the second word of interest then the following formula for measuring bigrams is produced:

$$BigramPMI(w_1,w_2) = \log_2 \frac{P(w_1,w_2)}{P(w_1)P(w_2)}$$

Similarly, we can extend the PMI formula for trigrams where w_1, w_2, w_3 represent three words for a trigram:

$$TrigramPMI(w_1,w_2,w_3) = \log_2 \frac{P(w_1,w_2,w_3)}{P(w_1)P(w_2)P(w_3)}$$

Following the pattern described above, the PMI formula can be extended for any size of N-gram such as quadgrams (4-gram) and pentograms (5-gram).

You might be wondering how this PMI formula can be utilized in practice on a problem in textual analysis. For an illustrative, practical example, we will look at the PMI calculation for the "don't know" bigram in a small corpus. We seek to determine how likely will the bigram phrase of "don't know" appear in the corpus when compared to the occurrences of the individual words "don't" and "know" considered independently in the text. In other words, how often does "don't know" appear in the text compared to all the times that "don't" and "know" are counted as separate and distinct units.

Consider the following text as our input: *"'I don't know', said Sally. 'I know that she went to the store. But I don't doubt that she came home right after. I don't know that it matters.'"*

For this example, we will count how many times the bigram "don't know" appears and how many times the words "don't" and "know" appear independently:

- "don't know": 2
- "don't": 3
- "know": 3

Note that when we calculate the total number of times an individual word like "know" appears independently, we include the times it appears next to the companion word "don't." In effect, when counting words in isolation we ignore the existence of any N-grams under investigation and view each word in the corpus as a separate unit. To determine the PMI for the "don't know" bigram we first need to determine the frequency of all the unigrams for individual words $P(\text{"don't"})$ and $P(\text{"know"})$.

In the above example text, there are a total of 29 words or unigrams. The probability of "don't" or $P(\text{"don't"})$ is therefore 3/29 or three occurrences out of the possibility of 29 words, which equals approximately 0.1, or 10%. The value for $P(\text{"know"})$ is also 3/29, so $P(\text{"know"}) \approx 0.1$, or 10%. In other words, the probability of randomly selecting a word from the corpus and retrieving "don't" is about one in ten. Similarly, the probability of randomly choosing a word from the corpus and obtaining "know" is about one in ten.

By comparison, $P(\text{"I"}) = 4/29 = 0.138$ and $P(\text{"Sally"}) = 1/29 = 0.034$. Intuitively, randomly picking the word "I" has a higher probability than any other word because it appears more than any other word and randomly picking "Sally" is at the low end of our probability range because it only appears once.

Given our calculation of $P(\text{"don't"})$ and $P(\text{"know"})$, we now determine $P(\text{"don't", "know"})$. In other words, what is the probability of randomly selecting the bigram "don't know" from all the other possible bigrams in the text. The bigrams are defined as every two-word combination. There are a total of 28 bigram combinations from the above text with the first few examples shown here:

- "I don't"
- "don't know"
- "know said"
- "said Sally"
- "Sally I"
- "I know"

Given 28 possible bigrams and the fact that "don't know" occurs twice in the text then $P(\text{"don't", "know"}) = 2/28 \approx 0.07$ or 7%. Recall the bigram PMI basic formula shown here:

$$BigramPMI\left(w_1, w_2\right) = \log_2 \frac{P\left(w_1, w_2\right)}{P\left(w_1\right)P\left(w_2\right)}$$

Substituting the words "don't" and "know" for w_1 and w_2 we derive the following:

$$BigramPMI\left(\text{"don't", "know"}\right) = \log_2 \frac{P\left(\text{"don't", "know"}\right)}{P\left(\text{"don't"}\right)P\left(\text{"know"}\right)}$$

Plugging in the probabilities for "don't" and "know" produces the following result:

$$BigramPMI\left(\text{"don't", "know"}\right) = \log_2 \frac{0.07}{0.1 * 0.1} = 2.74$$

The `bigram_pmi` function in Python listed below generates the same result where `unigram_frequency` is a dictionary of the frequencies of every unigram (i.e., each individual word) and `bigram_frequency` is a dictionary of the frequencies of every bigram (i.e., every combination of two consecutive words):

```
File: bigram_pmi.py:
def bigram_pmi(bigram_tuple, unigram_frequency, bigram_frequency):
    pw1 = unigram_frequency[bigram_tuple[0]]/float(sum(unigram_frequency.values()))
    pw2 = unigram_frequency[bigram_tuple[1]]/float(sum(unigram_frequency.values()))
    p_w1_w2 = bigram_frequency[bigram_tuple]/float(sum(bigram_frequency.values()))
    return math.log(p_w1_w2/float(pw1*pw2),2)
```

What does a result of 2.74 actually mean for BigramPMI ("don't", "know")? By itself it means nothing, but relative to other bigram PMI values the purpose of this computation starts to become clear. If we compute the BigramPMI of another bigram such as "I know" then we obtain a value of 1.32. Comparing those two bigram results we can deduce that the higher score for the bigram "don't know" seems to indicate that for this text this particular phrase is more important, or at least more prevalent, than the bigram "I know". If we now compute the PMI of every bigram for all the text and sort the results we would obtain a ranking of the relative importance of each bigram.

However, **proceed with caution when calculating N-gram scores**! If you do not filter the words and collocations that appear very infrequently, then you may assess collocations that only appear a few times or even once as the most important. For example, the bigram "it matters" only appears once in the text. Since the words "it" and "matters" also only appear once then the PMI function returns a high score for that bigram because "it" and "matters" only appear together. In this case, *PMI("it", "matters") = 4.9*, which places it in the final ranking list as one of the most important bigrams.

To overcome this vulnerability of the PMI function you simply filter out all words and collocations that rarely appear. For example, for an average sized novel of 60,000 words any N-grams occurring less than ten times (`filter_size=10`) should probably be dropped from consideration. Unfortunately, there is no magic number for filtering out collocations as this threshold depends on the size of the book being analyzed. Obviously, a book with only 10,000 words should have a smaller `filter_size` such as 3. Conversely, the `filter_size` of a 100,000-word book should be much larger, around 15.

At this juncture, we also want to point out that there are two main approaches to examining collocations. Let us look at an example involving bigrams again. Reviewing our original text, *"'I don't know', said Sally. 'I know that she went to the store. But I don't doubt that she came home right after. I don't know that it matters.'"*, we must decide if we wish to consider all possible bigrams or just bigrams that are actual ideas?

People tend to think in terms of discrete, definable thoughts, which translate into complete sentences. The first sentence in the text, *"'I don't know', said Sally."* ends with the word "Sally." Naturally it makes sense to extract the following bigrams from the first sentence: "I don't," "don't know," "know said," and "said Sally." The next sentence begins with *"I know that ..."*. In this case a bigram of "Sally I" built from "Sally", the last word of the first sentence and "I", the first word in the second sentence is semantically and structurally awkward.

Since the NLTK library does not recognize sentence boundaries when creating bigrams, we perform this step explicitly in the following code, which breaks the example text into sentences, removes all punctuation, makes all words lower case, and then extracts all the unigrams and bigrams for each sentence. The frequencies of the unigrams and bigrams are subsequently computed, then all bigrams that do not appear at least twice are filtered. Finally, bigram PMI values are calculated:

```
Partial file: full program in sentence_bigram_pmi.py:

input_text = "'I don't know', said Sally. 'I know that she went to the store. But I don't doubt
that she came home right after. I don't know that it matters.'"

...

filtered_bigrams = []
# remove any bigram's whose frequency is less than 2:

...

bigram_scores = {}
for b in filtered_bigrams:
    print(b)
bigram_scores[b] = bigram_pmi(b, unigram_frequency=unigrams_frequency, bigram_frequency=
bigrams_frequency)

# sort the results based on the score
results = sorted(bigram_scores, key=bigram_scores.get, reverse=True)

print("Most significant bigrams according to PMI:")
print(results)
```

Which produces the following output:

```
Most significant bigrams according to PMI:
[('that', 'she'), ('i', 'dont'), ('dont', 'know'), ('know', 'that')]
```

In other words, the top four bigrams in rank order are "that she," "I don't", "don't know", and "know that." However, a significant amount of coding was required to produce the previous result. The following is a much leaner program that more effectively leverages the NLTK package:

```
File: nltk_sentence_bigram_pmi.py:
import nltk
from nltk.collocations import *
import string

bigram_measures = nltk.collocations.BigramAssocMeasures()

text = "'I don't know', said Sally. 'I know that she went to the store. But I don't doubt that
she came home right after. I don't know that it matters.'"
text = text.lower()  # this makes characters lower case

# the following removes all punctuation and changes the text into individual words:
unigrams = nltk.word_tokenize(text.translate(str.maketrans('','',string.punctuation)))
finder = BigramCollocationFinder.from_words(unigrams)
finder.apply_freq_filter(2)  # filter out all bigrams that appear less than twice
print("Most significant bigrams according to NLTK PMI:")
print(finder.nbest(bigram_measures.pmi, 10))# the '10' means get the top 10 most significant
collocations
```

Which produces the following output identical to the prior program:

```
Most significant bigrams according to NLTK PMI:
[('that', 'she'), ('i', 'dont'), ('dont', 'know'), ('know', 'that')]
```

Both our custom code and the more concise version that extensively employs NLTK utilities produce the same results. It turns out that the extra computing required to tokenize along sentence boundaries is not necessary because the NLTK implementation of the PMI scoring metric automatically ignores any N-grams that cross sentence boundaries.

In addition, the NLTK library allows us to not only write code faster (because we don't have to write as much), but also to utilize NLTK's other built-in scoring functions. At the time of this writing, in addition to PMI, NLTK can also use the t-test, Chi-Squared test, and likelihood ratio methods. Likelihood ratio is a competing statistical method for determining goodness of fit, like the Chi-Squared test.

The four different metrics are demonstrated in the following code (assuming the above program is executed first):

```
Partial code: full program in nltk_sentence_bigram_pmi2.py:

print(finder.nbest(bigram_measures.pmi, 10))
print(finder.nbest(bigram_measures.student_t, 10))
print(finder.nbest(bigram_measures.chi_sq, 10))
print(finder.nbest(bigram_measures.likelihood_ratio, 10))
```

Which produces the following output:

```
[('that', 'she'), ('i', 'dont'), ('dont', 'know'), ('know', 'that')]
[('i', 'dont'), ('that', 'she'), ('dont', 'know'), ('know', 'that')]
[('i', 'dont'), ('that', 'she'), ('dont', 'know'), ('know', 'that')]
[('i', 'dont'), ('that', 'she'), ('dont', 'know'), ('know', 'that')]
```

9.9   Sentiment and Emotion

Sentiment and emotion analysis are some of the most useful NLP that data scientists perform. These strategies are intended to increase understanding of text and may be considered part of Natural Language Understanding (NLU).

We first address the difference between sentiment and emotion. Emotions are complex and varied. At any given moment you might feel dozens of emotions simultaneously. There are various psychological emotion theories that describe and classify emotions spanning a range from a few primary ones to over 100 possible categories.

One of the easiest psychological emotion theories to grasp is the six universal facial emotions. These six fundamental emotions can be observed in every human culture around the world regardless of language. This particular emotion categorization model has been popularized in Pixar's *Inside Out* movie in which the six emotions are characterized as follows: happy, sad, surprise, fear, anger, and disgust.

A psychologist named Dr. Paul Ekman and his team traveled the globe to both industrialized and non-industrialized countries and asked people to make faces when they felt certain emotions. For example, people could be asked to demonstrate the facial expression they might make upon hearing the sad news that their mother died. Dr. Ekman and his team then showed the resulting image to people around the globe. Universally people declared that person as sad.

When performing emotion analysis connected with textual expression the main motivation is to capture the emotions most likely being felt by the author. At the time of this writing, the practice and implementation of emotion analysis is in its infancy. First, there is no general consensus in the research community about what emotions are primary and should be utilized for analysis purposes. Questions such as "Should we use the six universal facial emotions?" or "If we include more emotions then which ones?" remain largely unresolved.

Another problem involves salience, or the degree of emotion. For example, if someone feels surprise and happiness, how do we quantify or measure the mixed levels of surprise and happiness? Are they extremely surprised and only mildly happy?

In the NLP community sentiment is considered to be based on emotion but is fundamentally ternary. In other words, there are only three general sentiments: positive, neutral, and negative. The idea underpinning sentiment analysis applied to text is determining if the author who wrote the text was feeling more positive, neutral, or more negative.

The following python code shows an example of the analysis of three sentences with the NLKT VADER (Valence Aware Dictionary and sEntiment Reasoner) library:

```
File: sentiment_analysis.py:
from nltk.sentiment.vader import SentimentIntensityAnalyzer
#import nltk
#nltk.download('vader_lexicon')  # this only has to be done once!)

corpus = ["I love to eat delicious pie on Sundays!",
        "My father died and I am horribly depressed.",
        "Watermelon grows on vines."]

analyzer = SentimentIntensityAnalyzer()

for sentence in corpus:
    print(f'The sentence: {sentence}\nThe score: {analyzer.polarity_scores(sentence)}\n')
```

Which produces the following output:

```
The sentence: I love to eat delicious pie on Sundays!
The score: {'neg': 0.0, 'neu': 0.379, 'pos': 0.621, 'compound': 0.8478}

The sentence: My father died and I am horribly depressed.
The score: {'neg': 0.72, 'neu': 0.28, 'pos': 0.0, 'compound': -0.8834}

The sentence: Watermelon grows on vines.
The score: {'neg': 0.0, 'neu': 1.0, 'pos': 0.0, 'compound': 0.0}
```

The above code produced scores for each of the three sentences. The `neg` key means negative, the `neu` key means neutral, the `pos` key means positive, and the `compound` key is an aggregate, single result. The negative, neutral, and positive results are simple to interpret and are normalized so they sum to 1.0. The first sentence is deciphered as more positive, the second sentence expresses a more negative situation, and the third sentence is considered declarative and neutral.

The `compound` result combines the values from `neg`, `neu`, and `pos` and is useful if a summary score is desired rather than three separate component scores. The `compound` value effectively quantifies the text as positive, neutral, or negative within the range −1.0 to 1.0 inclusive with −1.0 being completely negative, 0 being completely neutral, and 1.0 being completely positive.

Sentiment analysis is typically based on a dictionary of words with each entry labeled as either positive or negative. Words not included in the dictionary are considered neutral.

For example, if the designer that created the dictionary decided that "delicious" is positive then it is always considered positive, whereas if the word "depressed" were deemed negative then it is always considered negative.

Most sentiment analysis algorithms scan the corpus and count all the positive and negative words then offer a summative assessment indicating that the text is either positive, neutral, or negative. Alternatively, the VADER algorithm computes the normalized amount of positive, neutral, and negative sentiment.

At first glance, there are many obvious limitations to sentiment analysis. For example, what if there is disagreement with the creators of the sentiment dictionary such that the word "watermelon" should be labeled negative rather than neutral. Instead, what if "watermelon" should be positive, or what if "watermelon" should be positive in certain cases and negative other times depending on the context? In addition, what about the subtle nuances and tones in written language like sarcasm?

Sentiment analysis has only been around since the beginning of the current millennium and emotion analysis is even more recent. There is much activity in the research community on this subject and we are confident that many improvements in this discipline will be realized in the future.

Regardless of their current effectiveness, sentiment and emotion analysis are worthy tools to consider when attempting to understand the feelings an author may have intended to impart to their readers.

Exercises

1. What is the point of NLP?
2. How does Bag of Words work?
3. Why is TFIDF usually considered preferable to Bag of Words?
4. What is the rationale for removing stop words?
5. Should punctuation and capitalization be removed from the input of NLP algorithms? Why?
6. Why is feature independence in Naïve Bayes important?
7. When would you use stemming over lemmatization and vice versa?
8. WordNet is considered an amazing technological breakthrough. Why?
9. List three usage examples of WordNet.
10. What is so hard about NLU?
11. What is so hard about NLG?
12. What is the Turing test?
13. What is the difference between N-grams and collocations?
14. How can you use collocations in your data science work?
15. In your own words, what is the purpose of scoring collocations?
16. What is the difference between sentiment analysis and emotion analysis?
17. Provide at least three instances when you would use sentiment analysis and three separate instances when you would use emotion analysis.

Chapter 10
Time Series

A *time series* is an ordered sequence of data points equally spaced based on a designated time interval. An essential parameter for defining a time series is the *frequency*, which is the sample rate or number of equally spaced samples per unit of time. The time can be any unit such as nanoseconds, minutes, weeks, months, years, decades, or centuries. The temporal difference between any two consecutive points is called the *time step*. In data science we are primarily interested in both discovering patterns in an existing time series and forecasting or extrapolating for future trends.

Forecasting is a form of prediction that depends on time as a prime component of the analysis. In Chap. 5 we investigated the *Titanic* data. Our prediction examples with respect to the *Titanic* dataset did not include a time dimension but instead examined the set of features that could be leveraged to determine if a selected passenger survived or not. The key difference between prediction and forecasting is that forecasting is a method of prediction that requires the element of time.

There are many forecasting applications and the most immediately recognizable and popular ones are those related to financial markets and economic trends. For example, based on current activity and operations, will a company's revenue go up or down? By how much? Non-financial forecasting examples include weather conditions, the prevailing public sentiment about voting, the population of countries, energy consumption, and expectations regarding supply chain management.

The act of developing and proffering a forecast based on the past sustained behavior of a time series is in direct contrast to the theory expressed by the *random walk hypothesis (RWH)*, which states that a given time series, such as the stock market, exhibits only random price fluctuations at each successive step and cannot be reliably forecast. The fundamental idea of *random walk* is a time series starts at some predefined value such as 0, and at every subsequent step along the walk the next recorded value increases or decreases by 1 with equal probability, similar to the Bernoulli trial of flipping a fair coin. Is the time series under investigation truly random and thus generated by a random walk? If so, then forecasting is not feasible. However, if the time series is not random then there are various ways to forecast the succeeding set of values.

Similar to machine learning (see Chap. 8) in which the results of a trained model are compared to the base case of the majority class or the mean of a continuous measurement, all forecasting predictions can be compared to the *naïve approach* also known as the *naïve model*. The naïve approach is to simply forecast the current value into the next subsequent time step. All trends and other parts of time series are simply forecast as the currently reported values. For example, using the naïve approach we forecast tomorrow's high temperature to be the same as todays. We now have a base case against which to compare all other forecasting models.

Figure 10.1 displays an example time series of monthly US energy consumption from January 1973 to July 2021 in quadrillion BTU's (British Thermal Units). A BTU is a standard, universal energy measurement that translates well from one form of energy to another, and facilitates the comparison between different types of energy, such as nuclear energy and coal energy.

R. Ball, B. Rague, *The Beginner's Guide to Data Science*, https://doi.org/10.1007/978-3-031-07865-1_10

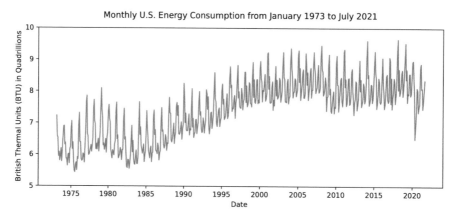

Fig. 10.1 Plot of U.S. energy consumption from January 1973 to 2021 in British Thermal Units (BTU) in quadrillions. (Image created from time_series_first_visualization.py and source data from eia.gov (U.S. Energy Information Administration))

Time series analysis is a major data science research tool in a variety of disciplines such as economics, finance, sociology, electronics, medicine, meteorology, zoology, and seismology. Regardless of the domain, time series definitions and terminology are essentially the same. We will discuss the general terms that are relevant across all time series data analysis then present a selection of popular algorithms used in forecasting.

Noise, *white noise*, or *residuals* are terms that describe the unpredictable behavior of a time series, comprising attendant segments of the series that cannot be adequately explained. These elements of the analysis can be considered error and do not help forecast future points in the time series.

A *level* or *level index* of a time series is the mean of the series.

A *trend* is the general, identifiable pattern the time series exhibits. For example, Fig. 10.2 plots Apple's stock price over a five-year period with the general trend overlayed. Despite the smaller fluctuations, the stock price reveals an overall increasing trend. Trend lines are not always simple linear regressions, but can be as sophisticated as necessary, a property illustrated in more detail below.

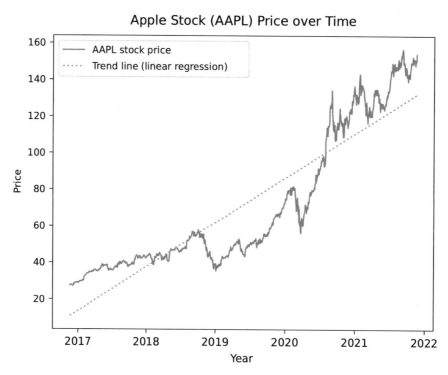

Fig. 10.2 Apple Stock (AAPL) trend line displayed. The trend line was created by a linear regression. (Image is created from apple_stock_five_years_visualization.py)

There are three primary types of trends: positive trend, negative trend, or no trend. Positive and negative trends are exactly as described in that the time series tends to incline (go up) or decline (go down) respectively. A time series with no or effectively zero trend will randomly fluctuate around a constant value.

A trend line can be described by a polynomial of any degree. For instance, the resulting trend line can be a simple line as highlighted in Fig. 10.2 or it can be a more complex polynomial such as a quadratic curve. Recall in Sect. 6.2, Fig. 6.6 presents four different polynomial trend lines for the time series data of an individual dieter, ranging from a first degree polynomial to a fourth degree polynomial.

Any degree of polynomial should suffice for your trend line. However, the higher the degree of polynomial utilized, the greater the probability of overfitting the trend line. For more on overfitting see Chap. 8.

Also, what is considered noise or error in a time series such as the one shown in Fig. 10.2 is relative to the viewer and the objectives of the analysis. One person might scan Fig. 10.2 and surmise that purchasing Apple stock in 2017 and ignoring the noisy ups and downs would have resulted in a considerable profit 5 years later – this is called the *buy and hold* method for stock market investing, essentially a long-term passive strategy.

However, another person might look at the same data and recognize the potential for buying and selling the same stock for even a greater profit. Consider *day trading*, the strategy employed by an investor who attempts to take advantage of the small stock price oscillations over a single day. In other words, for a given time series, the data points one person considers noise would conversely carry vital information for another individual.

Cycle refers to a discernible repeating pattern in the data. Common examples of cycles from mathematics are sine and cosine waves. All the associated cycle characteristics from mathematics are relevant in time series analysis. *Amplitude* is the unit measure from the *apex*, the top of a wave or its *crest*, to the *trough*, the bottom of a wave. *Frequency* is used to measure how often a wave repeats during a given period. Cycles with high frequency have many waves occurring during a given period while cycles with low frequency exhibit relatively fewer waves during that same period. *Phase* is the offset or shift of a wave from a given reference point.

Static patterns are patterns that do not change. One of the biggest challenges in data science when applying pure mathematics is the concept of infinity. Once a pattern has been identified theoretical mathematics presumes it will continue forever – a static pattern necessarily adheres to this assumption.

Concept drift or *drift* denotes a time series that is changing its prevailing pattern, the opposite of static. From the perspective of statistics, drift implies that the data is transitioning from one distribution to another. Intuitively, most time series long in duration will experience some version of drift, which simply means that times change and so too will the corresponding data.

There are various types of drift. *Sudden drift* occurs rapidly due to unforeseen events. *Recurrent drift* takes place on a regular basis such as seasonal weather. *Gradual drift* transpires slowly over time like most cases of inflation. Drift is a fundamental focus for forecasting which relies on the past to prognosticate the future.

10.1   Seasonality

Seasonality in a time series means there exists a predictable repeating pattern based on regular time intervals, a special case of recurrent drift. In other words, seasonality is a kind of cycle strictly synchronized to identifiable season variations. Temperature represents a classic example of seasonality in which the cycle of temperature increases and decreases depending on the seasons or time of year.

An example of a cycle that is not by definition seasonal would be a stock that noticeably oscillates in a repeated pattern, but the period of time covering a single cycle is not exclusively tied to discernible seasons in the year. Note a time series might possess both seasonality and other non-seasonal cycles.

For example, in Fig. 10.1 a seasonality trend that repeats consistent ups and downs each year is displayed. Figure 10.3 shows the same data plotted in Fig. 10.1, but for only a three-year time span to better visualize how the seasons of the year affect overall energy consumption. You can readily see in Fig. 10.3 the greatest amount of energy consumption is in winter around January and February and the least amount of energy utilization occurs during spring and fall around May and October.

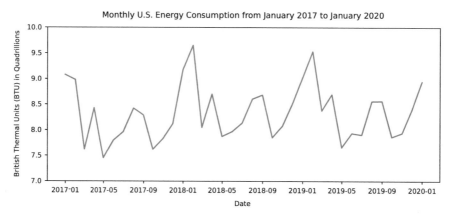

Fig. 10.3 Plot of U.S. energy consumption from January 2017 to 202 in British Thermal Units (BTU) in quadrillions to highlight the seasonality of energy consumption in the U.S. (Image created from time_series_second_visualization.py and source data from eia.gov (U.S. Energy Information Administration))

Seasonality does not have to conventionally track the calendar seasons of spring, summer, fall, and winter as in Figs. 10.1 and 10.3, but can be defined by any regularly repeating pattern in the year. For example, with baseball there is the regular (playing) season, spring training, and off season. Another example is the rain forest, which exhibits an alternating wet season and dry season. In all situations, seasonality implies a tight temporal link or coupling with the calendar year.

There are two types of seasonality: additive and multiplicative. Canonically speaking, for an increasing trend additive seasonality is seasonality **plus** trend and multiplicative seasonality is seasonality **times** trend. For clarity, Fig. 10.4 provides a visual example of additive compared to multiplicative seasonality. In each case the trend is displayed in the first row of graphs, the seasonality in the second row and the type (additive on the left or multiplicative on the right) in the third row.

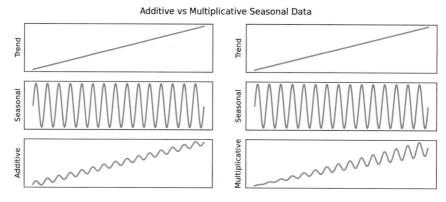

Fig. 10.4 Visual examples of additive vs multiplicative seasonality. (Image from time_series_additive_vs_multiplicative.py)

The difference between additive and multiplicative seasonality is evident in the third row of graphs. Namely, additive seasonality maintains a roughly constant amplitude (distance between the apex and trough of each cycle) regardless of the trend and multiplicative seasonality manifests a steady increase in the amplitude as the time series progresses.

Figure 10.4 is considered a textbook example in that the time series is perfectly rendered. In reality, an additive seasonality in which the amplitude is constant rarely occurs. Similarly, a multiplicative seasonality in which the amplitude increases uniformly in a predictable manner is also uncommon.

Different trends may also be categorized. The main types of trends are constant, linear, damped, and exponential, all of which are shown in Fig. 10.5 with their respective additive and multiplicative seasonalities.

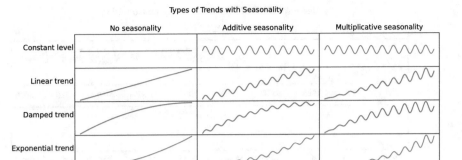

Fig. 10.5 Different types of trends shown with accompanying seasonality. (Image created from types_of_trends.py)

A *constant level trend* means the time series hovers around one single level. Although the time series may fluctuate, as shown with the additive and multiplicative examples, the mean of the time series remains constant.

A *linear trend* means the time series exhibits a general trend reducible by a linear regression. A positive linear trend, as shown in Fig. 10.5, generally increases whereas a negative linear trend generally decreases.

A *damped trend* means the time series begins on a linear trajectory but decays to a constant level. An example damped trend representation is diminishing returns in economics. An initial government stimulus may have a large galvanizing effect on the growth of the economy but every stimulus enacted thereafter has less and less positive effect until a nominal impact is ultimately realized.

An *exponential trend* is effectively the opposite of a damped trend in that it traces an exponentially increasing path as time progresses. An exponential trend typically describes population growth. Whether we consider the number of human inhabitants on earth over the last 10,000 years or count the bacteria in a petri dish, most population growth models will exhibit an exponential trajectory until some noteworthy intervention such as universal birth control for human growth or expiring food and space resources for bacterial growth. Exponential growth is often regarded as unsustainable in the long term.

Note that the linear, damped, and exponential trends can either be positive or negative. Figure 10.5 only depicts constant or positive trends. There are many other kinds of trends expressed in mathematical terms such as monotonic, logarithmic, and parabolic.

Figures 10.1 and 10.3 reveal an additive seasonality in U.S. energy consumption. Once the type of seasonality has been determined the `seasonal_decompose` function can be employed to automatically isolate the trend, seasonality, and residual error from a time series. The following code provides an example of using the `seasonal_decompose` function and produces the plots in Fig. 10.6:

```
From file: decompose_example.py:
import pandas as pd
import matplotlib.pyplot as plt
from statsmodels.tsa.seasonal import seasonal_decompose

df   =    pd.read_csv('US_energy_consumption_1973_to_Jul_2021.csv',   parse_dates=['Date'],
index_col=['Date'])

result = seasonal_decompose(df['total'], model='additive')
result.plot()
plt.suptitle('Decomposition of Total U.S. Energy Consumption')
plt.show()
```

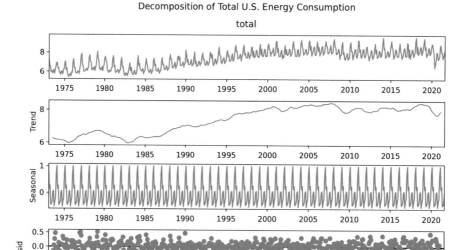

Fig. 10.6 Decomposition of U.S. Energy Consumption. The top graph shows the actual data for total consumption, the second graph displays the general trend line, the third renders the seasonal data, and the fourth plots the residual (error). (Image created from decompose_example.py)

Figure 10.6 displays the original time series in the top row, the overall extracted trend in the second row, the estimated seasonal cycles in the third row, and the residual error in the bottom row. The consistent amplitude in the seasonal data of Fig. 10.6 indicates the total U.S. energy consumption is a classic case of additive seasonality.

10.2   Time Invariant, Structural Breaks, and Piecewise Analysis

Time invariant means the time series response does not vary with time. For example, consider a restaurant patron who pays $1.00 in 1970 for a hamburger listed on the menu. If this price remains the same over time then this amount is time invariant. In reality inflation is a *time dependent* measure so that the $1.00 sufficient to purchase a hamburger in 1970 would not cover the cost of a present-day hamburger.

Structural breaks and *black swan events* are occurrences that deviate from the given trend and represent versions of sudden drift. A structural break reflects a sudden, instantaneous change in the ongoing trend. This conspicuous event might be predictable if coupled with external information. For example, a pharmaceutical company's stock price might remain fairly constant and flat around $10 a share while the company is researching a new drug to fight cancer. However, when the company reports a successful breakthrough in its research then the stock might quickly double or triple in price, a dramatic change which could be predicted based on this information even if the probability of a successful outcome was initially considered low.

Whereas a structural break is any deviation from the general trend, a black swan event is the special case of a rare and unpredictable aberration. In the wild, the black swan bird is nomadic and follows migration patterns that are hard to predict. In the world of time series, especially financial data, a black swan event is usually considered an event that has negative effects on the time series values of interest.

For example, the far-right portion of Fig. 10.1 illustrates the effect of a black swan event, namely the introduction of the coronavirus disease 2019 (COVID-19) to the US at the end of 2019 and the beginning of 2020. This was an unexpected, severe structural break that resulted in a significant drop in energy consumption brought about by a general national lockdown of most factories and major businesses.

Piecewise analysis of a time series is the practice of incorporating structural breaks of a time series in the analysis and effectively dividing the time series into disjoint local time frames. Conceptually, piecewise analysis is highly subjective. The decision of defining the optimal locations where the time series should be divided into different components is not deterministic. An old adage that applies in this case is that if you asked ten analysts to break a time series into meaningful successive components you would end up with ten completely different time series configurations. Figure 10.7 illustrates an example of the Apple stock price broken into three arbitrary component trend lines across a five-year window.

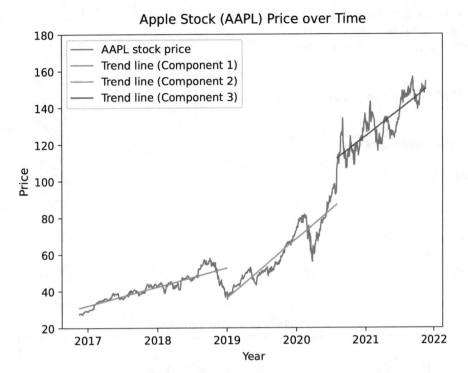

Fig. 10.7 Piecewise analysis of Apple stock price over 5 years. (Image created from apple_stock_five_years_visualization_piece_wise.py)

10.3 Stationarity, Autocorrelation, and Partial Autocorrelation

A time series may be considered *stationary* or *non-stationary*. A non-stationary time series has some time-dependent structure and does not exhibit constant variance over time.

There are two main types of stationary times series: strictly stationary and weak stationary (also known as covariance stationary). A *strictly stationary* time series is one in which different samples of the time series over the same duration will result in identical distributions. We mention strictly stationary time series more for completeness than for any practical value, since they are very rare beyond textbook examples.

A *weak stationary* or *covariance stationary* time series is one that manifests a constant mean, constant variance, and constant covariance between sample periods of equal duration, but in contrast to strictly stationary time series, the probability distribution can change over time. Note the covariance between any two sample points depends only on their relative position and not on their absolute position in the series. Recall that variance measures how a single variable deviates from the mean and covariance measures how two variables vary together with respect to their mean values, similar to correlation. However, in contrast to correlation, covariance is not standardized and measures only the direction of the linear relationship between variables whereas correlation is standardized (between -1 and 1) and measures both the strength and direction of the linear relationship between two variables. Correlation is a function of covariance.

To test for a weak stationary time series we use the *Augmented Dickey-Fuller (ADF)* test. The ADF test posits a null hypothesis H_o: "the time series is non-stationary" with an associated alternative hypothesis H_a: "the time series is stationary." The following code executes the ADF on each of the individual energy time series in the U.S. energy consumption dataset (US_energy_consumption_1973_to_Jul_2021.csv), generating the plot shown in Fig. 10.8:

```
From file: augmented_dickey_fuller_test_example.py:
from statsmodels.tsa.stattools import adfuller
import pandas as pd
import matplotlib.pyplot as plt

df     =    pd.read_csv('US_energy_consumption_1973_to_Jul_2021.csv',    parse_dates=['Date'],
index_col=['Date'])
```

```
for column in df.columns:
    p_value = adfuller(df[column])[1]
    print(f'{column}: {p_value:.2f}')

df.plot()
plt.title('US Energy Consumption by type')
plt.show()
```

Which produces the following output:

```
natural_gas: 0.99
coal: 0.88
crude_oil: 0.13
other: 0.15
nuclear: 0.02
renewables: 0.99
total: 0.57
```

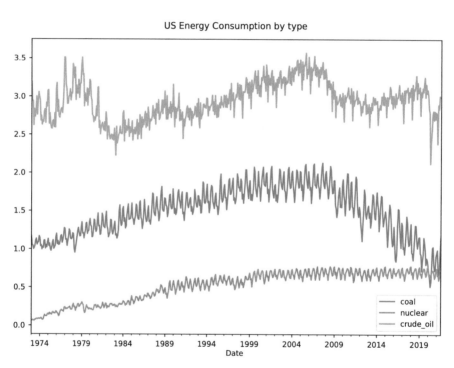

Fig. 10.8 Visualization of the different types of U.S. Energy Consumption between January 1973 and July 2021. (Image created from augmented_dickey_fuller_test_example.py)

Based on an alpha of 0.05, the only energy time series that is stationary (H_o rejected) is the nuclear consumption time series with p = 0.02. See Chap. 5 for more information on alpha, p-values, and statistical significance. As can be seen in Fig. 10.8, the nuclear time series (purple) is essentially a flat time series that has not significantly changed from 1973 to 2021.

Some autoregression models (see Sect. 10.4) require that a time series exhibit stationary behavior before being used. If you want to use autoregression models with a non-stationary time series then a transformation of the data series in necessary to make it stationary.

There are many techniques for converting a non-stationary time series into a stationary time series, which include the difference transform, log transform, power transform, moving window transform, and difference with log transform. The difference transform approach is the most common.

The *difference transform* stabilizes the mean of the time series by removing changes in the level of a time series, thereby reducing trend and shifting seasonality. Conceptually, the transform is straightforward: subtract the observation at time t-1 from the observation at time t, storing the result at time t. The following two lines of Python code illustrate two equivalent ways to perform this operation:

```
df['column_name'] - df['column_name'].shift()
df['column_name'].diff()
```

In either case, both `shift` and `diff` default to a lag of 1 period, but this can be controlled by passing a different integer parameter. For example, `diff(2)` differences the time series with a lag of 2 instead of 1. Unfortunately, due to the nature of differencing, the initial k values are sacrificed (assigned `NaN`) where k is the number of specified lag periods.

In some texts shift is labeled "backshift" and is represented as a capital B. Some references use lag as an equivalent to backshift.

Once the time series has been transformed and confirmed to be stationary then autoregression models can be applied. However, a forecasted value generated from a model utilizing a transformed data set must be inverted back to the original scale to make sense of this value. For a lag period equal to 1, this restoration can be accomplished by adding the differenced value at time t to the observation at t-1 to obtain the reconstituted value at time t. The `cumsum` function inverts or reverses the effect of the *diff* function. Note the `cumsum` function is able to reconstruct the original time series except for the initial value (or more if the lag is greater than 1). Also, because these value(s) are lost the level of the inverted time series does not align with the original and must be corrected by determining the difference in the mean between the original and inverted time series and adding this offset to the inverted time series.

The following Python code displays the U.S. total energy consumption, the differenced data, the inverted data, and the inverted data plus the corrected level. In addition, the code performs the ADF test on each of the four time series and visualizes the results in Fig. 10.9:

```
Partial code: Full program in differencing_example.py:

df['total_differenced'] = df['total'].diff()  # the periods or lag has a default of 1

df['cumsum'] = df['total_differenced'].cumsum()  # inverting the difference

level_diff = df['total'].mean() - df['cumsum'].mean()  # difference in level

df['cumsum_plus_level'] = df['cumsum'] + level_diff  # correcting the level
```

Which produces the following output:

```
ADF on total 0.5702166894487601
ADF on total_differenced 2.604534049025079e-09
ADF on inverted 0.572496598314925
ADF on inverted plus level 0.5724965983149495
```

The above results list the p-values of the ADF tests on the various time series. From this output we note the original time series is non-stationary (H_o not rejected) because the p-value is above 0.05, but the differenced result is stationary (H_o rejected) since the resulting p-value is below 0.05. The inverted and inverted with corrected level time series are both non-stationary (H_o not rejected) as expected. Note the ADF tests show slightly different p-value results for the inverted and inverted with corrected level time series compared to the original time series data because the initial value has been lost.

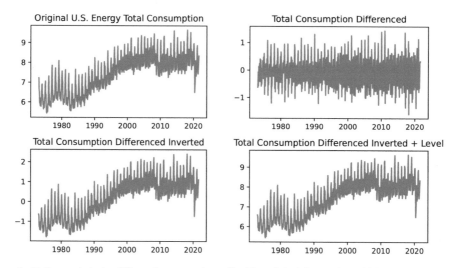

Fig. 10.9 Results from the Python code in the differencing_examply.py file. The original time series, which is non-stationary, is shown in the top-left corner. The result of differencing the original time series, which is stationary, is shown in the top-right corner. The inverted time series is shown in the bottom-left corner. Note that the scale is off because the level is wrong. The bottom-right corner shows the inverted time series with the level corrected. It now resembles the original time series, but with the first value removed. (Image created from differencing_example.py)

The *Autocorrelation Function (ACF)*, *autocorrelation*, or *serial correlation* determines the correlation or similarity measure between a given time series and a delayed or lagged version of the same time series. Autocorrelation estimates the connection or relationship between past observations and the most recent observations. In other words, if strong autocorrelation is present in a time series then discernible patterns exist such that future values in the time series can be predicted based on preceding observations located at specific lag intervals.

Figure 10.10 shows two autocorrelation examples using a sine wave time series in the top row and the total U.S. energy consumption time series in the bottom row along with accompanying correlograms. *Correlograms*, shown on the right side of Fig. 10.10, are visualizations of the correlation values of the time series with itself for given lags specified along the x-axis. Specifically, Fig. 10.10 plots how each of the latest k observations correlate to the most recent (last) observation.

The sine wave example displays the autocorrelation values spanning a range of 25 lags, which effectively provides a measure of the comparative significance of the most recent 25 observations in predicting the very last observation. As expected the autocorrelation value for a lag of 0 is equal to 1.00 – the sine wave is perfectly and positively correlated with a non-shifted version of itself – implying, somewhat tautologically, that the last observation is an ideal predictor of itself. Note from the correlogram the most recent 11 observations from the left of the autocorrelation figure show significant predictive power with each exceeding the confidence interval represented by the blue shaded region. The next 10 observations (now enclosed by the blue shaded area) do not provide significant predictive power, but the remaining four observations with lags from 22 to 25 emerge from the confidence interval, indicating usefulness in predicting the most recent value even though they are negatively correlated. In other words, for our specific sine wave example strong autocorrelation results point to predictive values that are at or near the most recent target value of 0.0.

For the U.S. energy consumption time series and autocorrelation displayed in the bottom row of Fig. 10.10, all the most recent 28 observations have predictive power when estimating the last observation because the autocorrelation values exceed the boundaries of the shaded area. In other words, significant autocorrelation is highly prevalent in this time series for lags up to and including 28 so knowing these past observations will help in predicting current values. Seasonality coupled with trends will usually yield high, periodic autocorrelations for shorter lag times as can be seen with the total U.S. energy consumption.

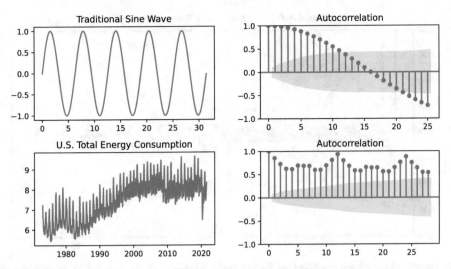

Fig. 10.10 Input on the left column and resulting visual results of an autocorrelation test – a correlogram – on the right column. The left column shows the original time series and the right column shows the results of the autocorrelation results. The individual observations in the correlogram with a projection to the X-axis indicate how positively or negatively correlated each of the latest observations are to the most recent one. The blue shaded region is the confidence interval with a default p-value of 0.05. Anything in the blue region represents an observation that has no significant correlation with the most recent value. (The image was created by autocorrelation_example.py)

Executing the code in `autocorrelation_example.py` produces Fig. 10.10, providing time series autocorrelation values for a given lag. Running the code will also calculate the Pearson correlation with a lag of 1, as shown in the following output (see Sect. 4.22 for more information about correlations):

```
Pearson correlation between the time series and itself with lag of 1 observation:
Sine wave correlation: 0.9950092281248635.
Total energy consumption correlation: 0.8522409065470394.
```

As expected, the above results confirm the sine wave is highly autocorrelated with a version of itself delayed by one time step, which makes sense given that this waveform exhibits a consistent, repeating pattern.

For the total U.S. energy consumption, the autocorrelation result of 0.85 for a time lag of 1 reveals a high degree of positive autocorrelation between consecutive time steps. This outcome is expected given the seasonality dependent on weather and general increasing trend of energy consumption over the past several decades. You can also use the *statsmodels.api.tsa. acf* function to produce additional autocorrelation information.

Another important correlation method instrumental in developing forecasting models for time series is the *Partial Autocorrelation Function (PACF)* or *partial autocorrelation*. The partial autocorrelation function is a variation of the autocorrelation function in that it emphasizes the direct correlation between two observations at a precise lag by removing the contribution of the intervening, shorter lags. If partial autocorrelation values are close to 0, then observations separated only by the specified lag are not correlated to one another. Conversely, partial autocorrelations with values close to 1 or −1 signal strong positive or negative correlations between the specific lagged observations of the time series.

Similar to the autocorrelation function, the plot generated by the partial autocorrelation function returns confidence intervals represented as blue shaded regions. If partial autocorrelation values exceed the boundaries defined by these confidence interval regions, then the observed partial autocorrelation values are statistically significant. Figure 10.11 displays a partial autocorrelation function plot for the total U.S. energy consumption.

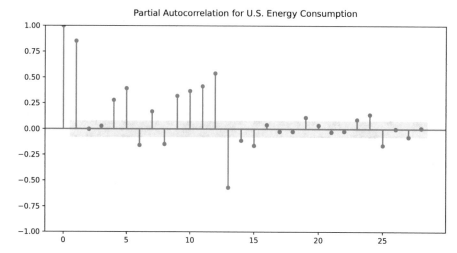

Fig. 10.11 Plot for the Partial Autocorrelation Function (PACF). The blue shaded region is the confidence interval with a default p-value of 0.05. Anything in the blue region represents an observation at the lag value defined by the x-axis that has no significant correlation with the most recent value. The lags 1, 4–15, 19, 23–25 are all statistically significant. (The image was created by partial_autocorrelation.py)

If the autocorrelation function helps assess the patterns and properties of a time series such as randomness and stationarity, the partial autocorrelation function is more useful for determining specific parameters for autoregression models (see Sect. 10.4). Figure 10.11 shows that lags 1 and 4–15, 19, and 23–25 are all statistically significant and might be considered for parameters in autoregression models. To further simplify the order of the resulting autoregression model, we could restrict the range of selected significant lag values to 1 and 4–13, or use only the immediately preceding value (lag = 1).

On a related note, the *Durbin-Watson test* is a well-known test for detecting a first-order autocorrelation in the residuals resulting from a regression analysis (not for a time series). One of the assumptions about linear regression models is that no correlation exists between consecutive residuals. If a discernible autocorrelation is found in the residuals of linear regression models then the predictors associated with these models are more likely to be incorrectly labeled as statistically significant. Instead of a range of 0–1, as in the Pearson correlation, the Durbin-Watson test statistic falls within a range of 0–4, where 2 indicates no autocorrelation, below 2 indicates a positive autocorrelation, and above 2 indicates a negative correlation.

10.4 Autoregression Models

Now that we have covered the fundamentals in discovering patterns in time series, we now turn to the important task of forecasting. Autoregression models are one of those topics that can occupy a significant portion of a statistics course. Even though the topic is quite extensive we provide an overview so the basic construction and implementation of these models are understood. However, do not let the sophistication of autoregression models hamper you because they are both very useful and essential in forecasting time series.

Autoregression models, also known as the *Box-Jenkins models*, are time series models that leverage observations from previous time steps as input to a regression equation for the purpose of forecasting the next value. Most of the background that you need about stationarity, autocorrelation, and partial autocorrelation to proceed with autoregression concepts is discussed in Sect. 10.3.

There are many different acronyms associated with autoregression models. We will briefly discuss these and their corresponding parameters. First, the *AR (Auto Regression)* model is based on the concept of autocorrelation in which the dependent predicted variable relies on the past values in the time series. AR takes a single argument, p, which defines the number of previous steps to use from the time series. This value p, known as the order of the autoregression model, is most effectively determined through analysis of the partial autocorrelation function.

The *MA (Moving Average)* model is a rolling average defined by a window of a given duration. One of the objectives of the MA model is to reduce noise. MA also takes a single argument, q, which specifies the number of samples within the time window for the moving average. This value q is best determined by the autocorrelation function.

Combining AR and MA results in the *ARMA (Autoregressive moving average)* model. The ARMA model assumes time series stationarity. If the time series does not exhibit the stationarity property then the available modeling option is called ARIMA, which is ARMA with the added *I (Integrated)*. The additional integrated model requires a single parameter, d, the order of integration. In other words, the value of d defines the number of difference operations required to achieve stationarity. If the time series does not employ differencing because it is stationary then $d = 0$. More concisely, ARMA(p,q) = ARIMA(p,0,q) where the middle parameter for the ARIMA model represents the value of d (0 in this case). Table 10.1 provides a summary of the three parameters described previously.

Table 10.1 Summary of parameter information for ARIMA

Model	Parameter
AR	p: number of autoregressive lags (use PACF to determine)
I	d: order of differencing required to make the time series stationary
MA	q: number of moving average lags (use ACF to determine)

If your data exhibits seasonal behavior such as the U.S. energy consumption dataset, then you should consider *SARIMA (Seasonal autoregressive integrated moving average)*. Table 10.2 lists the three main types of autoregression models.

Table 10.2 The three main types of autoregression models with their accompanying descriptions

Model Name	Model Description
ARMA	Autoregressive moving average
ARIMA	Autoregressive integrated moving average
SARIMA	Seasonal autoregressive integrated moving average

To achieve a successful autoregression analysis, you must first determine which model from Table 10.2 is the most appropriate. As mentioned above, the best choice for the U.S. energy consumption data is SARIMA because this model accounts for a seasonal component and does not require the time series to be stationary (recall the time series will be transformed to stationary via the I processing component). We will now use the results we obtained from Sect. 10.3 to determine the values of parameters p, d, and q for the SARIMA model applied to the U.S energy consumption dataset.

The preliminary step will focus on parameter d since this value defines the level of wrangling necessary to convert time series data from non-stationary to stationary. From our previous discussion of applying differencing operations and the augmented Dickey-Fuller test to the total U.S. energy consumption data we achieved statistically significant stationarity by differencing with a single lag of 1, so our $d = 1$.

Determining p involves examining the partial autocorrelation function results. Recall the PACF in Fig. 10.11 reveals that lags 1 and 4–13 are all statistically significant and might be considered for the value of p. Similarly, the ACF in Fig. 10.10 indicates there are many potential values for q. Although these plots are extremely useful for our analysis, determining precise values for p and q remains a difficult endeavor.

There are instances in which the results from the autocorrelation function and the partial autocorrelation function make the determination of values p and q straightforward, but our energy data example is complicated by the large consecutive series of significant lags generated by the autocorrelation function. At this critical juncture, the Akaike information criterion (AIC) can assist.

The *Akaike information criterion (AIC)* is an optimizer of the quality of statistical models for a given set of data. For our current purposes, AIC can be used to help determine the best values of p and q for our model. Given a value of d and a range of possible values for p and q, AIC estimates the quality of each model for each set of specified parameter values. The lower the AIC score the higher the quality.

We will use the `auto_arima` function from the *pmdarima* library to determine the optimal values of p and q for our model. The `auto_arima` function is similar to the `auto.arima` function in R. Note the `auto_arima` function can support a number of other optimizers besides AIC. We will pass several parameters to the `auto_arima` function: `d=1` – the value of d we found earlier, `start_p` – a suggested initial value for p, `start_q` – a suggested initial value for q, `seasonal=True` to indicate the data is seasonal, and `m=12` – to indicate the period of the season, in this case 12 time steps where 12 months equals a year. The following Python code demonstrates the use of AIC to determine the optimal parameters and produces Fig. 10.12:

```
Partial code: Full program in auto_arima_example.py:

model = auto_arima(train, start_p=1, start_q=1, seasonal=True, d=1, m=12)
print(model)
```

Which generates the following output:

```
ARIMA(0,1,2)(1,0,2)[12]
```

The first tuple in the displayed results indicate the best trend element values determined by the algorithm were $p = 0$, $d = 1$, and $q = 2$. The next tuple (1,0,2) defines the seasonal element values P, D, and Q (all uppercase). In other words, for the seasonal hyperparameters $P = 1$, $D = 0$, and $Q = 2$. The "[12]" indicates the single seasonal period m is equal to 12 time steps. For more detailed information on the model enter the command `print(model.summary())`. Figure 10.12 generated by the above code plots the predictions of the best selected model from the `auto_arima` function.

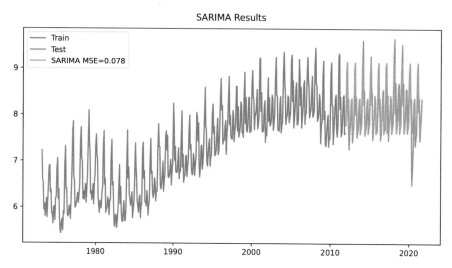

Fig. 10.12 Results from the auto_arima function. (Image created from auto_arima_example.py)

Figure 10.12 demonstrates excellent results with a MSE between forecast and test data of 0.078. When implementing walk forward validation (explained in detail in the next Sect. 10.5) even better results are achieved as shown in Fig. 10.13 where the MSE improves to 0.052.

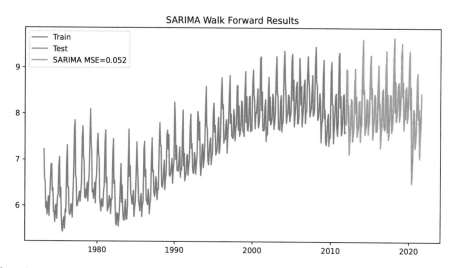

Fig. 10.13 Results from the `auto_arima` function with walk forward validation. (Image created from auto_arima_example_walk_forward.py)

An 'X', or exogenous variable, can be appended to our current models to produce prediction techniques ARMAX, ARIMAX, and SARIMAX, which are colloquially and collectively called the "MAX" models. An *exogenous variable* or input to our models essentially introduces terms from an external or independent time series drawn from outside the model under consideration. The opposite of an exogenous variable is an *endogenous variable*, effectively a dependent variable in the model. Viewing the summary details from the above code (using `print(model.summary())` will reveal that the selected model was actually a SARIMAX model.

There are various other autoregression models available. For example, we have been using a univariate time series for our examples up to this point. In other words, there is only one value associated with each time step in the time series. However, *VAR (Vector Autoregression)* is an autoregression model used for multivariate forecasting in which two or more time series variables may influence each other.

Research into the dynamic field of autoregression models is ongoing and new models are regularly being designed and released. Consequently, we now conclude this general overview of autoregression techniques with a sincere recommendation to independently pursue more detailed information about advanced autoregression models after completing the remainder of this chapter which covers competing forecasting paradigms.

10.5 Smoothing and Holt-Winters Method

Smoothing elicits general patterns and trends from time series data by removing the eccentricities and noise to obtain a clearer view of the underlying meaning communicated by the data. Figure 10.14 demonstrates different levels of smoothing applied to the total U.S. energy consumption data based on rolling average window duration. The top-left panel shows a plot of the original data. The top-right (6 months), bottom-left (12 months), and bottom-right (48 months) panels display the graphs resulting from applying a moving average of increasing length.

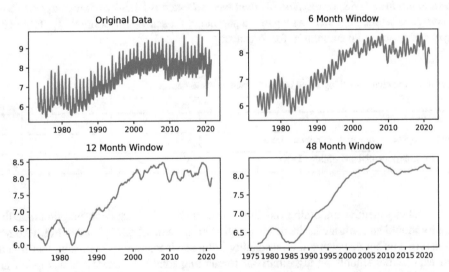

Fig. 10.14 Progressively increasing window size from 0 (original data) to a 48 months window to illustrate how simple moving averages work. (Image created from smoothing_example.py)

A *window* defines the number of observations used to generate a single smoothed result. Often averages across the samples included in the window are calculated, but any meaningful function, such as max, min, or sum can be applied. Figure 10.14 was created using Panda's `rolling` function. For example, to create the time series using a 6 observation sized window in the top-right corner (6 months window with 1 observation/month), we employed the following code statement: `df['total'].rolling(6).mean()`.

To compute the mean using a window of length 6 the first six observations, 1 through 6, are averaged to create the initial value for the smoothed time series. To compute the next resulting point in the smoothed dataset the observations 2 through 7 are averaged, followed by observations 3 through 8, and so on until the last point of the original time series is included in the rolling window.

The longer the window the smaller the resulting time series because the first few observations are sacrificed depending on the size of the window. For example, the bottom-right panel of Fig. 10.14 presents the smoothest of the four charts demonstrating a broad overall trend, but this plot also contains the least number of data points. Whereas the original data in the top-left panel begins in January 1973, the initial observation with the 48-month window applied in the bottom-right panel occurs 48 months (4 years) later on January 1977.

Smoothing can also be an effective tool for forecasting. One of the most popular forecasting techniques is exponential smoothing.

Exponential smoothing is a type of weighted smoothing technique that emphasizes the most recent observations for forecasting purposes while weighting successively older observations on an exponentially decaying curve.

There are three main types of exponential smoothing: simple (also known as single), double, and triple. *Simple Exponential Smoothing (SES)* is typically applied to smaller time series that exhibit irregularities, no seasonality, and no trend. Simple exponential smoothing utilizes a single parameter named alpha to determine the degree of decay in the weighting values assigned to previous observations. Alpha values are within the range (0,1]. An alpha closer to 1 implies a heavier emphasis on more recent observations.

Conversely, when alpha is closer to 0 then a larger span of past observations will influence current forecasts. An alpha equal to 1 means the prediction for the next time step will simply be the current value with all prior observations ignored, an approach referred to as *naïve forecasting* based on the naïve approach described earlier in this chapter. Regardless of the choice of alpha, the sum of all the weights assigned to prior observations is equal to 1 due to the geometric progression inherent in exponential weighting.

Double Exponential Smoothing (DES) or *Holt's linear smoothing* extends SES by adding support for trends. DES involves a forecast equation, a level equation, and a trend equation. DES calculations rely on an alpha parameter for level smoothing, as with SES, as well as a beta parameter for trend smoothing.

Triple Exponential Smoothing (TES) or the *Holt-Winters method* extends DES by adding support for seasonality. Along with the alpha and beta parameters, TES includes a gamma parameter for seasonality smoothing. Table 10.3 provides a summary of the three types of exponential smoothing for forecasting.

Table 10.3 Three types of exponential smoothing based on level, trend, and seasonality

	Level	Trend	Seasonality
Single exponential smoothing (SES)	✓		
Holt's linear smoothing or double exponential smoothing (DES)	✓	✓	
Holt-winters method or triple exponential smoothing (TES)	✓	✓	✓

To establish the correct exponential smoothing function for the total U.S. energy consumption data the existence of any trend and/or seasonality should be confirmed. The decomposition of this dataset as depicted in Fig. 10.6 reveals both a verifiable trend and seasonality. Further inspecting the seasonality influence leads to the conclusion that it is additive. Based on Table 10.3, the best exponential smoothing algorithm for forecasting total U.S. energy consumption is the Holt-Winters method. The following code example demonstrates the implementation of the Holt-Winters technique in Python, which produces Fig. 10.15:

```
Partial code: Full program in holt_winters_example.py:

model = ExponentialSmoothing(train, seasonal='add').fit()
pred = model.predict(start=test.index[0], end=test.index[-1])
```

Figure 10.15 indicates the Holt-Winters method does a fairly good job predicting the U.S. total energy consumption (MSE = 0.077). The predictions are in green and the actual observations are in orange. Overall, the method performs well with the exception of one major weakness readily apparent in the figure during the year 2020: the black swan event of the COVID-19 pandemic.

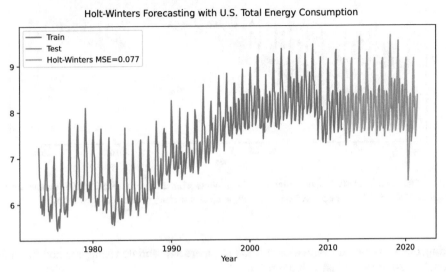

Fig. 10.15 Example of the Holt-Winters method applied to the U.S. total energy consumption dataset. (Image created from holt_winters_ example.py)

The results rendered in Fig. 10.15 would likely be improved with the walk-forward model validation method. One of the challenges with time series testing and validation is the rigid coupling between time and the fundamental behavior of the time series data used for training. For example, Fig. 10.15 displays the results of a Holt-Winters model, which accounts for trend and seasonality. However, upon careful inspection of the results, the crests and troughs of the forecast (green curve) are the same every year and do not vary over the length of the forecast. These outcomes make sense because the model can only forecast so far into the future and still maintain accuracy, though as already mentioned irregular future events are often overlooked.

The key idea behind *walk-forward model validation*, a time series cross-validation technique, is to generate the next prediction in the time series then update the model with the actual observation to then continue with the forecast for the next time step. Sometimes the model can be updated incrementally and sometimes the model will need to be recreated from scratch. For example, the model that produced the results from the SARIMA walk forward in Fig. 10.13 has functionality that easily allows for updating without requiring full-blown recreation and recalculation at each time step.

When we divide the U.S. energy consumption data using an 80/20 split as illustrated in Fig. 10.15, the forecast for the next 20 percent shows repetitive, consistent behavior. However, if we forecast the next time step then update the model with actual training data, we are testing the quality of the model from a more realistic standpoint.

The reason the walk-forward model is deemed realistic is that meaningful forecasts are typically short term and based on recent field (actual) measurements. For instance, consider forecasting the weather with the conventional 80/20 split rule. For 10 years of weather data, if we use the first 8 years of measurements to train our model then how reliable is the forecast going to be for the remaining subsequent 2 years?

Alternatively, the walk-forward model trains with 8 years of data, forecasts ahead for a short time period, updates the actual data and model accordingly, then repeats. The short forecasting time period could be 1 day, a week, or a month – whatever you like – but the model is then updated, or walks forward, with the most recent observed data before forecasting the next time period. Figure 10.16 plots the Holt-Winters function with the 80/20 traditional split in green (similar to Fig. 10.15) with a MSE (Mean Squared Error) of 0.077, and the same function leveraging the walk forward validation method in red with a MSE of 0.053, resulting in a 31% improvement in MSE.

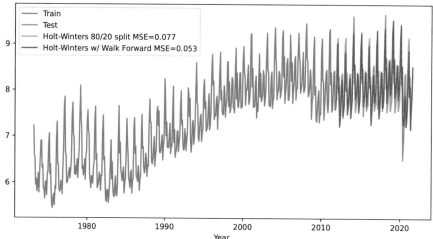

Fig. 10.16 A comparison of the traditional 80/20 split applying the Holt-Winters function with a MSE of 0.077 compared to the more appropriate walk-forward method with a MSE of 0.053. (Image created from holt_winters_example_with_walk_forward.py)

Generally speaking, the walk-forward validation method is a more appropriate testing method for time series when compared to the traditional 80/20 train and test split approach.

The walk-forward validation method is a clear favorite when compared to the k-fold validation method which is entirely inappropriate for time series applications. The severe consequence of utilizing the k-fold validation method on time series is the model could be trained with data from 2010 and subsequently tested with data from 1980. For more information on the k-fold validation method see Sect. 8.8.

Another specification associated with the walk-forward method is the selection of an *expanding window* or *sliding window*. In Fig. 10.16 we used an expanding window in which the training data in the model continued to grow with each additional observation.

Alternatively, a sliding window establishes a constant size of the training dataset that will inform the model and truncates old observations as new observations are introduced, similar to the decaying weighting coefficients applied in exponential smoothing. The fundamental idea for this strategy is that old data are not as useful as new data. For instance, if U.S. energy consumption data were available as far back as 1900 to the present, would information on consumption in the year 1900 truly assist with present-day forecasting? The most likely answer to this question is "Probably not."

10.6   Time Series with Neural Networks

Neural networks exhibit significant promise when applied to forecasting. As mentioned in Sect. 8.6, there are many types of neural networks. Each neural network occupies particular niches. Although generic MLP (Multi-Level Perceptron) neural networks often work well for time series problems, the LSTM (Long short-term memory) recurrent neural network traditionally demonstrates better accuracy.

Regardless of the kind of neural network employed, there are two main strategies to consider when leveraging these algorithms for time series analysis: the time step method (also known as step-by-step method) or the windowing method (also known as sliding window method). The *time step method* passes the time step of the x-axis as input along with the value of the y-axis as target. For example, for the U.S. energy consumption data, the first month is January 1973, or month 0, and the total energy consumed was 7.226.

To illustrate further, if we were to train our neural network with the first five time steps of the total energy data we would have X = [0, 1, 2, 3, 4] and y = [7.226, 6.595, 6.524, 5.944, 6.075]. In other words, the time step method passes an indexed numerical list of the time steps with the accompanying series values.

In contrast, the *windowing method* passes the time series values as the X training data and the next value as the y target data. Windowing can be performed with various sliding window sizes. For example, with the energy consumption values [7.226, 6.595, 6.524, 5.944, 6.075] and a window size equal to 2, the first (X,y) pair would be ([7.226, 6.595], 6.524), followed by ([6.595, 6.524], 5.944), and ([6.524, 5.944], 6.075). To create sliding windows either a for-loop or numpy's `sliding_window_view` function can be utilized.

Suppose we specify the following training data: [7.226, 6.595, 6.524, 5.944, 6.075] and a window size of 2. The following code will produce the X and y values necessary to train a neural network:

```
File: sliding_window_example.py:
from numpy.lib.stride_tricks import sliding_window_view

data = [7.226, 6.595, 6.524, 5.944, 6.075]
window_size = 2
X = sliding_window_view(data[:-1], window_size)  # the -1 leaves the last value for the y value
y = data[window_size:]

print(f'Original data = {data}')
print(f'X = {X}')
print(f'y = {y}')
```

Which produces the following output:

```
Original data = [7.226, 6.595, 6.524, 5.944, 6.075]
X = [[7.226 6.595]
 [6.595 6.524]
 [6.524 5.944]]
y = [6.524, 5.944, 6.075]
```

A sliding window size can be any positive integer that does not exceed the size of the data. Determining the best window size is an optimization problem that is data dependent. One window size might return optimal results for one particular dataset, but evoke less optimal results with a different dataset. Figure 10.17 compares examples of using the time step method on the left and the windowing method with three different window sizes on the right. The code for Fig. 10.17 is in `mlp_example.py`.

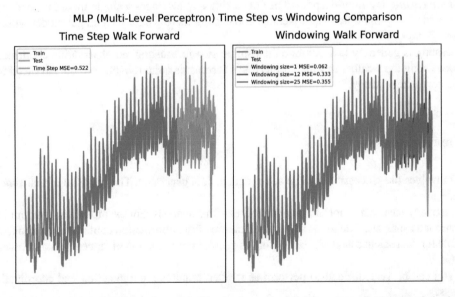

Fig. 10.17 Comparison of results with the MLP neural network. The graphs compare the time step method to the windowing method (or sliding window method). The walk-forward model is applied in both cases. The best results are with windowing and walk forward. (Image created from mlp_example.py)

Figure 10.17 demonstrates the windowing method is generally better than the time step method, and the size of the window has a significant impact on the accuracy of the resulting model. A quick glance between the left side and right side results of Fig. 10.17 confirms that windowing exhibits better accuracy. From a quantitative perspective, contrasting the MSE (Mean Squared Error) between the time step method (0.522) and the best MSE from the windowing method (0.062) reveals approximately an order of magnitude improvement in accuracy!

For time series analysis, there are many different neural networks you might research to achieve better accuracy. As mentioned above, recurrent LSTM is often referenced in the literature as well as many other options including stacked LSTM, bidirectional LSTM, and CNN LSTM.

Many of these neural networks are not available with the *sklearn* library. However, other Python library options exist such as the *keras* library. We leave the exploration of other exciting and innovative machine learning libraries to the reader. Researching the various neural network possibilities will take considerable time, but the return on this investment may be additional significant accuracy for your time series models.

Finally, we address *multi-step forecasting* vs *single-step forecasting*. At this point we have employed single-step forecasting with our walk forward validation. Single-step forecasting is forecasting the next datapoint in the time series. For example, a single-step forecast applied to the U.S. energy consumption model involves an estimation of the next step in time, which translates to the next month.

Multi-step forecasting is evaluating more than just the next time step. For example, with the energy consumption data, we might perform a 12-step or full year forecast. Multi-step forecasting is often utilized in weather models in which a three-day or five-day forecast is typical.

Examining the code we used for autoregression and exponential smoothing forecasting will reveal that both of these models allow any amount of multi-step forecasting by simply passing a parameter such as `n_periods=7`. In contrast, specifying multi-step values as keyword parameters is not available for more general purpose neural networks.

Instead, there are many customized multi-step forecasting techniques with neural networks. For example, *recursive multi-step forecasting* is performed by feeding the current prediction from the model as input to compute the next subsequent prediction. For instance, we predict tomorrow's weather and use that estimate as the input (or part of the input in the case of windowing) to generate the forecast for the day after tomorrow.

Direct multi-step forecasting involves creating a different model for each time step. For instance, with direct 5-step weather forecasting there would be five distinct models, one for each step. Essentially, model 1 would forecast for tomorrow, model 2 would forecast for the day after tomorrow, and so on.

Multiple output forecasting is a multi-step form of forecasting that leverages a single model to produce multiple predictions. For instance, with 5-step multiple output weather forecasting there would be a single model which outputs a list of forecasts for the next 5 days.

Multi-step forecasting is generally less accurate than single-step forecasting, an observation confirmed by the common experience with forecasting the weather in which tomorrow's forecast is quite reliable, but the advanced forecast for 5 days from now might be completely wrong.

10.7 Real-Time Analysis

Real-time analysis involves the processing of time series data as it is generated. The opposite of real-time analysis is *asynchronous analysis*.

The investigation of any past data is not a component of real-time analysis. For example, all the examples in this book are classified as asynchronous analysis. Due to the static, fixed nature of the information contained in a book, current or updated data are not available for data science analysis except for the coarse approximation of "recent" data afforded by creating new versions of the book.

Real-time analysis can be very difficult to perform as it often requires a combination and coordination of automated, synchronous processes.

In addition, from the perspective of the data scientist real-time analysis requires a different mindset. Most researchers who analyze data do so with the presumption there is copious free time to clean and fit the data to the appropriate model, and subsequently forecast the result. The prediction results can then be evaluated by running a multitude of tests to determine how well the model performs.

Even for a relatively short data collection timeframe, say a day, the investigator can at least be certain that the data is either not changing or at least not shifting significantly enough to influence the overall analysis.

Consider two different types of investing models for comparison. Even though company prices in the stock market continue to fluctuate when the stock market is open, the period of time for normal trading spans only six and a half hours for the New York Stock Exchange (NYSE). These circumstances mean brokers and trading companies have most of the 24 hours in a given day to perform asynchronous analysis on the trades for that day and decide the best strategy for the upcoming day. In addition, the NYSE is closed on weekends and various holidays providing even more time for asynchronous analysis.

Contrast the environment of trading on the NYSE to crypto currency markets that are always open and active. Crypto currency, of which Bitcoin is one of the more recognized examples, can be traded all day every day. A moment never emerges when the data can be analyzed without changing. If a crypto currency buyer takes an hour to mull over the decision to purchase some currency, a recent positive profitable trend may have dramatically reversed during that time.

There are advantages and drawbacks to real-time analysis. For example, on the positive side of the ledger real-time analysis allows eccentricities to be discovered instantly, fraud can be detected and interdicted straightaway, and the best sales data are available immediately.

However, the relentless inflow of new data requires more powerful computers, significantly greater Internet bandwidth capacity, and most importantly, a new mindset for the analysts working with this continuous stream of information. In addition, real-time analysis systems are often much more difficult to debug than traditional systems.

Going back to the previous example of trading Bitcoin, consider either an automated system that buys and sells Bitcoin or, if the user wishes to be a more active participant in the process, consider a semi-automated system that generates an alert upon every major price change, prompting the user to either buy or sell based on the forecast.

In this particular real-time analysis scenario of buying and selling Bitcoin the trader could conceivably make substantially more money than traditional stock buying and selling. Conversely, a considerable amount of money could be lost in a very short period of time if the accuracy of the automated system degrades, such as when the crypto currency price pattern is rapidly fluctuating but the model is working with stale data.

Real-time systems are becoming more commonplace and are an established tool in disciplines such as finance, supply chain management and inventory systems, weather, and seismology (earthquakes).

10.8 Stock Market

Invariably, people interested in money and wealth accumulation who learn about time series analysis techniques gravitate naturally toward the stock market.

Can you capitalize on time series data science techniques applied to the stock market? Yes. In fact, the name of the subfield of data science that pertains to the stock market is called technical analysis.

In the stock market there are two main investing strategies rooted in different analyses. The more common analysis technique is called *fundamental analysis*. This approach crafts investment plans primarily from a business perspective and examines the financial health and prospects of the company behind the stock. For example, before investing in Apple, a large technology company, investors may consider the current CEO, the p/e ratio, the quarterly returns, the average dividends, and how the competition is performing. In other words, fundamental analysis is about assessing the stock as an extension of the company. These concepts underpinning fundamental analysis are rigorously taught in business schools and will not be discussed here.

On the other hand, *technical analysis* is about the mathematical and associated visual trends in stocks. When a trader (one who buys and sells stocks) performs technical analysis they are not concerned with who is running the company or the profits generated by the company. The trader is simply scrutinizing the price patterns of a particular stock. For example, does the stock AAPL (Apple's stock symbol) appear to be trending up or down? If so, when will the trend likely plateau or reverse?

Technical analysts are mostly concerned with *market sentiment* - a term that describes the attempt to understand the mass psychology of people who buy and sell stocks. The basic tenet associated with market sentiment is the stock market represents millions of people who buy and sell stocks. To technical analysts it is more important to understand why people buy and sell a particular stock than the detailed operations of the company behind the stock. After all, if people are **thinking** that Apple will likely perform well, regardless of how well the company actually functions, then there will likely be a concomitant widespread inclination to buy the stock. Therefore, a technical analyst who can anticipate the trading behavior of a vast number of people will consequently be more successful.

Is one technique superior to the other? Time typically decides that question. However, there are many people who have profited significantly by using either technique and at the same time many people who have lost considerable amounts of money employing these very same techniques. Since both techniques are seeking to model the stock market, hybrid techniques that effectively combine both fundamental analysis and technical analysis will likely emerge as valuable companions to those participating in the wild ride called the stock market.

One important principle to consider with respect to the stock market is the efficient market hypothesis (EMH). The *efficient market hypothesis (EMH)* is an economic theory introduced by Eugene Fama in 1965 which proposes that financial markets, like the stock market, accurately and instantaneously incorporate information about any security (e.g., a company's stock) into the current price of that security.

If this hypothesis is truly credible, actively trading market securities based on *historical* data is not an effective strategy to generate abnormally high returns, that is, consistent returns above those produced by the market while simultaneously assuming less risk than the market average. In other words, the EMH states that if you want higher returns over and above the average market returns you will need to take higher risk than average. However, the higher the risk the greater the chance you will lose money.

We encourage readers interested in applying data science techniques to stock market analysis to rigorously learn about these techniques and understand that it is impossible to consistently and correctly forecast market trends and price points.

As a disclaimer, we mention here that the application of the time series data techniques described in this book to the stock market is solely at the discretion of the reader, and the results obtained are not the responsibility of the authors.

10.9   Facebook Prophet

Facebook Prophet or *fbprophet* is a library from Facebook that is becoming increasingly popular primarily because it is robust, easy to use, and exhibits generally good forecasting performance.

Prophet is based on an additive model where yearly, weekly, daily, and holiday seasonality is integrated into the forecast. Prophet works particularly well on time series with strong seasonality and also detects trend changes using piecewise linear or logistic growth techniques.

One thing to note about Prophet is the required naming convention in which the time-based x-axis must be labelled with *ds* and the y-axis value must be labeled *y*. The following Python code provides an example of implementing Prophet by reading in the U.S. energy consumption data, relabeling the data, fitting the model to the entire dataset, obtaining predictions for the next 4 years, and generating Prophet-specific visualizations as shown in Fig. 10.18:

```
File: prophet_example.py:
import numpy as np
from prophet import Prophet
import matplotlib.pyplot as plt
import pandas as pd

df = pd.read_csv('US_energy_consumption_1973_to_Jul_2021.csv', parse_dates=['Date'])

# prophet requires that the x-axis be 'ds' and the y-axis be 'y'
df.rename(columns={'Date': 'ds', 'total': 'y'}, inplace=True)
df = df[['ds', 'y']]

m = Prophet()
m.fit(df)
future = m.make_future_dataframe(periods=1460)   # predict for the next 4 years
future.tail()
forecast = m.predict(future)

fig1 = m.plot(forecast)
fig2 = m.plot_components(forecast)
plt.show()
```

(a)

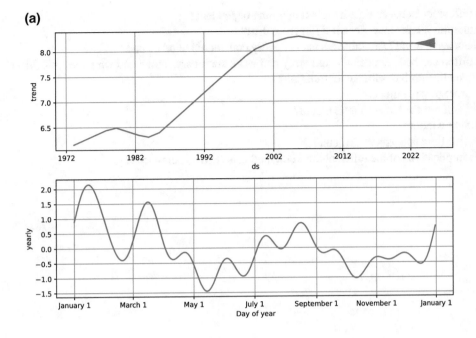

(b)

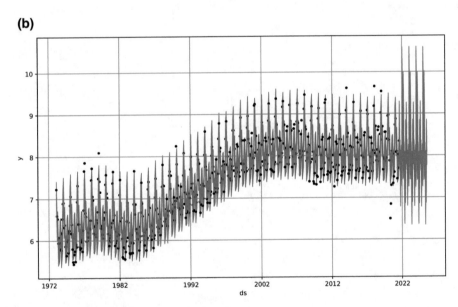

Fig. 10.18 Visual output from Facebook Prophet. (**a**) Visual trend and yearly seasonality. (**b**) Original data indicated by black dots, fitted curve, and forecast for the next 4 years. (Images created from prophet_example.py)

Exercises

1. Provide a comprehensive description of time series.
2. What is the difference between prediction and forecasting?
3. List at least two purposes for smoothing a time series.
4. Explain noise, level, trend, cycle, and seasonality.
5. What is drift? What are the different types of drift?
6. Why is the concept of drift so important in terms of forecasting?
7. What is the difference between additive and multiplicative seasonality?
8. What is piecewise analysis?

9. What is the difference between time invariant and time dependent?
10. What is a structural break? How does it differ from drift?
11. What is a black swan event? Are these events generally considered good or bad?
12. Explain the difference between strictly stationary and weak stationary. When and why does this difference matter?
13. Why should you be familiar with autocorrelation?
14. What is an autoregressive model?
15. When should you use the Holt-Winters model?
16. What is a sliding window?
17. What makes real-time analysis so difficult?
18. What are the implications of the efficient market hypothesis (EMH) being true?

Chapter 11
Final Product

"Begin with the end in mind" is both sound and timeless advice. No matter how noble, potentially profitable, or seemingly wonderful a project may be, if you lose sight of the end goal you are likely to fail.

The key to success in any data science project is to maintain the discipline to keep moving toward your end goal while simultaneously balancing an awareness to adapt to changing conditions as necessary. By *discipline* we mean sustaining the self-control to avoid distractions imposed by external or internal factors. *External factors* often include uncontrollable outside forces that upend your plans, which can range from family and friend commitments to a snowstorm or hurricane. Achieving some satisfactory measure of life-work harmony requires discipline.

Internal factors comprise the personal diversions that can be controlled to some degree; some examples include an overwrought curiosity, perfectionism, or bad health. While curiosity is a vital and fundamental trait for all data scientists, excessive curiosity will likely sidetrack you from the main focus and impetus of the project. There are countless projects that have been derailed because the team was inexplicably drawn into marginal issues and interesting curiosities, subsequently losing sight of the primary project goal.

Perfectionism is also a potential evil that may convince you that only the absolute best product and results are acceptable. While it is certainly important and commendable to perform at your best, you can ultimately distract yourself with nominal enhancements to the data science project such as incrementally improving machine learning models, designing the fastest database schema, or other unprofitable modifications that do not increase the overall quality of the end product.

In addition, compromising one's health by not devoting time to exercise and/or relaxation and ensuring proper nutrition could also derail your data science project. In the end, if you do not take care of yourself then you may be your own weakest asset.

The *willingness to course-correct* is also essential. Although it is important to maintain the discipline to keep an eye on the end goal of your project, you must be wise enough to realize when the end goal is not achievable and make a decision either to terminate the project and move on or to modify the end goal with one that is realistic and attainable.

How do you know when you should terminate the project? Identifying the time to discontinue a faltering project is extremely difficult. Consider the history of the Concorde passenger jet. The United Kingdom and France invested billions developing the airplane and never fully recovered their losses. Years of deficits in the Concorde research and development program coupled with a devastating plane crash finally took its toll and the governments decided to terminate the project.

Should the governments have terminated the Concorde project earlier? Should they have channeled additional money into the project to recoup their losses? There are no easy answers to these questions, only guesses whose impacts are only clarified through hindsight. However, one strategy is certain: a project destined to terminate with losses should be concluded sooner rather than later. If at any time you suspect your project is doomed to failure then which is better: to fail with some losses during the early stages or fail with substantial losses at the end? Because of the numerous challenges associated with this particular scenario, a strong-willed person is needed to speak up and suggest that a project should be terminated.

© The Author(s), under exclusive license to Springer Nature Switzerland AG 2022
R. Ball, B. Rague, *The Beginner's Guide to Data Science*, https://doi.org/10.1007/978-3-031-07865-1_11

Alternatively, adjusting or changing the course of a project **might** be better, especially if you evaluate the current path as unproductive and diverge to a potentially more profitable plan. Generally speaking, the later in the project timeline that goals are updated and reworked, the more expensive the project becomes.

However, by relying on the good principles and practices related to exploratory data analysis, you will likely ask many more meaningful and relevant questions to assess the validity and purpose of a data science project. In other words, once you understand the data you are working with it may be completely natural and justified to change the direction of the project.

Ultimately, no matter the avenue you pursue to design and implement your project, when you reach the culmination of your hard work you will need to reckon with the following question: **Did you keep your focus on making your organization profitable or did you chase after many shiny distractions?** As mentioned in Chap. 1, the profitability of your organization is paramount, since failing to maintain solvency may usher in the end of your projects, your position, or even your organization.

11.1 Presentation

Invariably at the end of your project, or most likely several times during milestone reports to your superiors and stakeholders, you will need to provide presentations. Presenting highly technical data to others from varied backgrounds (e.g., administrative, technical, novice), is inherently very challenging.

When presenting your data, start with a story. People love stories. Explain how the data represent people's experiences and how the data reflect their stories.

However, do not fall into the trap of telling **your** story. Although your audience may like or admire you, your presentation is about the story *behind the data*, not the personal struggles you endured to gather and understand the data. For example, if you devoted six months to uncover how to improve the test grades of elementary school children, then it's in your best interests to focus on those children, not on your own story and adventures of applying six months of your life to unearth the actionable insights.

Remember if you share non-relatable facts in a perfunctory manner no one will listen to you. Even if your data are by all accounts absolutely amazing and could ultimately change the world, if you cannot effectively communicate your message then your project was a wasted effort. For instance, if you found a cure for cancer, but you could not explain the cure to others so that it could be understood and developed independently, then you have applied your time and talents unsuccessfully.

When you present know your audience, especially their level of expertise and interest. They most likely are not concerned about detailed computational concepts such as gini, information gain, feature elimination, or neural networks. However, be cognizant of unnecessarily "dumbing down" your presentation: professionals do not like to be patronized or demeaned. Do not assume your audience members are incompetent.

As with the execution of the overall project described above, for your presentation focus on the big picture! What is the big picture? Recall why you originally started this project.

Sometimes minor deviations from the initial data science question may be fine, but typically they can usurp valuable time. Keep your presentation simple and focus on why you started the project in the first place.

One of the best ways to fail at a presentation is to improvise or ab-lib. Conversely, one of the best ways to succeed at a presentation is to prepare.

If you are nervous or unsure then practice your presentation in front of others with whom you feel comfortable such as friends or family. Presenting to a mirror is also a good technique to overcome any hesitation.

Many professional speakers also suggest recording your presentations so you can later analyze what you did well and what you can do to improve. There are many available online videos you can watch of professional speakers. Find inspirational speakers you like to listen to and mimic their methods of presenting.

Many modern presentations are delivered with slides such as Microsoft's PowerPoint. Slides have some benefits in that they help you stay focused on topic and stick to prepared material. On the other hand, a strict slide mindset can sometimes limit flexibility in entertaining questions or interesting comments from the audience. One compromise is to provide a formal slide presentation at the beginning then offer a less structured demonstration or casual dialog with your audience toward the end.

If you decide to use presentation slides we have two additional cautions: Do not list too much information on a single slide and do not read from your slides. Including too much information on a slide (slide overload) makes it difficult for your audience to digest the key points contained on the slide. Similar to the software engineering concept of high cohesion (Sect. 11.3), every slide should address a single, self-contained idea. If you are presenting several concepts then provide an overview slide followed by individual slides for each of the separate ideas.

Do not read from your slides. Reading your slides does not honor the time invested by your audience members, who are seeking elaborations and colorful descriptions about the information posted on the slides. The wording on each slide should offer a quick overview of what you want to express. You should refer to your slides, not read them. If you are planning to read to your audience then save everyone time and prerecord your audio, then post it publicly so that interested persons can listen to your voice at their leisure.

Deliver your presentation with enthusiasm. If you display any apathy towards your own topic of discussion your audience will mentally shut down and their minds will wander. However, if you exhibit passion for your work your audience will sense the merit of the information you are disseminating. Even if you feel awkward and do not present well the passion and enthusiasm you express will keep your audience engaged.

Impart your enthusiasm through your physical presence. The best professional speakers are all animated people that make good use of facial expressions, eye contact, gestures, and body language. Use your entire body, but do not get too carried away with animated behavior and act cartoonish.

Use your voice like you do your body. In other words, do not drone on in a monotone voice, but be engaging by capitalizing on modulations in pitch and intonation. The key is to be enthusiastically engaged with your mind, body, and voice.

Relax and enjoy yourself. You are presenting your work, so try to enjoy the results of your labors. If you are apprehensive or defensive then you will not be able to effectively communicate your message.

Finally, use visualizations to carry your message forward. We have used numerous visualizations throughout this book for the purpose of helping you understand the material. People are generally visual creatures that rely on their sight to understand material more so than touch, taste, or sound. What can be effectively explained in a single glance by a diagram, chart, or plot can sometimes be difficult to describe with words. The next section of this book provides more details about visualizations.

11.2 Information Visualization Theory Basics

Whether you are creating visuals for presentations, exploring the data, or creating a visual-based product, this section will help you with the basics of information visualization.

Information visualization is the method of representing data in a visual and meaningful way so a viewer can better grasp the underlying purpose and structure of the data. Information visualization, like almost all the topics in this book, is an extensive and complex field on its own, encompassing psychology, user experience, computer graphics, and art.

Information visualization usually incorporates some form of interaction to achieve maximum insight. In other words, for an individual to best understand and gain knowledge from a dataset they must interact with it personally. For more about the difference between data, information, and knowledge see Sect. 3.1.

Interaction is not always possible. For example, the visualizations in this book are not by definition interactive. The library or tool you use to create visualizations will depend on whether interaction is an integral part of the design.

For example, the vast majority of visualizations created for this book were produced using the *Matplotlib* library, which provides effective tools to generate a host of static, non-interactive visualizations.

Alternatively, *Plotly* is a popular tool used with many languages. At the time of this writing, *Plotly* can be integrated with Python, R, Julia, Javascript, F#, MATLAB, and Dash. Regardless of the language, *Plotly* creates interactive browser-based visualizations. When exploring data it is appropriate to use interactive visualizations such as *Plotly*. However, when presenting the data and summary results either take screenshots of *Plotly* visualizations or use static visualization tools like *Matplotlib*.

There are a host of Python visualization libraries, such as *Plotly, Matplotlib*, Seaborn, *GGplot, Altair, Bokeh, Pygal*, and *Geoplotlib* (especially for maps).

One type of non-interactive product that has become popular is an infographic. An *infographic* is a static presentation that combines images, charts, graphs, and text to produce a holistic perspective of the data. An infographic is a stand-alone product that can be viewed and understood in isolation without any supplemental support or commentary from the creator.

Infographics are usually poster-like documents that share a small story or provide a particular perspective about some data. For example, consider an infographic that illustrates how the use of cars has changed over the last century. Your infographic would emphasize visual representations with minimal text. Most infographics flow from top to bottom in a manner that helps the reader draw a specific conclusion about the data.

In information visualization there are various fundamental interactive techniques available. *Multiple views* literally allow for multiple perspectives of the same dataset, which can be accomplished by displaying different parts of the data – such as showing the first half of a dataset in one window and the second half in another window – or aspects of the same data can be shown in different panels – such as a bar chart beside a pie chart.

Small multiples arrange many small, thumb-sized visualizations sorted in a meaningful way. For example, consider a heatmap of a country where the colors on the map indicate the higher and lower population density centers. A single visualization would demonstrate a global view of the location of these centers. Now consider a series of small maps that depict the evolution of these population density centers over time. For instance, there might be ten small, thumb-sized maps organized in a row showing the progression of time over a given century, one for each decade revealing how the population centers have shifted and developed. All ten of the small maps are individually labeled and are collectively considered a single visualization.

LATCH (Location, Alphabet, Time, Category, or Hierarchy) encapsulates the philosophy that, although there is an infinite amount of data available collectively from the past, present and future, there are only a few ways that people can reasonably modularize it. In other words, data can only be organized by location (such as maps), alphabetically (or similar arrangements such as numerical order), by time, by category, or by hierarchy. In the end, all visualizations should adhere to one or more of those five organizational methods.

Visualizations can sometimes lead to misperceptions about the actual relationship between different values contained in the dataset. For example, let's suppose a person creates an infographic that demonstrates how inflation has affected the dairy industry. For this example, in the year 2000 a gallon of milk cost approximately $2.80 in the United States. How much money is required to purchase the same amount of milk at a later time?

The designer might select an image of a cow to represent the price of a gallon of milk for any particular year and then choose to display different sized cows over the years based on inflation. However, sometimes the designer might not accurately map the underlying data to the correct cow size, which could be the result of laziness or an outright intention to exaggerate or deceive. Consider the prices of a gallon of milk in Table 11.1.

Table 11.1 Average price of milk in the United States for the years 2000, 2010, and 2020 with inflation indicated[a]

Year	Gallon of milk	Inflation since 2000
2000	$ 2.80	–
2010	$ 3.47	24%
2020	$ 3.86	38%

[a]These prices for a gallon of milk are based on real US milk inflation and are not made up

If the person were to depict the size of the cows correctly the cow from 2010 should be 24% larger than the cow from 2000 and the 2020 cow should be 38% larger than the 2000 cow. It might be tempting to make the 2010 cow only 20% larger than the 2000 cow and render the 2020 cow 40% or even 50% larger than the 2000 cow to emphasize a point or drive home the idea of runaway inflation. Ensuring accurate representations of data in our visualizations is crucial.

The *lie factor* is a formula quantifying the relationship between the size of effect displayed in a graphic or visual and the size of effect in the actual data. The formula for the lie factor follows:

$$Lie\,Factor = \frac{Size\ of\ effect\ shown\ in\ visual}{Size\ of\ effect\ in\ data}$$

where

$$Size\ of\ Effect = \frac{|Second\ value - First\ value|}{First\ value}$$

In other words, the lie factor is the size of an effect rendered in a visualization divided by the actual size of the effect in the data on which the visualization is based.

The lie factor reminds us to make certain our visualizations are not suggesting relationships about data that exaggerate the truth. History is littered with thousands of examples in which newspapers and media outlets have employed the size and scale of visuals to misrepresent the truth and mislead consumers.

We now address the practice and implementation of interaction in visualizations. *Brushing* is selecting a subset of the data items with an input device such as a mouse or finger on a touchscreen. This action usually results in highlighting selected data in some meaningful way or otherwise fading the non-highlighted data. For example, if you were presented a scatterplot with 100 points and you brushed (selected) 10 of them then the 10 points might become bolder and therefore easier to see.

Alternatively, the 10 points might visually remain the same and the other 90 points suddenly become more transparent. Either way, brushing enables users to select and highlight specific data from the overall dataset.

Linking seeks to connect multiple views. Brushing the 10 points in the aforementioned scatterplot in one view would result in the same data being brushed in all other views. Linking is so intertwined with brushing that the technique is called *linking and brushing*.

Linking and brushing is a common technique available in modern visualization tools. For example, a user might display a map and bar chart separately with each showing average home prices for communities within a particular region. The map lists the average price for every community geographically and the bar chart provides a binning of the sorted average prices. The user might find it convenient to brush the top average price bin displayed in the bar chart to view the associated communities highlighted in the linked map. This use case allows a user to leverage the strengths of different visualizations to attain better insights about the underlying data.

For another interactive technique, consider a context-free map visualization that allows you to increase and decrease magnification and explore any direction (e.g., east, west, north, and south). This traditional method of map interaction is called *pan and zoom*. To pan means to move in any direction without changing magnification. For example, to pan a map visualization would be similar to flying in a plane at 10,000 feet and traversing any direction without changing altitude. Zooming means changing the magnification. For example, a plane that zooms out (decreases magnification) climbs to a higher altitude.

The issue with pan and zoom is the lack of overview. *Overview* refers to context, which is a vital component in visualizations and cannot be overemphasized. Without context many insights will be lost.

Consider a pan and zoom interactive map in which you have zoomed all the way in and are presented with a few street names and house numbers. If you do not immediately recognize the streets and houses, what will you do? You will most likely zoom out to establish a frame of reference, essentially the city/state/country you are viewing.

This action suggests the *visual information-seeking mantra* (also known as *Schniederman's mantra*): "Overview first, zoom and filter, then details-on-demand." The essential design concept embodied by this mantra is to leverage a person's natural thinking processes of first securing the contextual overview of the data, then zooming and filtering the data as necessary, and finally gathering the appropriate details on demand.

Overview plus detail is a common visualization technique in which the overall context is available to the viewer at all times while details are simultaneously displayed. This technique is often accomplished by providing two views, one of the detail and one of the overview. For example, consider a map of the United States in which the visualization furnishes a small map of the entire country in the bottom right corner of the display while the majority of the display is devoted to details of the current map location.

For example, when you zoom into a specific area of the country, say California, your main window displays details about California while the small overview map in the bottom right corner shows the specific part of the country currently being viewed. This is usually accomplished by overlaying a rectangle of the detailed area on the small map, which provides context about the region currently under investigation.

Focus plus context is a less common technique that enables viewers to see the object of primary interest presented in full detail while at the same time providing an overview. Focus plus context helps solve the problem of insufficient screen space and is also known as *detail-in-context*, and *detail-and-context*.

The *fisheye view* is one of the more common examples of focus plus context. Fisheye views magnify the center of the field of view or screen with a continuous and sometimes precipitous decrease in magnification toward the edges. Static fisheye views are commonly utilized in bus routes, subway routes, and similar maps that enlarge your current location and allow you to see the broader transportation network at the same time. Unfortunately, the enlargement of your current location fundamentally distorts the map, but has the benefit of maintaining the overview at all times within a single view.

11.3   Software Engineering

During the design and implementation of your data science project you are likely to create software. This section provides a brief overview of some of the professional principles and practices of developing and managing code called software engineering.

Software engineering comprises various methodical, disciplined, measurable approaches to the development, operation, and maintenance of software. If your project creates software as an end product (e.g., a recommendation engine, or a forecasting tool) the basic software engineering concepts in this section will assist you.

In computer science there is often a fundamental difference between the responsibilities assigned to *programmers* and those assumed by *software engineers*. A programmer is often the title of an individual who can program a computer but has little formal training and expertise. Sometimes programmers are given the pejorative name *code monkey* – an overworked, underpaid, and unappreciated person who programs only the most basic coding tasks. Programmers and code monkeys may be bright people but their skillsets or their tasks are often characterized as mundane and basic.

A software engineer is generally considered to be a highly-trained and skilled professional. If anyone can become a programmer by learning coding basics from online sites and videos then a software engineer has been through rigorous training, often at universities or boot camps, to become proficient at both the rigor and skillsets necessary to create the requirements (the specifications) necessary for successful software projects, to work with a team, to design the project based on these requirements, and to complete the project on time and within budget.

Like many topics in this book, software engineering is a complex discipline with its own accompanying literature. However, we cover many of the basic principles related to software engineering within the context of data science projects.

Software engineering is concerned not only with developing programs, but also with ensuring the end product satisfies the needs of the client. Software engineering focuses on validation ("doing the right thing") and verification ("doing the thing right"). Consequently, one of the most important skills for a software engineer is *communication* – the ability to interact with the client to determine precisely what they require. It is rare for a client to be able to easily and precisely specify the requirements for their project. Often a software engineer will hold many meetings with the client, typically involving mockups and prototypes, before reliable and valid project specifications are clearly defined.

Software design should achieve several objectives, but in general, the following priority list applies to most software engineering projects:

- Accessibility
- Dependability
- Efficiency
- Maintainability
- Usability

Accessibility is developing usable software for as many people as possible. For example, a large percentage of the population is colorblind with the most common type being red-green color blindness, which means those individuals have a difficult time distinguishing between the colors of red and green. If your visualizations rely heavily on a viewer's capacity to distinguish between red and green then you have an accessibility problem. Other common software accessibility issues include audio enhancements for those hard of hearing and visual acuity support for those not able to see well.

Dependability is a measure of the software's likelihood of failure. A very common illustration of the lack of software dependability for new programmers is the inevitable program crash when an unexpected input is entered. For instance, when software is expecting a number as input, but the user inadvertently enters a non-numeric input such as "42k" (as opposed to the intended "42") then poorly written programs will crash (halt execution). A similar common mistake is code that accepts a non-existent file name, attempts to read from that file, and subsequently crashes. Dependability is such a significant concern in software development an entire subfield of testing has emerged. There are many ideas and concepts related to testing. For example, *unit testing* creates individual tests for every small module (unit) of a program.

Efficiency is concerned with software performance, specifically in terms of speed, memory usage, and energy. There are many computer science courses such as data structures and algorithms analysis that emphasize the design of faster algorithms that require less memory. Reducing energy usage is a newer design metric related to sustainability but is becoming more prevalent as more people use mobile devices.

Maintainability is the long-term management and upkeep of software after its initial creation and release. After the software has been deployed, how easy will it be for other software engineers to modify and update the code base? A software engineering proverb associated with maintainability is "If people do not ask for updates to the software then they are not using it." As a result, you should create all your software with the hope that support personnel will continuously apply future modifications as needed without requiring your time or assistance.

Usability is ensuring an easy and intuitive design such that any user can rapidly figure out how to operate your software. The field of *Human-Computer Interaction (HCI)* focuses on developing usable software. Whereas efficiency prioritizes increased speed as well as decreased memory and energy usage, usability is concerned with speeding up the interaction times between the user and the software application. For some perspective, Donald Norman, one of the founders of HCI, claims that if a user has to consult a manual to use your software then you designed it incorrectly.

Although we cannot discuss every important software engineering principle, we will examine a few of the most important ones relevant to developing successful data science projects.

KISS (Keep It Simple, Stupid) or KISS (*Keep It Super Simple*) is exactly that: keep it simple. There are many ideas that will increase the speed and complexity of algorithms. However, in general, it is much better to implement simple ideas that are easily understood and maintained than extravagant ones that are difficult to understand and implement. For example, if you create a fast and efficient piece of code that no one else understands because of its complexity then when another person seeks to update or modify that part of the program they may have to rewrite a substantial amount of code because the complexity confounded them and it had to be replaced.

Good KISS practices include following code styles, using comments, and modularizing programs. If your code does not make logical sense to colleagues and peers then consider a redesign because most likely someone else will be required to rewrite the code in the future.

DRY (Don't Repeat Yourself) addresses the issue of repeating the same code within a program. Many programmers will write the same block of code in many places. The main purpose for functions and libraries is to leverage code reuse, so follow this principle of code reuse in your programs. If the same code is duplicated in several locations move that code to its own function or library so that it resides in only one place. This practice allows for more efficient debugging and code maintenance.

Defining a block of code that is used in multiple places leads to the design concepts of coupling and cohesion. *Coupling* is the amount of interdependence between modules (functions and libraries). Good software has **low coupling**. For instance, if your software invokes a function that calculates the distance formula then you should be able to easily replace that function with another function that performs the same calculation with no resulting adverse effects to program operation. Function replacement might be necessary because the older version was slow or required excessive memory, but in either case the inputs and outputs to the function are the same.

An automobile analogy of low coupling is being able to replace one car engine with another without affecting the rest of the vehicle. The number of belts and connection points between the engine and the rest of the car would remain the same, but the new engine might be more fuel efficient or provide more horsepower.

Cohesion is the degree to which elements of a function or library are related and directed towards achieving a single conceptual objective. The distance formula function example from before should be included in a library with only other related functions such as those that compute similar mathematical metrics. A good software design will exhibit **high cohesion**.

An example of low cohesion is a library or file that contains unrelated functions, which might include a function for calculating the distance formula, another for opening a file, and a third for retrieving user input at the beginning of a program.

Returning to the car analogy, a quality car design would exhibit low coupling so a mechanic could easily replace one engine with another and high cohesion so that all the principles of creating power are encapsulated in one single engine. A bad car design would have high coupling which prohibits engine replacement because of the byzantine interconnections between engine and car body and low cohesion in that power is produced from several places such as three or four independent power cells distributed throughout the car.

Version control is a key component for all software engineering projects that allows both individuals and teams to track revisions and updates to software, return to previous versions of a program, and merge changes between multiple developers. For a single person working alone version control allows reverting to an earlier file to safeguard against newly introduced errors and accidental deletions. A team of collaborating developers can also resurrect prior versions, but also work simultaneously on the same file while maintaining a reliable, deterministic method for merging the code from each individual programmer into a single working program. There are many different version control systems available such as Git for local version control and GitHub for distributed online version control. Other popular systems include CVS and SVN.

We conclude this section by referencing two important postulates in software engineering: the mythical man-month and no silver bullet. Both these concepts were originally popularized by Frederick Brooks in the 1970's.

The *mythical man-month* encompasses two important principles. The first is the proposition that you can measure software development productivity in terms of the man-month, a hypothetical unit of time a software engineer (regardless of gender) contributes to a project. The concept can be logically divided into smaller units such as man-week and man-hour.

The man-month metric can be applied effectively for isolated simple tasks. For example, how much time is required to mow your lawn? After ten mowing sessions, we could calculate the average time expended and obtain a fairly accurate estimate of the number of man-hours it takes to complete the task. This concept works surprisingly well for many straightforward hands-on labor tasks. Unfortunately, it does not work well for software development projects.

Estimating the number of man-hours to complete software is very difficult due to many different factors. First, some software tasks are simple and can be completed quickly. Second, complex software typically requires a significant amount of time that cannot be estimated. Third, complex software projects that comprise many interconnected activities and subtasks necessarily involve the additional overhead of communication between developers, project managers, software architects, and various teams. As a result, the time necessary to satisfactorily complete a software project does not increase linearly with the complexity of the project.

A second proposition about the mythical man month is adding manpower to an overdue software project actually results in a later delivery date. This appears counterintuitive. However, let's return to the lawn mowing task. If you have completed a quarter of the mowing task and suddenly five more people with mowers show up to assist then you will be required to pause mowing, determine where people can be the most efficient, and worry about additional safety concerns because there are so many additional mowers working the yard. Most likely, with the addition of the extra five people it will take you longer to finish you lawn because of the additional communication and safety issues involved.

Software projects are similar in that if you add additional software engineers during the later stages of an existing project to accelerate progress you will most likely push back the delivery and deployment date even further. The additional overhead of extra management, extra communication, extra resources, time spent explaining original project goals, placing the additional people and so on will likely lead to an even later project.

The misleading concept of man-hours also lures people to the erroneous belief that given a project initially estimated to require say, nine months, then hiring nine people necessarily translates to completing the work in exactly one month. This interpretation is conceptually possible, but typically does not apply for software development projects. For instance, if a grounds crew manager divided a baseball field into nine distinct sections and clearly defined the exact mowing responsibilities for nine workers well in advance, then theoretically the baseball field could be mowed nine times faster.

However, many software projects are not as starkly parallel as this mowing example, but instead involve complex sequential workflows. A more instructive example is attempting to birth a single baby with nine women in one month. This line of reasoning illustrates the absurdity of our assumption: If one woman can birth a single baby in nine months then why can't nine women birth a single baby in one month, or eighteen women birth a single baby in half a month?

The expression *no silver bullet* originates from the mythical werewolf purported to be deathly allergic to silver. Essentially, shooting a werewolf with a silver bullet anywhere in its body will precipitate its immediate demise. The silver bullet presents itself as the panacea to all werewolf threats. Unfortunately, there is no blanket solution to software problems. In other words, there is no single mythical solution that will solve all your software errors and design challenges. For example, if the success of a prior project was facilitated by the introduction of some tool, such as an IDE, that particular solution cannot be guaranteed to speed up your current project.

For instance, if a parallelization toolkit was applied to greatly speed up one project it might have only minimal effect on a subsequent, unrelated project. The reason is the design of the first project may have included primarily independent modules but your current project requires extensive communication and interaction between different components on a regular basis making parallelization a less effective solution.

In summary, there is no silver bullet – there is no single all-encompassing software solution that will consistently resolve all your software engineering challenges. As a data scientist, you must constantly innovate and discover different solutions to new emerging problems. The good news is the lack of a silver bullet means there is a tremendous demand for your many talents and abilities in the exciting field of data science.

There are many other software engineering principles that we encourage you to research as you pursue your goal of life-long learning.

Exercises

1. What does "Begin with the end in mind" mean for you in the context of data science?
2. What are the differences between the internal and external factors that might distract you from your task? How can you avoid them and stay focused?
3. When is it best to consider a different project path and change focus?
4. How do you know when to terminate a project? Why is that such a difficult decision?
5. For you personally, what is the hardest part about giving a presentation?
6. Why are stakeholders not interested in your personal adventures in exploring data?
7. What is the most important part of any presentation?
8. How can you improve your own presentation abilities?
9. What is the purpose behind information visualization?
10. What is the difference between a static visualization and an interactive visualization?
11. When would you create an infographic? When would it be inappropriate?
12. What is the difference between multiple views and small multiples?
13. The LATCH rule dictates there are five ways to organize data. Assume that each technique stands for a concept, such that the "A" stands for alphabet, but really is a type or ordering, like ordering numbers from highest to lowest. Find at least one more way to organize visualizations that is not part of the LATCH rule.
14. In your own words, what is the lie factor? Why do you think it is so prevalent in newspapers and media outlets?
15. Is the visual information-seeking mantra outdated? Can it be improved?
16. What is the difference between a programmer and a software engineer?
17. From the list of main objectives of software engineering (accessibility, dependability, efficiency, maintainability, usability), which one do you think is the most important?
18. What is the mythical man-month?
19. What does it mean that there is no silver bullet?

Printed in the United States
by Baker & Taylor Publisher Services